T0296497

Spatial Pattern Analysis in Plant Ecology

The predictability of the physical arrangement of plants, at whatever scale it is viewed, is referred to as their spatial pattern. Spatial pattern is a crucial aspect of vegetation which has important implications not only for the plants themselves, but also for other organisms which interact with plants, such as herbivores and pollinators, or those animals for which plants provide a habitat. This book describes and evaluates methods for detecting and quantifying a variety of characteristics of spatial pattern. As well as discussing the concepts on which these techniques are based, examples from real field studies and worked examples are included, which, together with numerous line figures, help guide the reader through the text. The result is a book that will be of value to graduate students and research workers in the fields of vegetation science, conservation biology and applied ecology.

Mark R. T. Dale is Professor of Biological Sciences at the University of Alberta, Edmonton, Canada.

CAMBRIDGE STUDIES IN ECOLOGY

This series presents balanced, comprehensive, up-to-date, and critical reviews of selected topics within ecology, both botanical and zoological. The Series is aimed at advanced final-year undergraduates, graduate students, researchers, and university teachers, as well as ecologists in industry and government research.

It encompasses a wide range of approaches and spatial, temporal, and taxonomic scales in ecology, experimental, behavioural and evolutionary studies. The emphasis throughout is on ecology related to the real world of plants and animals in the field rather than on purely theoretical abstractions and mathematical models. Some books in the Series attempt to challenge existing ecological paradigms and present new concepts, empirical or theoretical models, and testable hypotheses. Others attempt to explore new approaches and present syntheses on topics of considerable importance ecologically which cut across the conventional but artificial boundaries within the science of ecology.

Spatial Pattern Analysis in Plant Ecology

MARK R.T. DALE
University of Alberta, Edmonton, Canada

CAMBRIDGE UNIVERSITY PRESS
Cambridge, New York, Melbourne, Madrid, Cape Town, Singapore, São Paulo

Cambridge University Press
The Edinburgh Building, Cambridge CB2 8RU, UK

Published in the United States of America by Cambridge University Press, New York

www.cambridge.org
Information on this title: www.cambridge.org/9780521452274

First published 1999
First paperback edition 2000
Third printing 2006

A catalogue record for this publication is available from the British Library

Library of Congress Cataloguing in Publication data
Dale, Mark R. T. (Mark Randall Thomas), 1951–
Spatial pattern analysis in plant ecology / Mark R. T. Dale.
p. cm. – (Cambridge studies in ecology)
Includes bibliographical references (p.) and index.
ISBN 0 521 45227 9
1. Plant ecology – Statistical methods. 2. Spatial analysis
(Statistics) I. Title. II. Series.
QK901.D27 1998
581.7′07′27–dc21 98-15360 CIP

ISBN 978-0-521-45227-4 hardback
ISBN 978-0-521-79437-4 paperback

Transferred to digital printing 2007

Contents

Preface *page* ix

1 Concepts of spatial pattern 1
Introduction 1
Pattern and process 1
Causes of spatial pattern and its development 6
Concepts of spatial pattern 12
Concluding remarks 29

2 Sampling 31
Introduction 31
Sampling for pattern in a fixed frame of reference 32
Sampling for pattern relative to other plants 45
Location of sampling 48
Concluding remarks 49

3 Basic methods for one dimension and one species 50
Introduction 50
Data 50
Blocked quadrat variance 56
Local quadrat variances 58
Paired quadrat variance 71
New local variance 78
Combined analysis 83
Semivariogram and fractal dimension 90
Spectral analysis 91
Other methods 95
Concluding remarks 98

4 Spatial pattern of two species 100
Introduction 100
At most one species per point 101
Several species per point 103
Blocked quadrat covariance (BQC) 104
Paired quadrat covariance (PQC) and conditional probability 106
Two- and three-term local quadrat covariance (TTLQC and 3TLQC) 113
Comparison of methods 116
Extensions of covariance analysis 120
Other approaches 121
Relative pattern: species association 123
Concluding remarks 124

5 Multispecies pattern 125
Introduction 125
Multiscale ordination 128
Semivariogram and fractal dimension 136
Methods based on correspondence analysis 137
Euclidean distance 139
Comments 142
Spectral analysis 143
Other field results 143
Species associations 147
Concluding remarks 165

6 Two-dimensional analysis of spatial pattern 168
Introduction 168
Blocked quadrat variance 169
Spatial autocorrelation and paired quadrat variance 169
Two-dimensional spectral analysis 174
Two-dimensional local quadrat variances 177
Four-term local quadrat variance 178
Random paired quadrat frequency 186
Variogram 190
Covariation 192
Paired quadrat covariance (PQC) 193
Four-term local quadrat covariance 195
Plant–environment correlation 198
Cross-variogram 198
Landscape metrics 200

Other methods 202
Concluding remarks 204

7 Point patterns 206
Introduction 206
Univariate point patterns 207
Anisotropy 227
Bivariate point patterns 231
Multispecies point pattern and quantitative attributes 237
Concluding remarks 241

8 Pattern on an environmental gradient 242
Introduction 242
Continuous presence/absence data 248
Quadrats: presence/absence data 270
Density data 272
Concluding remarks 275

9 Conclusions and future directions 277
Summary of recommendations 277
What next? 279
Three dimensions 279
Relation to spatial structure of physical factors 286
Obvious extensions 288
Temporal aspects of spatial pattern analysis 288
Wavelets 290
Questions and hypotheses 293
Concluding remarks 296

Bibliography 297
Glossary of abbreviations 314
List of plant species 319
Index 325

Preface

This book is designed to help the reader understand the concepts and methods of spatial pattern analysis. The book is divided into three sections of three Chapters each. The first three Chapters lay the foundations of the material by discussing the basic concepts, considerations for the acquisition of data, and the basic methods for a single species in one dimension, concentrating on data from strings of contiguous quadrats. The middle third of the book describes extensions of the basic methods to the analysis of two species, of multiple species and of data used to investigate two-dimensional patterns. The last three Chapters describe different aspects of spatial pattern analysis: point pattern data, pattern on environmental gradients and future extensions of pattern analysis.

The book is written in the first person plural throughout, not as an affectation, but because the material presented here is not the work of just one person, but of a whole group of people who have contributed to the overall research program. That group includes students and associates whose names will be obvious from the citations: Dan MacIsaac, Dave Blundon, Elizabeth John, Maria Zbigniewicz, Rob Powell, Colin Young, and so on. Other students and researchers have allowed us to use their data for illustrative purposes and these include John Stadt and Michael Hunt Jones.

The book does not present the material with a thoroughly consistent notation. This was a deliberate decision, based on the reasoning that a book-wide notation would be forced to be elaborate and thus eventually clumsy. Therefore, it is possible that the variable 'x' can take on different meanings in different parts of the book. The meanings should be clear within their contexts.

There is a certain amount of redundancy in the material; some equations, for instance, appear more than once. Again, the choice was deliberate and based on convenience, to reduce the amount of flipping between pages and Chapters to find the required information.

This volume represents work that is very much 'in progress'. It describes material in a rapidly developing field of research. There is much contained here that really only came to light in the writing of the book itself, and there is obviously much more to be discovered. The emphasis in the description is methodological, in part because that aspect of the subject has the greatest need for exposition in order to encourage researchers to tackle the subject and, in part, because too few studies have been done of many spatial pattern phenomena to allow satisfactory generalizations to be made. Because spatial pattern analysis has close links with other areas of plant ecology, the book could easily have been expanded to include more plant community ecology, more on theories in vegetation science, more on multivariate analysis techniques, etc. The effort was made, however, to concentrate very much on the main topic, but without ignoring the important connections to other areas.

We gratefully acknowledge the technical assistance of Megan Lappi, Nadia Sas, Elaine Gordon, Jiangfen Zhang, Sara Suddaby, Dianne Wong and Marko Mah. P. A. Keddy, N. C. Kenkel, C. J. Krebs, R. Turkington, J. Birks, J. A. Wiens, and an anonymous reviewer provided helpful criticism and suggestions on earlier versions of the Chapters.

1 · *Concepts of spatial pattern*

Introduction

The natural world is a patchy place. The patchiness manifests itself in many ways and over a wide range of scales, from the arrangement of continents and oceans to the alternation of the solid grains of beach sand and the spaces between them. Plants in the natural world also are patchy at a great range of scales from the global distributions of biomes to the arrangements of trichomes and stomata on the surface of a leaf. When the patchiness has a certain amount of predictability so that it can be described quantitatively, we call it spatial pattern. Although the concept of pattern is often associated with nonrandomness, in some cases we will want to allow the possibility of random pattern, because true randomness does permit a certain amount of prediction. As an illustration of spatial pattern, Figure 1.1 presents an example from the literature, a map of the patches of *Calluna vulgaris* (heather) in a 10 m × 20 m plot in central Sweden (redrawn from Diggle 1981). A transect through the vegetation, such as the one illustrated in the lower part of the figure, reveals a fairly regular alternation of patches of high density and gaps between them.

Pattern and process

The impetus to study spatial pattern in plant communities comes from the view that in order to understand plant communities, we should describe and quantify their characteristics, both spatial and temporal, and then relate these observed characteristics to underlying processes such as establishment, growth, competition, reproduction, senescence, and mortality. A large proportion of the studies described in this book have been profoundly influenced by A. S. Watt and his famous paper 'Pattern and process in the plant community' (1947). The influence of Watt is the view of the community as a mosaic of phases at different stages in a

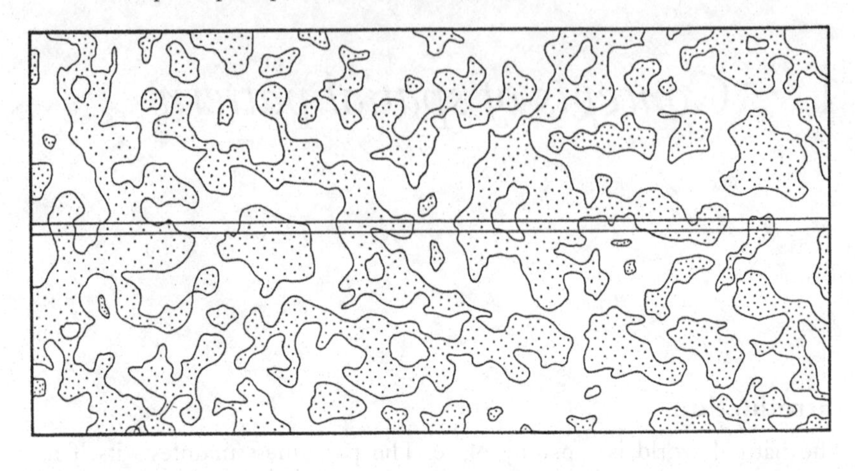

Figure 1.1 An example of spatial pattern: the upper part is a map of patches (shaded) of *Calluna vulgaris* (heather) in a 10 m × 20 m plot. The patches of high density are stippled. The lower part is the transect through the map as indicated; it reveals a more or less regular alternation of patches of high density and gaps between them (redrawn from Diggle 1981).

similar cycle of events, driven by the same processes. The spatial pattern of this mosaic can be used to generate hypotheses about the underlying processes or to suggest the mechanisms that have given rise to it. Whittaker and Levin (1977) expanded the mosaic concept by relating intracommunity patterns to microsite differences and successional mosaics to the responses following disturbance. In a world in which most vegetation systems have not been studied in any detail, the description and analysis of spatial relationships within them is a first step to understanding them.

A central point of discussion in plant ecology has, then, been the relationship between the processes that occur in vegetation such as growth, competition, or senescence, and the spatial pattern that is observed (Watt 1947; Lepš 1990a). A similar discussion has taken place in the broader discipline of ecology in which 'pattern' is interpreted not only spatially but in reference to all the observable characteristics of a system; however, the question is the same, i.e., to what extent can process be inferred from pattern? (Cale *et al.* 1989).

Although early studies of spatial pattern in plant communities were based on the belief that past process could be deduced from pattern, it is now generally agreed that it cannot, strictly speaking, be done (Shipley &

Keddy 1987; Lepš 1990a). Because spatial pattern is the result of past process, however, it can be used to test some hypotheses about process, even if it does not provide complete knowledge. For example, a change in the arrangement of individual plants over time that includes an increase in the distance between surviving individuals is not compatible with positive interactions among them (Lepš 1990a). In addition, the clear and objective description of spatial pattern is an important part of generating hypotheses about how controlling biological or environmental processes work (Ford & Renshaw 1984).

Spatial pattern is a crucial aspect of natural vegetation because it affects future processes, both of the plants themselves and of a range of other organisms with which they interact. The spatial scale at which pattern is seen to affect process goes from the neighborhood of an individual *Arabidopsis thaliana* plant, a few centimeters or less (Silander & Pacala 1985), to the scale of landscapes, where it may affect biodiversity and ecosystem functions (Turner 1989). Natural vegetation is sometimes viewed as a mosaic of patches of different kinds (cf. Burton & Bazzaz 1995) and the size and spacing of those patches are important characteristics of the vegetation.

In general, vegetation provides animals with their food, directly or indirectly, and also, to a large extent, the physical environment in which their activities take place. There is increasing awareness of the importance of evaluating and quantifying habitat complexity or structure in studies of how mobile organisms interact with their environment (McCoy & Bell 1991). Doak *et al.* (1992) summarize the findings of many researchers looking at the interaction of plant patches with animals, showing that patchiness, patch size, density, and isolation can affect herbivores, their predators, parasitoids, pollination, population density and so on in a variety of ways. For example, Wiens & Milne (1989) found that *Eleodes* beetles in a semi-arid grassland respond to the patch structure of their habitat in a nonrandom fashion, avoiding areas with a spatial structure of intermediate complexity. Usher *et al.* (1982) found that the distribution of plants in an Antarctic moss-turf community had important effects on spatial distribution in communities of soil arthropods. It is clear that, in many systems, the spatial pattern of vegetation is an important part of habitat structure.

Given an average vegetation density, animals of different sizes and mobilities will be affected differently depending on whether that density arises from small gaps alternating with small patches, or large gaps alternating with large patches. This kind of knowledge in one particular range of spatial scales is central to management decisions in forestry. Different organisms are helped or harmed by the differences between single tree cutting, the cutting of small patches, or large-scale clearcutting (cf. Kimmins 1992).

Spatial pattern also has an effect on plant–herbivore interactions. A study of the biennial herb *Pastinaca sativa* and its specialized herbivore *Depressaria pastinacella* found that plants in patches were more susceptible to attack than isolated plants of the same size (Thompson 1978). In the forests of northern Ontario, there are periodic outbreaks of tent caterpillar (*Malacosoma disstria*) which feed principally on trembling aspen (*Populus tremuloides*); fragmentation of the forested areas increases the duration of the caterpillar population highs (Roland 1993). Kareiva (1987) found that increased host plant patchiness (*Solidago canadensis*) caused less stable dynamics in populations of its herbivore (the aphid *Uroleucon nigrotuberculatum*) because of the search and aggregation behavior of the predator at the next trophic level (the ladybird *Coccinella septempunctata*). Kareiva (1985) studied the effects of host plant patch size on flea beetle populations and found that patch size affected processes such as emigration rate to the extent that there may be a critical patch-size below which herbivore populations cannot be maintained. He also found that the herbivore's discrimination between patch quality ('lush' *vs.* 'stunted') depended on the distance between patches (Kareiva 1982). Colonization of neighboring patches will often be influenced by the distance between the patches. Bach (1984, 1988a,b) also found that patch size affected herbivore population densities which responded nonlinearly with intermediate-sized patches having the highest density. It is not only patch size, but also patch density that has an effect (directly or indirectly) on herbivores (Reeve 1987; Cappuccino 1988). Other studies (e.g., Sih & Baltus 1987; Sowig 1989) have shown that patch size affects flower visits and pollination by different species of bee. The influence was sufficiently strong in catnip (*Nepeta cataria* L.) that it affected seed set, which was lower in smaller patches.

The general conclusion from these studies is that patch size, patch spacing, and patch density, all of which are elements of the plants' spatial pattern, have important influences on their herbivores (and the herbivores' predators) and pollinators. It is probably equally true that these characteristics of patchiness affect the plants and their interactions also, although fewer studies have been done with that focus. In her study of squash plants and their herbivores, Bach (1988a,b) found that patch size did affect both the growth and the longevity of the plants themselves.

Because the plants of one species can have a positive or negative effect on the occurrence and spatial arrangement of another species, one important effect of spatial pattern is its affect on other plants. It is well known that gaps in a forest canopy are very important for the establishment of new individuals or the release of suppressed saplings (Platt & Strong 1989; Leemans 1990; among many). The spatial pattern in one group of plants may affect the pattern of another group; for instance, Shmida & Whittaker (1981) found that the spatial arrangement of shrubs in California shrub communities had a strong effect on the herb species, with some species being found primarily under the shrubs' canopies and others found mainly in the openings between. Maubon *et al.* (1995) describe a dynamic interacting mosaic of bilberry (*Vaccinium myrtillus*) and spruce (*Picea abies*) in the Alps, in which the established bilberry makes soil conditions unfavorable for spruce recruitment and the spruce trees make conditions less favorable for the bilberry by shading.

In summary, the spatial pattern of plants has important effects on the interactions between plants, between plants and other organisms such as herbivores, and between other organisms such as herbivores and their predators. The impact of the spatial pattern of the plants may be felt directly, as in the provision of biomass, or indirectly through its modification of microclimates. We should probably expand our list of organisms affected to include mycorrhizae and other fungi, decomposers and detritivores, and a variety of microorganisms, but little research has been done on how these groups are affected by the spatial pattern of plants.

In some kinds of vegetation, the spatial pattern is very obvious. In arctic and alpine regions, 'patterned ground' of geometric shapes of sorted stones is a common phenomenon resulting from frost action and it has clear effects on the spatial pattern of the vegetation (Washburn 1980). Areas that are no longer under climatic conditions that form these patterns may have 'fossil' patterned ground which continues to affect vegetation (Embleton & King 1975). The action of freezing and thawing may also contribute to the development of hummocks, of step features on sloping ground, solifluction lobes and so on (Washburn 1980), all of which may affect spatial pattern of plants. In boreal regions, a common feature at a somewhat larger scale is the patterned fen or string bog in which strings of slightly higher elevation alternate with pools or flarks (Glaser *et al.* 1981).

In other cases, the spatial pattern may be more subtle and detectable only by analysis; for example, in areas of *Agrostis/Festuca* sward chosen for their visual homogeneity, it was found that several of the important species had marked spatial pattern at the same scale (Kershaw 1958, 1959a,b). In a study of the banner-tailed kangaroo rat (*Dipodomys spectabilis*), Amarasekare (1994) found that its habitat could not be considered as consisting of discrete patches, some occupied and some not, but that the differences between occupied areas and the surrounding unoccupied habitat were quantitative and could be detected statistically. Even tended lawns, which may look uniform, have spatial pattern in the form of fine-scale community structure (Watkins & Wilson 1992).

Causes of spatial pattern and its development

It will become clear from the examples described in this book that the arrangement of plants in natural vegetation is usually not random and in fact there are usually several scales of spatial pattern present. This fact alone suggests that there is a range of factors that cause spatial pattern, and these can be classified into three broad categories: (1) morphological factors, based on the size and growth pattern of the plants; (2) environmental factors that are themselves spatially heterogeneous; and (3) phytosociological factors that permit the spatial arrangement of one species to affect the occurrence of plants of another species through their interaction (cf. Kershaw 1964, Chapter 7).

Some of the classic examples of spatial pattern determined by morphological factors, as described in Kershaw (1964), are from clonally growing plants, such as *Eriophorum angustifolium* and *Trifolium repens*, in which the first three scales of pattern are related to first- and second-order branching and to the entire stolon or rhizome system. In a study of pattern development on proglacial deposits in the Canadian Rockies, we found that the smallest scale of pattern was related to the sizes of the clonally growing patches of *Dryas drummondii* (Dale & MacIsaac 1989). Mahdi & Law (1987) concluded that the spatial organization of a limestone grassland community was probably the result of the pattern of clonal growth of the individual species. Kershaw (1964) provides other examples, but it must be remembered that while morphology may determine the size of a patch for one particular scale of pattern, the scale is also affected by the sizes of the gaps between them, which may be determined by other factors.

A large number of studies have found a relationship between the spatial pattern of plants and spatial heterogeneity in an (abiotic) environmental factor. Such factors include soil depth (Kershaw 1959a,b), topography (Greig-Smith 1961a), soil nutrients (Galiano 1985), positions of subsurface rocks (Usher 1983), and so on. Maslov (1989) concluded from a study of forest plants in Russia that environmental heterogeneity was the major factor determining pattern for vascular plants; interestingly, however, that did not appear to be the case for bryophytes.

We have already mentioned that a common feature of arctic and alpine landscapes is what is called 'patterned ground'. Washburn (1980) provides an interesting and thorough discussion of this phenomenon, as well as some excellent pictures. Patterned ground actually takes a variety of forms, including circles, polygons and stripes and these can be classified further as sorted or nonsorted depending on whether there is a trend in particle size across the feature or whether particle size is more or less uniform. Because they result from frost action, the pattern elements can affect where plants grow. For instance, in a study of the development of sorted polygons in Norway, Ballantyne & Matthews (1983) found that plants colonized only the margins of the polygons first, where the substrate was more stable. Heilbronn & Walton (1984) studied striped ground on the island of South Georgia and found that colonization by grass plants was more successful on the unsorted parts of the pattern. They also suggest that the presence of the plants can contribute to the persistence of step features on sloping patterned ground.

Polygonal features can develop also on soils and mud as a result of desiccation (Termier & Termier 1963). For instance, Harris (1990) describes polygons on the saline soil of the Slims River delta at Kluane in the Yukon and illustrates the fact that the vegetation tends to grow along the margins of the polygon cracks. Termier & Termier (1963) suggest that the polygonal markings on some sandstones are the result of similar processes.

It is clear from many studies that the variability of environmental factors will have a direct effect on the growth and spatial pattern of plants. Sources of underlying spatial topographical heterogeneity that may be reflected in spatial pattern in vegetation include features such as pillow lava, the developing cracks and grikes in a limestone pavement; eskers, moraines, and striations resulting from past glaciation; drainage channels, gullies, meanders and braided streams; ancient dunes, beach fronts and reef ridges. The list is too long to permit a complete listing of examples and so we will mention just one from the literature: Whittaker & Levin (1977) describe the climax pattern on coastal ridges in California which have redwood (*Sequoia sempervirens*) forests on the terrace slopes, pigmy cyprus (*Cupressus pygmaea*) in the centers of the terraces and bishop pine

(*Pinus muricata*) and rhododendron (*Rhododendron macrophyllum*) on the old beach deposits on the terrace crests (their Figure 5). The spatial pattern observed in the vegetation is the result of the interaction of the topography, the processes of soil formation and the vegetation itself.

Another category of environmental factor that will cause spatial pattern in vegetation is disturbance. Crawley (1986) comments that a great many of the spatial patterns observed in plant communities reflect recovery from disturbances that occurred at different times in the past. At the landscape level, potentially widespread disturbances such as fire can have an obvious effect on spatial organization (Turner & Bratton 1987). Fire can also have a much more local effect in maintaining the spacing of savanna trees or in segregating tree cohorts of different ages (Cooper 1961). At a smaller scale, the gaps left by the falling of individual trees can have a profound effect on the growth and regeneration of the vegetation, causing spatial pattern (Kanzaki 1984; Veblen 1992 and references therein).

The importance of disturbance and regeneration in vegetation has been generalized into the 'mosaic-cycle' concept of ecosystems (Remmert 1991). In this view, vegetation is a mosaic of patches, with different patches being at different stages of a temporal cycle of aging, decay or destruction and rejuvenation. There is an obvious parallel with Watt's (1947) description of building, mature and degenerate phases of cyclic succession, but the difference is that Remmert (1991) suggests that the mosaic cycle model is valid for most ecosystems, if not all.

As a particular example of a kind of cyclic process, Sprugel (1976) describes the phenomenon of wave regeneration in high-altitude fir forests in the Northeastern U.S.A. Each wave consists of a strip of old dying trees under which there is vigorous regeneration with a progression of trees of increasing age and size until the next region of mature and dying trees is reached. The waves are on the order of a hundred meters across and move in the same direction as the prevailing wind. As mature upwind trees die, the trees immediately leeward are exposed more directly to the effects of the wind which increases mortality. As the canopy thins and opens, recruitment can then take place.

Animals also are agents of disturbance in a variety of ways, including trampling and browsing. Even more obvious effects on patchiness can be produced by digging animals such as moles, or from the burrows of herbivores such as rabbits, gophers, or ground squirrels (Peart 1989). Similar patchiness may arise from the effects of termite mounds (Mordelet *et al.* 1996), or localized dung or urine deposition. Umbanhowar (1992) examined four patch types in northern mixed prairie (ant nests, mammal earth mounds, bison wallows and dry prairie potholes), and found that the different patch types supported different groups of plant species. In a similar system, Steinauer & Collins (1995) found that the small-scale patch structure was significantly affected by urine deposition, which increased or decreased species diversity within the patch.

The interactions of plants may also give rise to spatial pattern in natural communities. For example, Kenkel (1988a) attributes the local highly regular dispersion of trees in an even-aged pure stand of jack pine (*Pinus banksiana*) to competition for soil resources and light. In populations of knapweed, *Centaurea diffusa*, which is monocarpic, Powell (1990) found that spatial pattern is created by three processes: recruitment, rosette mortality (which increases dispersion), and post-reproductive mortality (which decreases dispersion). Intraspecific competition may have a secondary effect on other species: in studying the spatial pattern in a mire, Kenkel (1988b) found that the hummock-hollow complex arises from the accumulation of *Sphagnum* species about the branches of the shrub *Chamaedaphne calyculata* which creates the hummocks, and therefore the spacing of the hummocks reflects past intraspecific competition in *Chamaedaphne*.

Interspecific competition may also be a force in determining spatial pattern; for instance, the exclusion of *Sphagnum fuscum* to dryer hummock sites by other *Sphagnum* species (Rydin 1986; Gignac & Vitt 1990). In addition to negative effects, plants can drive spatial pattern by positive interaction, such as the provision of more favorable sites for recruitment, a phenomenon referred to as nucleation when it occurs during primary succession (Yarranton & Morrison 1974; Day & Wright 1989; Blundon *et al.* 1993). For instance, in primary succession in the Canadian Rockies, we found that at one site, *Hedysarum mackenzii* acts as a center for further colonization whereas at a second site, 200 km away, it is *Dryas drummondii* that is a center for nucleation (Blundon *et al.* 1993). It is no coincidence that both species have the ability to fix nitrogen, a limiting resource under those conditions, and the input of nitrogen may be an important factor in the nucleation we observed.

The way in which pattern develops depends very much on the factors that are creating the pattern. It is easy to imagine spatial pattern becoming more pronounced with time as small differences in substrate structure or chemistry are expressed by increasing differences in the plants that grow on it, or as the levels of soil nutrients themselves change in response to successional development (cf. Symonides & Wierzchowska 1990). A more extreme case is the development of strong spatial pattern on a substrate that was originally relatively homogeneous, such as the development of strings and flarks (pools) in a patterned wetland, driven by the interaction between the biological properties of the plants and the physical properties of the peat they create and the flow of water (Glaser *et al.* 1981; Swanson & Grigal 1988). In that particular instance, the pattern that is produced is strongly anisotropic with the lengths of the strings running across the direction of water flow.

Interestingly, arid regions can have somewhat similar landscape features with bands of vegetation alternating with stripes of bare ground. This phenomenon is known from Australia, Mexico, and several regions of Africa, in some parts of which it has the picturesque name of *brousse tigrée* (Figure 1.2). It occurs on gently sloping sites where the sheet runoff of water is slowed by the upslope edge of the vegetation stripe where the resulting better moisture regime facilitates plant establishment. The advantage of the upslope edge is mirrored by the disadvantage of water shortage and drought at the downslope edge and the stripes migrate up the slope (White 1971; Montaña 1992; Thiéry *et al.* 1995). It seems logical to assume that the spacing between the stripes is determined by the balance between the amount of precipitation received and the amount of moisture needed for successful regeneration. The parallel between this system of vegetation stripes and the stripes of wave regenerating fir forests (mentioned above) is striking, with abiotic stress being an important factor at the trailing edge of the stripe in both systems.

In many cases, such as those just described, the development or intensification of spatial pattern in plant communities is the result of what Wilson & Agnew (1992) describe as 'positive-feedback switches' in vegetation. These are mechanisms by which small differences between patches are magnified by the interaction of the plants with particular environmental factors. The list of environmental factors that can be involved is long and includes water, nutrients, light, fire, allelopathy, and herbivores. The switches can act temporally to accelerate or delay change and they can act spatially to produce sharp vegetation boundaries or stable mosaics of distinct patches in a previously more uniform environment (Wilson & Agnew 1992). Since these mosaics can be at a range of scales, from the individual plant to the landscape, these switches can play an important role in the development of spatial pattern.

It is also easy to imagine a situation in which initial differences due to substrate heterogeneity are blurred and eventually erased as the biotic factors of the vegetation itself come to dominate the system. Sterling *et al.*

Figure 1.2 Aerial view of brousse tigrée in an arid landscape in Niger (drawn from part of Figure 1 in Thiéry *et al.* 1995). The vegetation is dark and bare areas between are light. The area shown is about 1 km × 1.3 km.

(1984) studied pastures of different ages since plowing. After 7 years, there was a strong relationship between the vegetation and the microtopography, with some species tending to be found in small drainage channels and others on the higher and drier parts. On older sites (25–30 years), however, that relationship had vanished and the plants did not seem to respond to small differences in microtopography.

Some ecologists have suggested that spatial pattern follows a predictable development sequence during succession. Kershaw (1959a,b) and Greig-Smith (1961a) proposed that, in the initial stages of succession, pattern should merely become more intense at the same scales, as the density in patches increases. With the plants continuing to grow and colonization proceeding, some scales of pattern should be lost as the patches coalesce (Figure 1.3). New larger scales of pattern may develop as patches coalesce, eliminating small gaps, or as patches die, making larger gaps. This view of spatial pattern seems to suggest that in climax vegetation, any pattern that persists is irregular and low in intensity. In our studies of spatial pattern development during primary succession, however, we did not find that the pattern became more irregular (Dale & MacIsaac 1989; Dale & Blundon 1990).

Concepts of spatial pattern

In the first parts of this chapter, we have discussed the importance of spatial pattern and its relationship to population and community processes. Because the techniques and research described in this book are based on a set of related concepts, we will now describe and discuss those concepts in greater detail.

Spatial pattern

Spatial pattern is the arrangement of points, of plants or other organisms, or of patches of organisms in space which exhibits a certain amount of predictability. In many instances, this predictability will take the form of periodicity of some kind, such as groves of trees alternating with open grasslands across a landscape. We might want to insist that spatial pattern is nonrandomness in spatial arrangement, which then permits prediction, but some authors allow the possibility of random pattern (Ludwig & Reynolds 1988). Of course, true randomness does allow a certain amount of predictability, even if it is probabilistic. For example, if points are inde-

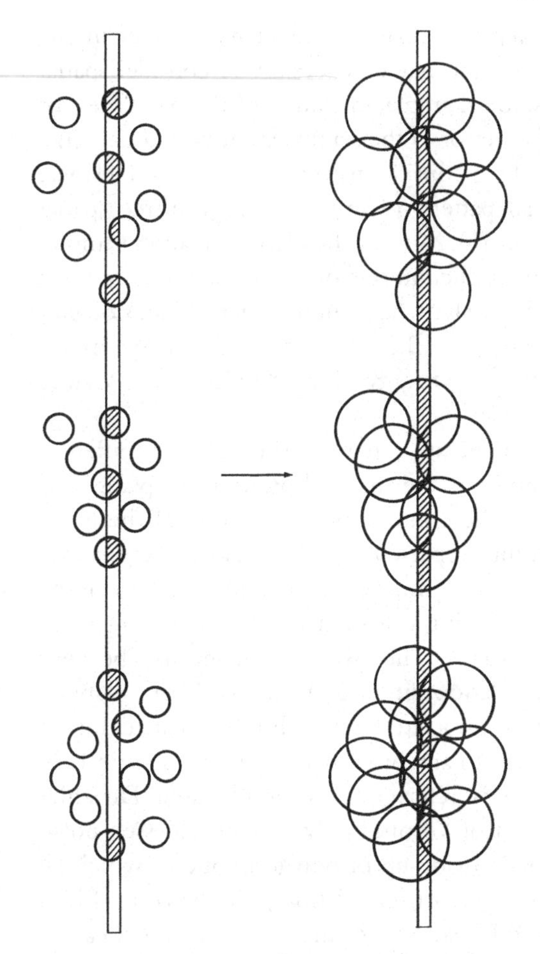

Figure 1.3 On the left, small patches occur in clusters and the transect shown would detect two scales of pattern. On the right, the patches have grown so that the only gaps are those between the original clusters. The transect will only detect the larger scale of pattern. As patches coalesce, the smaller scale of pattern is lost. As in Figure 1.1, the shaded area shows where the transects intersect the patches.

pendently and randomly placed in the plane, we can predict that the number of points in a set of samples of fixed area will follow, approximately, a Poisson frequency distribution.

For most of the following discussion of spatial pattern, it will be assumed that pattern exists mainly or essentially in two dimensions so that the region under study can be treated as a plane surface. In many instances, however, the two-dimensional pattern may be studied in only one dimension at a time. In other instances, it may be necessary to consider spatial pattern in three dimensions; for example, in studies of the patchiness of phytoplankton in a body of water or of the arrangement of branches in a forest canopy. It is even possible to consider pattern in higher dimensions, such as the four-dimensional pattern of leaf phenology (three spatial dimensions and time as the fourth). Most of the discussion and examples here will be from phenomena studied in one or two dimensions, but it is possible that the objects themselves have noninteger dimensionality which may be better described by the concepts of fractal geometry described below (Palmer 1988; Sugihara & May 1990; Kenkel & Walker 1993).

The spatial arrangement of plants can be treated in two different ways. The first treatment considers the mapped positions of plants and deals with only two elements, the continuous background of the plane itself and dimensionless points representing the plants. The second approach is to treat the plane as a mosaic of discrete nonoverlapping continuous domains, each of which is classified as belonging to a particular type or phase (Matérn 1979). The two treatments may be very closely related. One simple relationship is to begin with the mapped positions of points and then to associate with each point that region of the plane that is closer to it than to any other mapped point (Figure 1.4). The result of this procedure is referred to variously as a Dirichlet tessellation of the plane, the set of Voronoi polygons, or Thiessen polygons. (A tessellation is a mosaic made up of polygons; one in which all the polygons are triangles is called a triangulation.) Okabe *et al.* (1992) provide an extensive and useful treatment of the theory and application of Voronoi tessellations.

The simplest form of spatial pattern would be the alternation of regions in which the density of a particular species was high (patches) with regions of low density (gaps). In point patterns, it may not be easy to delimit the regions of high or low density. In the mosaic treatment of pattern, a simple patch–gap pattern can be thought of as a two-phase mosaic, in which patches alternate with gaps. It is not necessary for a region that is recognized as a patch to be internally homogeneous; in fact, in real situations, we do not expect it to be, but recognize the possibility of a hierarchical mosaic of patches within patches over a range of scales (cf. Kotliar & Wiens 1990). For instance, many textbooks (e.g. Ricklefs 1990; Silvertown & Lovett Doust 1993) include the familiar

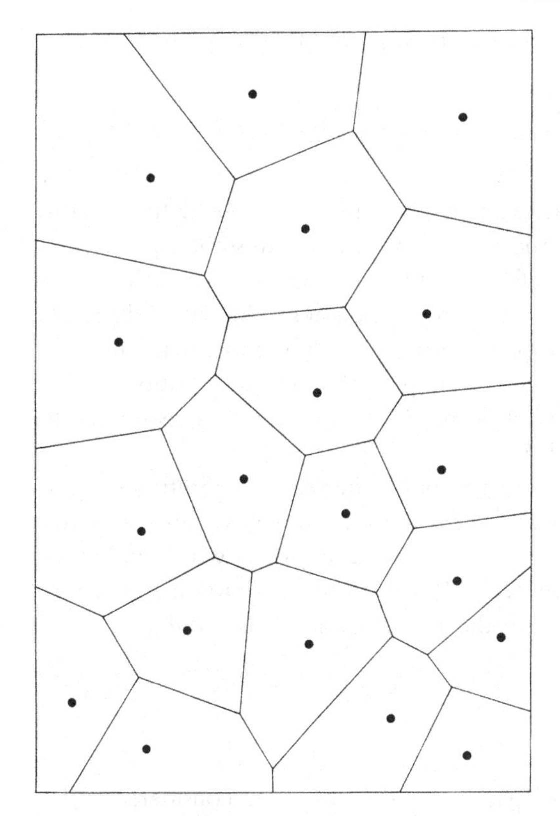

Figure 1.4 One relationship between a point pattern and a tessellation of the plane: Dirichlet domains associated with points. A point's domain is all parts of the plane closer to it than to any other point.

figure of the spatial distribution of *Clematis fremontii* in Missouri, which shows its geographical range, its distribution within a region, a cluster of glades within a region, patches within a glade, and individual plants within a patch.

If more than one species or kind of plant is being considered, it is obviously possible to have a mosaic for three or more phases. For example, if the joint pattern of two species was being investigated, regions might be classified into four categories: both species at high density, species 1 high and species 2 low, species 1 low and species 2 high, and both species at low density. When many species are being considered simultaneously, this simplistic approach may not be appropriate, as the number of phases would rise rapidly with the number of species, and a different approach to summarizing multispecies density will be necessary (Chapter 5).

Scale

The scale of spatial pattern in a two-phase mosaic can be defined as the average distance between the centers of adjacent dissimilar phases. An equivalent definition is to refer to half the average distance between the centers of similar phases that are separated by a single domain of the alternate phase. It is possible for spatial pattern to exhibit more than one scale even in a two-phase mosaic, for example, when the distribution of distances between the centers of domains of the same phase is obviously bimodal, as part of Figure 1.3 illustrates.

For a mosaic of more than two phases, the second definition of scale needs to be modified slightly to refer to half the average distance between the centers of domains of the same phase between which no other domains of the same phase occur. Based on this definition, it is clearly possible for different phases in the same mosaic to have different scales (Figure 1.5).

Intensity

Another property of spatial pattern that needs to be considered is the pattern's intensity, which in simple two-phase pattern is the degree of contrast between the dense and sparse areas. Dale & MacIsaac (1989) define intensity as the difference in density between the gap phase and the patch phase of such a pattern; if the gap phase has zero density and the patches and gaps are the same size, the intensity will be the average density in the patches. When the patch size is not equal to the gap size, the intensity is the patch density that would give the observed variance in pattern of the observed scale in which patch and gap size were equal. This concept is illustrated in Figure 1.6. Other authors have a different concept of intensity, defining it as a property that would remain constant under random thinning (cf. Hill 1973; Pielou 1977a, p.182); that is, if half the plants in each patch were removed at random, the resulting patch-gap pattern would have the same intensity (Figure 1.7). Another way of saying the same thing is to say that in those authors' view, rare species can have patterns as intense as those of common species. For the approach to the study and analysis of spatial pattern that is presented here, however, it is most straightforward to define intensity by reference to density differences.

Figure 1.5 Different phases in the mosaic have different scales of pattern. The two darker phases are less common and have larger scales of pattern (imagine this part of the mosaic repeated in all directions).

The concept of pattern intensity, as we have just defined it, will have to be modified for multiphase or multispecies pattern, but we will use the definition based on density difference as a basis (see Chapters 4 and 5).

Types of single-species pattern

In discussing single-species pattern, it is convenient to use a variety of terms to describe its characteristics, particularly for artificial examples. If the patches are of a constant size and the gaps are of a constant size, then the pattern is referred to as regular; if the patches and/or gaps vary in size, then the pattern is irregular. If the average patch size is equal to the

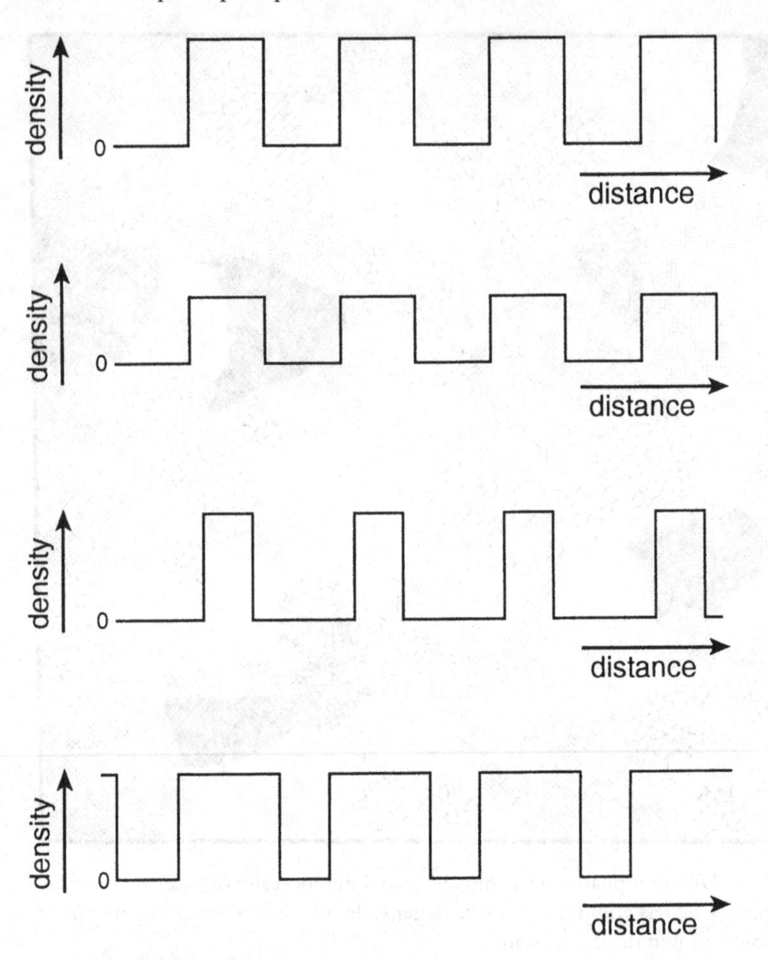

Figure 1.6 Intensity. Here are four graphs of density, x, as a function of distance along a transect, t. The top pattern has the greatest intensity of the four which all have the same scale. In the second graph, intensity is less because the density in the occupied quadrats is less. In the third, intensity is less because the patches are smaller than the gaps and similarly in the fourth, because the gaps are smaller.

average gap size, then the pattern is balanced; otherwise, it is unbalanced. If the scale of the pattern is more or less constant along the length of the transect, the pattern can be said to be stationary, with the alternative being pattern with a trend in scale. In real data, pattern may be more or less stationary over short distances, but display trends when greater distances are considered (Matérn 1986). It is also possible for sections of

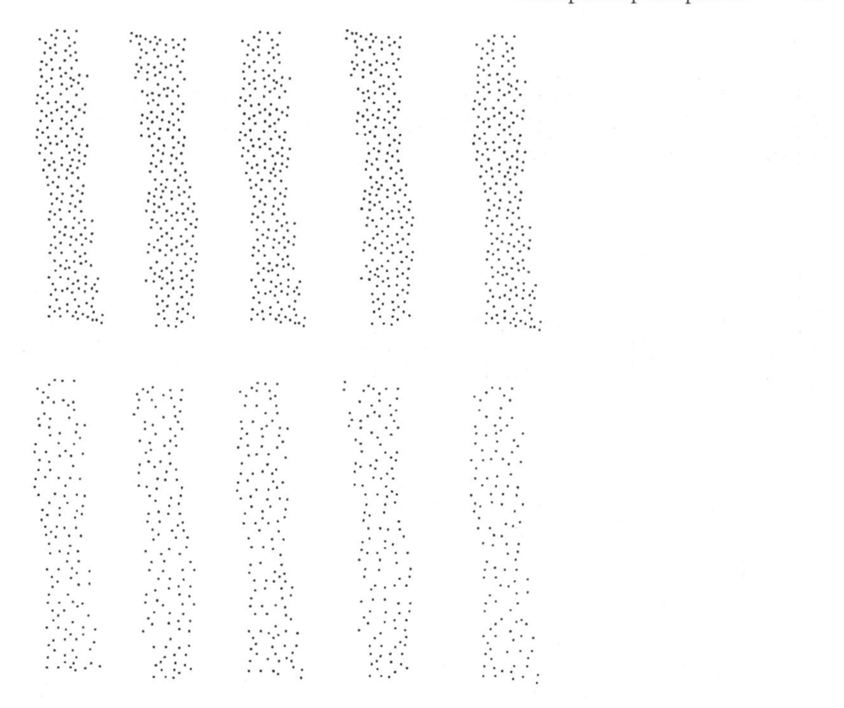

Figure 1.7 Two maps representing stems of plants that grow in stripes. The lower part of the Figure is derived from the upper part by thinning. In our definition of intensity, thinning reduces the intensity of the pattern.

pattern to alternate with larger scale gaps or patches, resulting in an interrupted pattern which essentially has two scales. These descriptions are illustrated in Figure 1.8.

Dispersion

Dispersion is a concept closely related to that of spatial pattern, and refers specifically to the arrangement of points in a plane. Pielou (1977a) notes an important distinction: 'dispersal' is the process such as the movement of individual organisms, whereas 'dispersion' is the spatial arrangement that results.

The null model of dispersion assumes that the points occur independently of each other, so that all regions of the same size have the same probability of containing a given number of points (Figure 1.9a). This kind of dispersion is usually referred to as a random pattern, or because

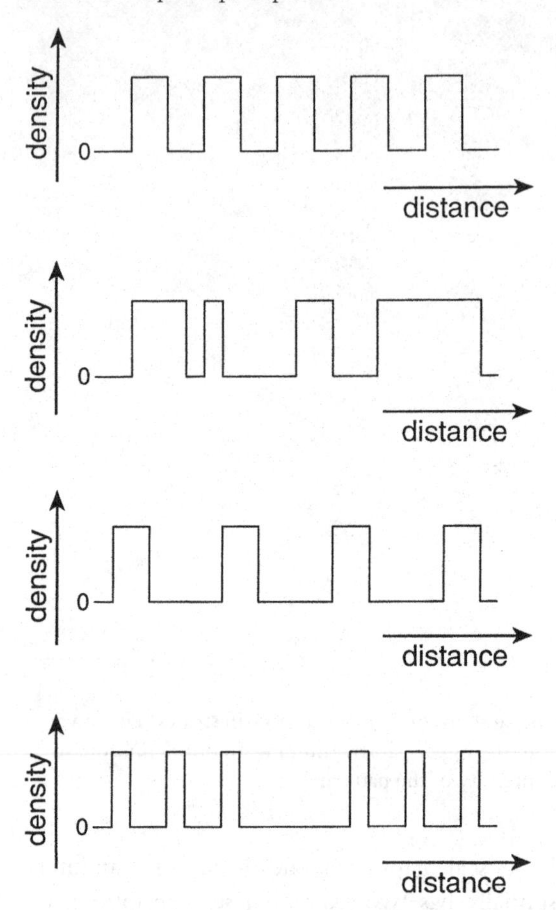

Figure 1.8 Descriptors of pattern. These graphs show density as a function of distance along a transect. The top one shows regular balanced pattern in which the patch and gap sizes are constant. The second is an irregular pattern in which patch and gap sizes are variable but may have the same average. The third is an unbalanced pattern in which the patch and gap sizes are consistently unequal. The fourth is an interrupted pattern in which several patch-gap alternations are separated by long gaps; such an arrangement gives rise to two scales of pattern.

the number of points in a given area follows the Poisson distribution, as a 'Poisson forest' (Keuls *et al.* 1963; cf. Upton & Fingleton 1985). This dispersion is also referred to as complete spatial randomness (CSR, Diggle 1983).

There are two main alternatives to the null model. The first includes the cases in which the points are clumped or underdispersed, such that the presence of one point increases the probability of finding another in its vicinity (Figure 1.9b). This dispersion pattern is also referred to as contagious and sometimes as 'aggregated' but it is better if terms such as aggregated and segregated are reserved for the description of the relationship between plants of two different kinds (see below). The second alternative includes those cases in which the points are overdispersed, such that a point's presence reduces the probability of finding another nearby (Figure 1.9c). Some texts refer to the overdispersed pattern as 'regular', but that term has connotations of the points being arranged in a geometric lattice of some kind, and should be reserved for that situation.

All three patterns of dispersion can be observed in real examples, and a range of causal mechanisms can be invoked. Because of the variety of interactions between organisms, we do not expect their positions to be truly independent of each other, but it is possible that their dispersion can appear to be indistinguishable from the random dispersion (Skellam 1952; Grieg-Smith 1979). Clumped patterns can result from environmental heterogeneity so that organisms of the same species are found close together in areas of favorable conditions. Many biological processes such as the vegetative production of ramets will lead to clumped patterns

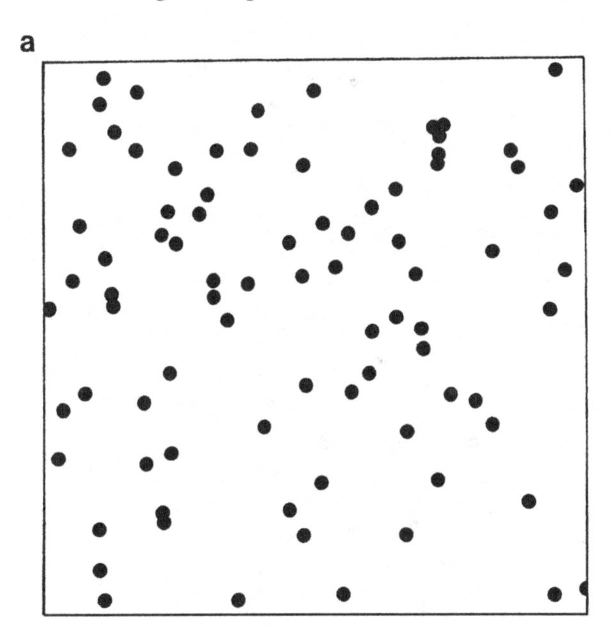

Figure 1.9 Dispersion. *a* Random, in which the points occur in the plane independently of each other. *b* Clumped, the presence of one point increases the probability of finding another nearby. *c* Overdispersed, the presence of one point decreases the probability of finding another nearby.

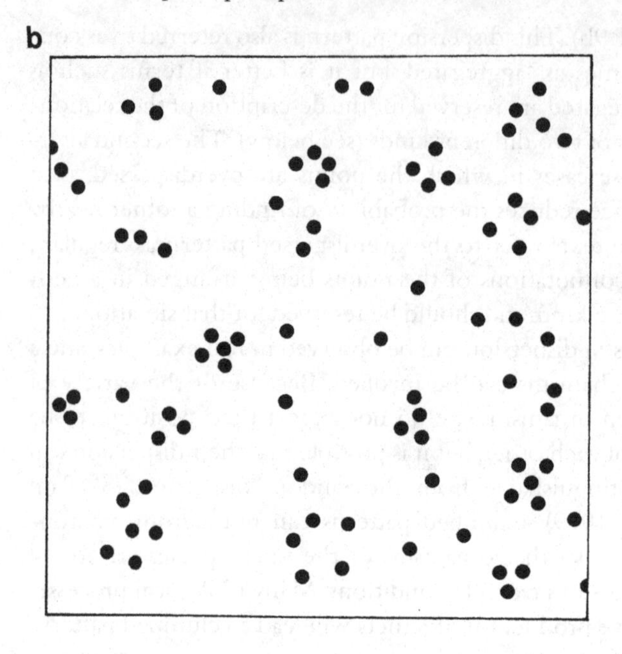

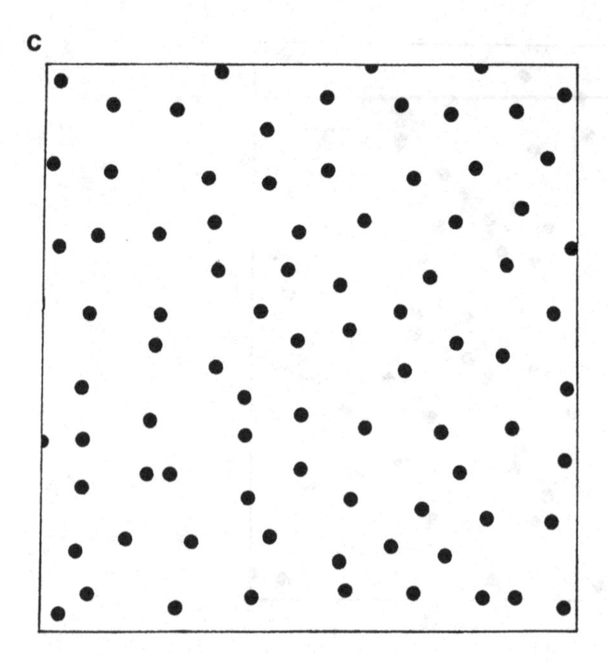

Figure 1.9 *cont.*

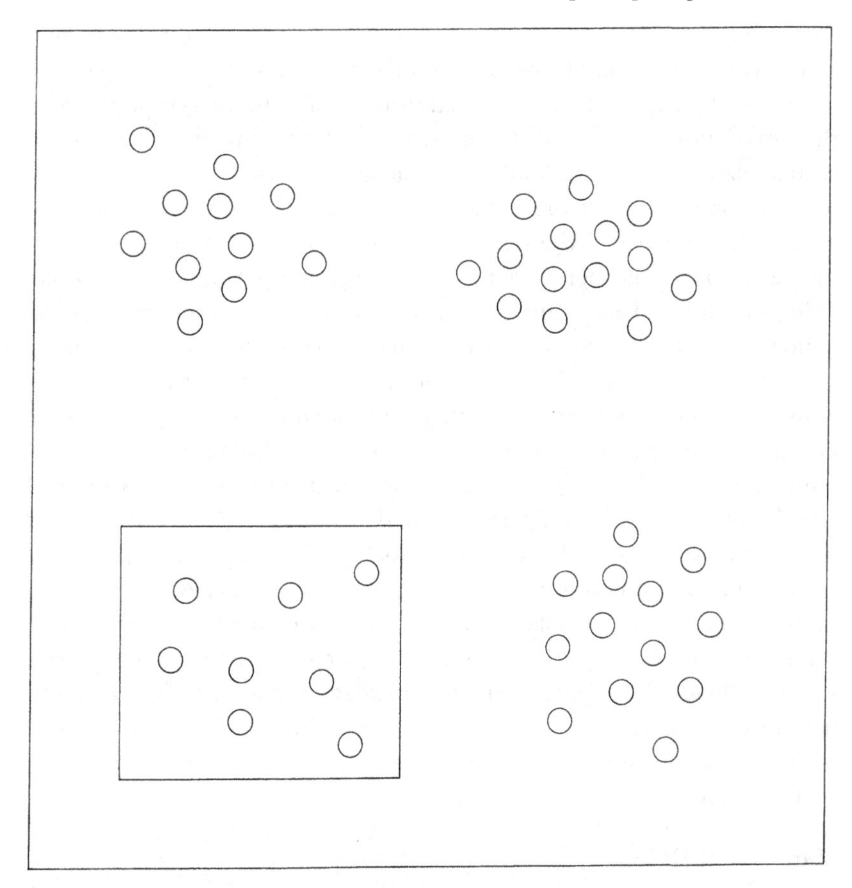

Figure 1.10 The interaction of dispersion pattern and scale. The circles in the small square seem overdispersed, but in the large square they appear clumped.

and some inhibition processes such as competition may lead to the over-dispersion of positions.

It is important to realize that the perception of the dispersion of plants depends on scale; if only a single grove of trees is considered, they may appear overdispersed (Figure 1.10, inner square), but when several are considered, the plants are seen as being clumped (Figure 1.10, outer square).

Spatial autocorrelation

When a plant population or community is sampled, the samples have a spatial relationship with each other. To a certain extent, samples that are close to each other are more likely to be similar. For example, if vegetation is sampled using a transect of small contiguous quadrats, adjacent quadrats are likely to be more similar than those at greater spacing. This lack of independence is referred to as spatial autocorrelation because the correlation occurs within the data set itself and arises because of spatial relationships. One result of spatial autocorrelation is that statistical tests performed give more apparently significant results than the data actually justify because the number of truly independent observations is smaller than the number used in the test (Legendre 1993; Thomson *et al.* 1996). It must be kept in mind, however, that it is the same lack of independence that provides the spatial predictability that is the essential characteristic of pattern. In spatial pattern, while the similarity of samples at first decreases with distance, with greater distance similarity rises again because the elements of the pattern are repeated. Under these circumstances, we can think of the scale of spatial pattern as the distance at which similarity ceases to decrease with distance.

An interesting example of spatial autocorrelation in plant populations comes from a competition experiment using the annual *Kochia scoparia*. Franco & Harper (1988) planted the seedling along density gradients and found that the plants on the outside of the arrays were the largest. These largest individuals suppressed their immediate neighbors, allowing their second-order neighbors to be larger than otherwise. Thus, the data's autocorrelation was positive for plants at even separations and negative at odd separations.

Markov models

One approach to the spatial dependence in vegetation data is to consider transects or strings of contiguous quadrats in which the presence or absence of a species is noted. Such data can be treated as a sequences of 1's (presence) and 0's (absence), and may be represented by a two-state (0 and 1) stochastic model. If there is no spatial dependence, each element of the sequence is independent of those next to it, but spatial dependence may make each element depend on the state of the m elements preceding it. If the state of the element depends only on the m elements preceding it and not on the entire history of the process, it is a Markov process of order m. For example, in a Markov process of order 1, each element depends only on the one immediately preceding, but is independent of all others. In Chapter 5, we shall look at the order of Markov models appropriate for the description of multispecies pattern.

Markov models of a different sort have been used in the study of the spatial pattern of plants treated as points in the plane. The underlying idea is that if the plants are observed to be overdispersed, a simple explanation is that each plant inhibits the establishment or success of other plants in its immediate vicinity. The most direct way of modelling the situation would be to have 'hard' inhibition such that no plant can occur within some specified distance, δ, of another plant (see Diggle 1983, section 4.8). A more flexible approach is to have the probability of a plant existing at a given point in the plane decline with the number of plants within radius δ of the point (Ripley & Kelly 1977). This approach can be treated with a two-dimensional Markov model. Kenkel (1993) found that the spatial pattern of the clonal herb *Aralia nudicaulis* fit such a Markov point-process model very well. Point pattern models will be examined in greater detail in Chapter 7.

This kind of two-dimensional Markov modelling is, in concept, similar to the use of cellular automata to examine spatial processes. An example of cellular automata would be a grid of cells, each of which can exist at one of a number of discrete states. The state of a cell is defined by rules that depend on the states of neighboring cells and, from starting configuration, the rules are applied iteratively. What is of interest, in the context of spatial pattern, is that starting from random configurations, the governing rules can cause pattern to emerge from the randomness (see Green 1990). The difference between cellular automata and Markov models of grids is that the cellular automata rules are deterministic rather than stochastic. Ratz (1995) used a probabilistic approach that was nevertheless based on the cellular automata approach to model the long-term spatial patterns created by fire in a fire-dominated system such as the boreal forest.

Association

The term 'association' will not be used here to refer to a vegetation unit or commonly occurring grouping of species, but rather to describe the tendency of the plants of different species to be found in close proximity more often than expected (positive association) or less often than expected (negative association). Associations, positive or negative, can also be classified according to their cause: ecological coincidence refers to cases in which the plants of different species grow close together or far apart because of similar or divergent ecological requirements or capabilities. For instance, the lichens *Rhizocarpon eupetraeoides* and *Umbilicaria vellea* are found together on alpine rock surfaces because they are both tolerant of desiccation and temperature fluctuations but tend to inhabit steeply sloping surfaces perhaps because they are intolerant of snow cover; they exhibit positive association at the scale of a boulder face (John 1989). On the other hand, in the same community, *Rhizocarpon superficiale* and *R. bolanderi* are negatively associated at the scale of a rock face because of different microhabitat correlations (John & Dale 1989), with *R. superficiale* tending to be found on the top edges of boulders because of its intolerance of high temperatures (Coxson & Kershaw 1983.) At the scale of the landscape, however, the two species are positively associated because they both tend to be found on rockslides and other exposed rock surfaces, not in forests or alpine meadows. *Typha latifolia* and *Typha domingensis* are positively associated at the scale of a lakeshore, but *T. latifolia* is excluded to deeper water by its congener's competition, so that the species are negatively associated at the scale of neighboring plants (Grace 1987).

At the plant neighborhood scale, plants of early successional sites will be positively associated with each other because they are good dispersers and shade intolerant, but they will be negatively associated with late successional species, which are usually shade tolerant in the regeneration phase and perhaps better competitors. At the landscape scale, the association between these two groups of species would depend on the size of the disturbed areas in which the early successional species are found. Ecological coincidence, whether positive or negative, is expected to bring about a symmetric association between two species: if species A is found to be associated with B, B is expected to be associated in the same way with species A.

The other cause of association can be referred to as influence, where the plants of one species modify the environment to the extent that they have a direct effect on the occurrence of the other species. In the case of epiphytes, it is clear that the presence of the host makes the presence of the epiphyte possible; for example, the red alga *Polysiphonia lanosa* grows almost exclusively on the brown rockweed *Ascophyllum nodosum* (Lewis 1964). In such a case, the association might be considered to be asymmetric with *Polysiphonia* being positively associated with *Ascophyllum*, but not the other way around. Clearly demonstrated examples of positive influence are not easy to find, but the influence of 'nurse plants' that make the regeneration of cacti possible (small cacti may overheat if fully exposed to sun) is a good example (Valiente-Banuet & Ezcurra 1991; Arriaga *et al.* 1993; among many). In a review of the subject, Bertness & Callaway (1994) conclude that 'positive interactions during succession and recruitment ... are unusually common characteristic forces in harsh environments'.

In looking for examples of negative influence, it is clear that plants affected by allelopathy are negatively associated with the plants producing the allelopathic chemicals, but not the other way around. Allelopathy is difficult to demonstrate conclusively, but one case seems to be the effect of the shrub *Adenostoma fasciculatum* on annual herbs in California drylands (cf. Crawley 1986). Another source of negative influence that may be symmetric is competition for resources such as soil moisture or nutrients. Competition for light, on the other hand, is expected to be asymmetric, with shorter plants being more adversely affected.

The association of species is generally treated as a pairwise phenomenon, and the network of these pairwise associations is often referred to as the phytosociological structure of a plant community (Dale 1985). In some instances, the presence or absence of a third species can affect the relationship between a pair of species, and this possibility leads to the consideration of multispecies association, where the frequencies of various combinations of species presences and absences are examined (Dale *et al.* 1991). This topic will be discussed at greater length in Chapter 5.

The importance of species association to spatial pattern is that the spatial pattern of one species can affect the spatial pattern of the species associated with it (whether positively or negatively) and thus affect the whole vegetation. If one or a few species are particularly important in the structure and function of a plant community, the spatial pattern of those species may be amplified by their effects on other species.

Another way of thinking about species association is to recognize that it *is* spatial pattern defined relative to the positions of plants of particular species rather than relative to a system of strict spatial coordinates. Knowing that species B is associated with species A may be as useful in predicting the presence of that species as knowing that it exhibits patchiness at a scale of 5 m.

In textbooks, the interactions between species are often codified according to the effects on the interacting species: competition is a −/− interaction because it has a negative effect on both species; mutualism is +/+; predation (including herbivory) is +/−; amensalism (i.e., an interaction that has a negative effect on one population and no discernible effect on the other) is −/0; and commensalism is +/0. From the above discussion of the factors involved in the association of species, it is clear that the interactions between species in natural vegetation cannot be classified in this simple way. Not only are the interactions not pairwise, but they may also depend on the spatial scale considered.

Fractals

Most of us have grown up in a strictly Euclidean culture; we were taught that there are dimensionless points, lines of one dimension, planes of two dimensions, and so on. In fact, if you review the comments in the first sections of this chapter, you will see that the material is phrased in exactly that kind of language. Mandelbrot (1982) is credited with introducing the concept of a fractal, a phenomenon that has fractional dimension rather than that of a whole number like 0, 1, or 2. Since their introduction, fractals have been taken up by a wide range of artists and researchers, and they have found application in a variety of scientific fields. It is not appropriate to give a technical treatment of fractals here; there are many expositions available, for example Schroeder (1991). Instead, we will give a short introduction to the topic of the relevance of fractals to ecology, referring the reader to the recent review of the subject by Kenkel & Walker (1993). Their thesis is that 'concepts derived from fractal theory are fundamental to the understanding of scale-related phenomena in ecology ...' (Kenkel & Walker 1993).

A simple example of the concept of a fractal is to consider curves on a plane. A straight line on the plane has a dimension of 1.0, as does a parabola. One explanation for the dimension of the parabola being 1.0 is that very small subsections of the curve can be treated as if they were straight lines. If, however, we have a curve so complex that it has spatial structure at all scales, so that no subsection is sufficiently small that we can treat it as a straight line, the curve has a dimension that is greater than 1.0 and less than 2.0. For example, the Koch curve (Figure 1.11), which has each line divided into three with an equilateral triangle erected on the middle third, iteratively *ad infinitum*, is a fractal object of fractional dimension $\mathfrak{D} = 1.26$ (see Sugihara & May 1990). It illustrates a common property of fractals, that of self-similarity at an infinite number of scales. Many natural objects are sufficiently complex in their geometry that they have fractional dimension. For instance, Morse *et al.* (1985) found that a spruce branch has a fractal dimension of between 2.4 and 2.8. One of the plates in Schroeder (1991) used to illustrate real-world fractals is a picture of red algae growing on a rock surface, with patches of a range of sizes, some of them coalescing. The relationship between fractals and spatial pattern is evident; Palmer (1988), for example, used a fractal approach to examine spatial pattern of vegetation along a transect. We will discuss that application in detail in Chapter 5.

Some authors have suggested that the fractal nature of biological struc-

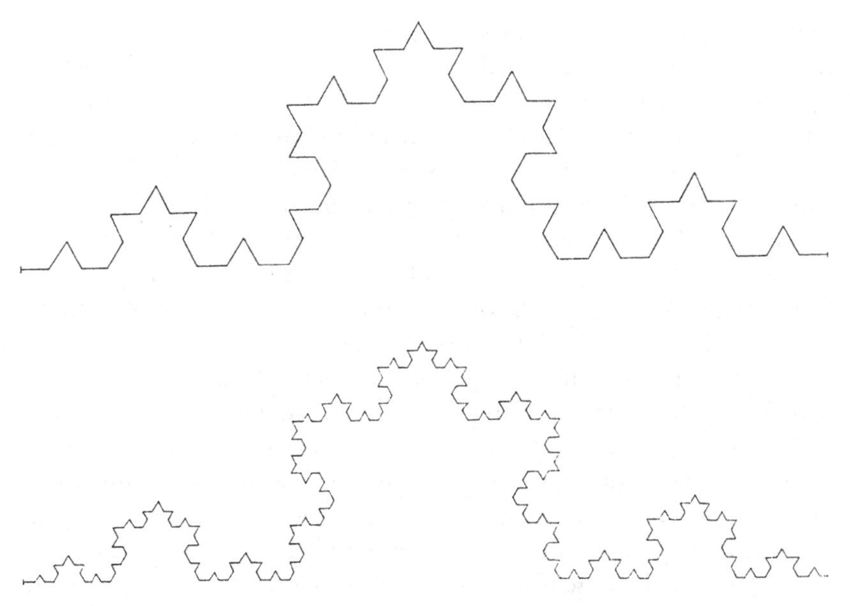

Figure 1.11 Koch curves made by the iterative construction of an equilateral triangle on the center third of each line: the third and fourth iterations. The iterations continue indefinitely so that each section of the curve is the same as the whole diagram, and so on to smaller scales too fine to illustrate.

tures is 'self-evident' (e.g. Zeide & Gresham 1991). We need to distinguish between two related but different characteristics of fractals: fractional dimension and self-similarity over a wide range of scales. We may need to separate these two features in our thinking and realize that many of the biological phenomena that we study, like the spatial pattern in vegetation, may exhibit fractional dimension without extended self-similarity. The laws of biological scaling (once we know them better) may explain why most biological fractals will show trends in $\mathfrak{D}$ with changing scale (cf. Palmer 1988; Sugihara & May 1990). In their concluding remarks, Kenkel & Walker (1993) suggest that while ecologists are still at an early stage of realizing the potential application of fractal geometry to testable hypotheses, the concept has important implications for many ecological processes.

Concluding remarks

Plants are patchy in their spatial distribution and the patchiness is usually evident at several different scales. At each scale, the spatial pattern has several important characteristics, including the size, density and spacing of the patches. Some hypotheses about past processes can be tested based on observed spatial pattern, but pattern is probably most important for its influence on future processes, both the interactions between plants such as competition and their interactions with other organisms such as herbivores and their predators, pollinators, pathogens and so on.

The spatial pattern we observe can be the result of the interaction between a number of factors including climate, topography, past disturbance, predation, competition and other interactions with neighboring plants. The important interactions between plants cannot be classified into simple categories like competition and mutualism; the processes in natural vegetation seem to be much more complicated.

One topic that needs to be explored in greater detail is the relationship between all kinds of temporal cycles and the spatial patterns they may produce. We have discussed several ways in which the freeze-thaw cycle in arctic areas can give rise to patterned substrate which affects the positions of plants. There are many other temporal cycles yet to be considered in this context. The advance and retreat of glaciers can give rise to pattern in features like recessional moraines. Alternating sedimentation regimes can give rise to spatial pattern in the soil's parent material (e.g., particle size or chemical composition), and thus in the vegetation that grows on it. The seasonal cycle of weather may affect the spatial structure of eroding rock or drainage channels and the seasonal cycle of snow accumulation can produce obvious spatial pattern with vegetation differences between areas where the snow melts early and snow beds where it lies for a long time. Cyclic changes in herbivore densities, such as the famous snowshoe hare or lemming cycles, can affect the spatial pattern of their food plants, depending on the herbivore's behavior and density-dependent feeding response. Forest insects and pathogens that display less regular outbreaks may also affect plant spatial patterns either very locally or at the landscape scale. For example, if small patches of trees have prolonged defoliation compared to large patches, as in the tent caterpillar example (see 'Pattern and process'), small patches will be selected against and the scale of pattern will increase. The daily cycles of tides have a very strong influence on spatial pattern of algae and other seashore plants, especially in the intertidal zone where obvious zonation often develops in response to the desiccation gradient. The relationship between temporal and spatial pattern is a fascinating area that deserves much more research effort in the future.

2 · *Sampling*

Introduction

The most important criteria that determine how sampling will proceed in the study of spatial pattern are the question being asked and the scale at which we wish to answer it. Secondly, the kind of analysis that is required to answer the question must be considered because particular methods of analysis require certain kinds of data. Then, the sampling method will be determined by the interaction of a number of factors including the morphology, size and density of the plants of interest; the topography, accessibility, and area of the study site; the availability of time, money, technological and field assistance. It will be influenced fundamentally by whether the spatial pattern is to be treated as the arrangement of points in continuous space or as a mosaic of domains. We must also consider how much disturbance the sampling technique will cause, because we will want to minimize disturbance in long-term studies or in ecologically sensitive areas.

The methods used will also depend very much on whether the focus is on the spatial pattern of plants relative to a fixed frame of reference, on the elucidation of a community's response to an environmental gradient, or on the spatial arrangement of plants relative to other plants (species association). Kenkel *et al.* (1989) make the important point that the considerations for sampling design that are traditionally emphasized in statistics textbooks may not apply in studies of spatial pattern, because they are designed to provide efficient and unbiassed estimates of parameters such as mean cover or diversity. This distinction is obvious when we consider questions of quadrat size, for instance: a precise estimate of mean cover is best obtained when the quadrat size is chosen to minimize variability among samples, but to investigate pattern, we want to see all that variability and will choose a quadrat size that will maximize the variance among samples (Kenkel & Podani 1991). Similarly, for parameter estimation, it is usually recommended that the samples are randomly placed; for spatial pattern analysis, a simple spatial relationship among the samples is desirable and so nonrandom regularly spaced or contiguous samples are appropriate. While some methods of analysis can use data from randomly placed samples, others require a specific sampling design such as contiguous samples.

In this chapter, we will look first at sampling relative to a fixed frame of reference using points, various arrangements of quadrats, lines, mapped mosaics, and the special techniques for sampling on environmental gradients. Next, we will describe methods for studying pattern relative to other plants. This will be followed by discussion of general considerations related to sampling.

Sampling for pattern in a fixed frame of reference

Points

In some circumstances, plants can be treated as dimensionless points in a plane. The practicality of this approach will depend on the morphology and spacing of the plants, because reducing the spatial relations of, for example, twining vines to single points would entail the loss of much of the spatial information. The technique has been used often for forest trees which have a similar erect structure and a single central columnar stem; the center of the stem can be treated as the tree's location, and may well represent the place at which its seed germinated (Williamson 1975; Kenkel 1988a; Szwagrzyk & Czerwczak 1993; among many). Herbaceous plants that grow as more-or-less compact rosettes with a discernible center are also suitable subjects for this approach (Silander & Pacala 1985; Powell 1990), as are the erect stems of clonally growing plants such as bramble (*Rubus fruticosus*) or goldenrod (*Solidago canadensis*) (Hutchings 1979; Dale & Powell 1994; among many). In these cases, the spatial information about the plants is simplified to the positions of dimensionless points in a plane.

Spatial point patterns are usually studied by mapping the points' positions, which can be accomplished in a variety of ways. Powell (1990) mapped knapweed (*Centaurea diffusa*) seedlings onto acetate sheets supported on a sliding thick glass table that was mounted on rails. The accuracy of the mapping was ensured by a Plexiglass sighting tube with three sets of cross hairs. We mapped *Solidago* plants using an (x,y) coordinate system with a fixed rail at ground level, set square and meter sticks (Dale

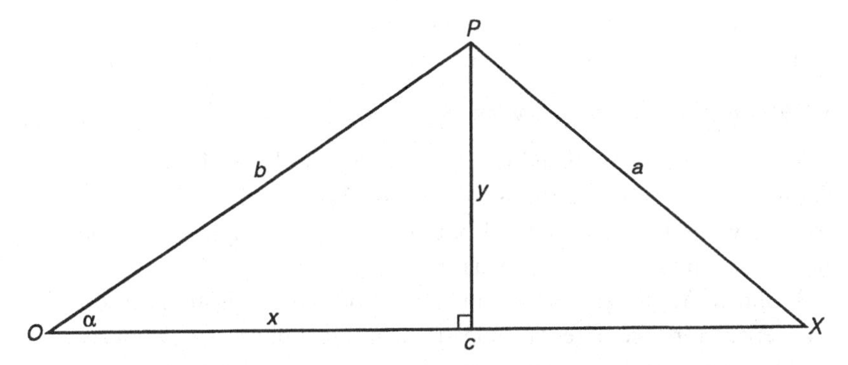

Figure 2.1 Line *OX* is the baseline between the two reference points, length *c*. Lines *a* and *b* give the measured distances from the two reference points to point *P*. Angle *POX* is α. The text describes three methods for recovering the (x,y) coordinates of *P*.

& Powell 1994). That particular system worked under the circumstances because we were able to clip the senescing stalks off at ground level as they were mapped, allowing clear access to the remaining stems. Maps of plant positions can also be obtained, sometimes more easily, by digitizing the information from photographs (Galiano 1982b). In open vegetation, mapping by triangulation based on angles is possible, but of low accuracy (Mosby 1959). A more accurate technique is to use the measured distances from two or more reference points from which (x,y) coordinates can then be recovered.

Given the distances from two fixed reference points with known locations, the position of any plant can be recovered from its distances from the reference points in several ways. Given a measured baseline *OX* of length *c* and the distances *a* and *b* from its endpoints to the plant *P* (Figure 2.1), the cosine method solves for the Cartesian coordinates (x,y) of the plant by calculating angle *POX*:

$$\alpha = \cos^{-1}[(b^2 + c^2 - a^2)/2bc].$$

Then

$$x = b \cos \alpha \text{ and } y = b \sin \alpha.$$

The Heron's formula method calculates the half perimeter of the whole triangle: $s = (a + b + c)/2$. The area of the triangle is $cy/2$ and also $\sqrt{s(s-a)(s-b)(s-c)}$ by Heron's formula. Therefore: $y = 2\sqrt{s(s-a)(s-b)(s-c)}/c$ and $x = \sqrt{b^2 - y^2}$ by Pythagoras' formula.

Beals (1960) provides a third alternative based on the areas of the two smaller triangles which is algebraicly related to the cosine method:

$x=(c^2+b^2-a^2)/2c$ and $y=\sqrt{b^2-x^2}$.

For example, if the baseline is 125cm long and $a=91$ and $b=71$ (cf. Figure 2.1), all three methods give $x=49.54$cm and $y=50.86$cm. The cosine law method is often used, but our experience suggests that the two alternatives are less sensitive to error.

Rohlf & Archie (1978) provide a method by which the positions are recovered more accurately by a least-squares technique. Mapping may be facilitated by subdividing the study area into smaller plots which are mapped individually and then the data can be combined for analysis. For instance, Gill (1975) mapped the trees in a 40m × 40m area by subdividing it into 5m × 5m plots; at high tree densities, these were further divided into 2.5m × 2.5m quadrats.

Having obtained the mapped positions of the plants, a variety of methods can be used to assess the spatial pattern; these are described in Chapter 7. In some cases, such as for the investigation of anisotropy, it may be most appropriate to map the plants in circular plots rather than in square or rectangular quadrats, so that all directions are treated equally. For instance, when Szwagrzyk & Czerwczak (1993) mapped the positions of trees in natural European forests to study possible effects of competition on spatial pattern, they used circular plots with radius 50m.

Quadrats

Other sampling methods frequently used for the study of spatial pattern involve the use of samples of a fixed area, usually in the form of quadrats. Some methods require the use of contiguous quadrats that are laid out in one-dimensional transects or two-dimensional grids. It is traditional to use square quadrats in grids, which has the apparent disadvantage of forcing an x-direction and a y-direction on the sampling. The forced orientation of sampling is not always a problem because there are cases in which it is known before sampling starts that the vegetation has a particular directional characteristic; for instance, in patterned wetlands the lengths of the strings and flarks are at right angles to the direction of the flow of water. Some authors have used several parallel and, in some cases, contiguous strings of quadrats. For example, Anderson (1967) used small square quadrats in narrow grids, ranging in size from 6 × 48 to 2 × 1024, and Bouxin & Gautier (1982) present data from a 4 × 120 grid of

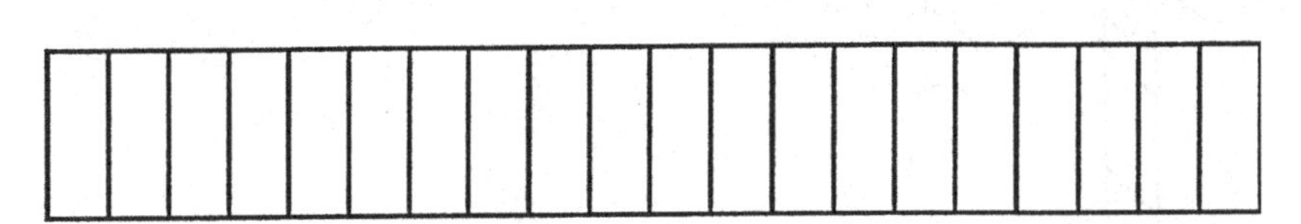

Figure 2.2 Changing the ratio of quadrat area to quadrat spacing where the longer side of a rectangular quadrat is oriented along or across the direction of a transect. In the two parts of the Figure, the spacing of the quadrats is the same, but the areas sampled by the two shapes of quadrat are different.

contiguous 0.25 m × 0.25 m quadrats. One problem with this sort of sampling scheme is that is has a strong directional bias, which might affect the interpretation of the results.

As alternatives to square quadrats, we can imagine using a lattice of triangular or hexagonal samples but the advantage gained seems small, compared to the extra labor and difficulty in laying out such lattices in the field. An even more extreme suggestion is to use 'quadrats' that form an aperiodic tiling. The disadvantages of such a scheme may be greater but its benefit is that there is no directionality in the sampling scheme itself to interact with any directionality of pattern in the vegetation. Whatever the shape and arrangement of the sampling units, in each quadrat of the grid, stems are counted, or density measured or cover estimated, in order to quantify the density of the plant(s) of interest in the quadrat.

Quadrats set out in strings or transects are also traditionally square, but there seems to be no reason for them not to be rectangular provided that they are small enough that the variation among quadrats is maximized and variation within is mimimized (cf. Dale *et al.* 1993). The use of rectangular quadrats with the longer side oriented along or across the direction of the transect permits adjustment of the ratio of quadrat area to quadrat spacing (cf. Figure 2.2). Kenkel *et al.* (1989) warn against the use of 'elongated' plots that might include more than one clump of plants and the same concern may apply to quadrats; therefore, extreme departures from equal sides probably should be avoided. Greig-Smith (1983) also cautions against the use of sample units that are very elongated because of problems associated with a high edge-to-area ratio.

The use of contiguous quadrats represents a high intensity of sampling, relative to distance, and it is tempting to reduce the effort required

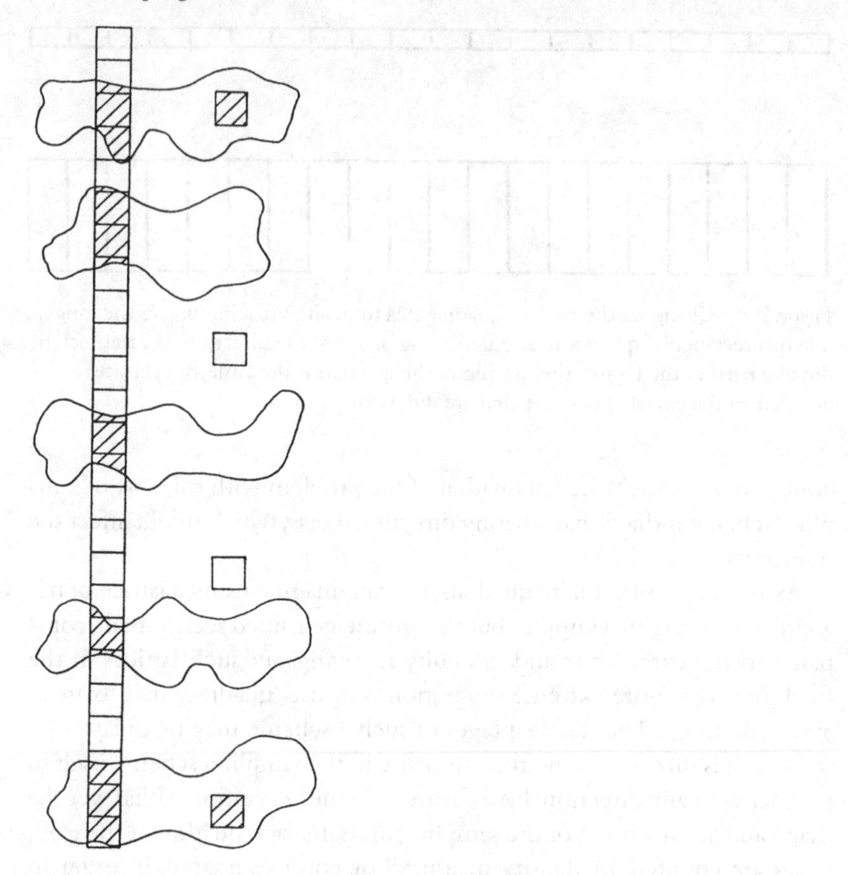

Figure 2.3 Regularly spaced quadrats may give misleading results when their spacing interacts with the spacing of patches. The contiguous quadrats on the left detect all the patches and gaps; the spaced quadrats on the right do not. In this case, an apparent two-quadrat gap is divided by an unperceived patch.

by leaving spaces between the samples. This is a case where the intended method of analysis will affect the choice of sampling, because many of the techniques of spatial pattern analysis require that the quadrats are contiguous. Other arrangements, such as regular spacing, are not generally appropriate and may give misleading results due to an interaction between the scale of pattern of the vegetation and the spacing of the quadrats. In Figure 2.3, the central two quadrats, which are empty, seem to indicate a large gap but they are actually separated by a patch. Similarly, two or three high-density samples will appear in the data as a single patch even when they are separated by an unperceived gap.

In spite of these problems, in some studies there may be no alternative to using regularly spaced samples. For instance, in studying the spatial characteristics of mycological flora or chemical properties of soil, it is not practical or desirable to excavate long trenches and regularly spaced cores are used instead. For instance, Jackson & Caldwell (1993) used spaced samples to examine the patterns heterogeneity of soil properties around perennial plants in a sagebrush steppe. Regularly spaced quadrats may also be the best compromise when the area under study is too large to be sampled fully; for example, Legendre & Fortin (1989) studied spatial pattern in forest trees using a regular array of 10 m × 20 m quadrats separated by 50 m. Finally, on the topic of contiguous and spaced samples, we should note that, while some methods of analysis require contiguous samples, there is a set of methods derived from geostatistics conceived specifically to deal with spaced samples such as rock cores.

Many of the sampling techniques used to study density or to test for randomness, such as scattered quadrats or sampled point locations, are not usually suitable for spatial pattern analysis because, while they can signal departure from random dispersion, they do not permit an investigation of the scale(s) of pattern or other characteristics of that departure. It is important to remember, however, that even randomly placed samples do have a spatial relationship among them which could be investigated using some of the spatial autocorrelation methods; for example, irregularly spaced samples of amphipod abundance are analyzed in that way by Jumars *et al.* (1977). Again, techniques from geostatistics were designed to deal with the spatial relations among irregularly positioned samples.

Having mentioned the possibilities of grids or transects of quadrats, the question is then whether it is better to sample in one or two dimensions. The answer to this question depends on both theoretical and practical considerations that are best discussed in reference to examples. Consider a study site that has a known physical gradient, such as a consistent slope; in that case it is sensible to question whether spatial pattern is similar across the slope and up and down it. There, a grid of quadrats with sides oriented along and across the slope would make it possible to study not only the overall two-dimensional spatial pattern, but also pattern in the two principal directions separately. In fact, as we will discuss in Chapter 5, it is possible to use that kind of data to look at the nature of anisotropy in the pattern. If the number of quadrats that can be studied is restricted by practical considerations, a system of transects oriented across and along the gradient, like a trellis, would be a sensible compromise (Figure 2.4). For instance, Yarranton & Green (1966) used a combination of vertical and horizontal transects to study the spatial pattern of a lichen community

Figure 2.4 A trellis system of quadrats is oriented across and along the gradient. There are fewer quadrats than a full rectangular grid of the same dimension, but the data will be more informative than if the transects ran in only one direction.

growing on a cliff face. Similarly, in a study of spatial pattern in an African savanna grassland, Carter & O'Connor (1991) used two sets of transects, one running up- and down-slope and one along the contour. The same considerations apply when the gradient is biological rather than physical; for example, where a clonally growing plant such as *Solidago* or *Aster* is invading an abandoned field, there is a natural orientation for grids or transects parallel with and at right angles to the direction of advance (cf. Dale & Powell 1994).

An apparently similar situation with a different solution is found in studies of the development of spatial pattern following glacial retreat. Here, we would be interested in the pattern of vegetation on surfaces of the same age. A sampling grid would, therefore, be inappropriate and instead transects at right angles to the glacier's retreat should be used (Dale & MacIsaac 1989; Dale & Blundon 1991).

The most difficult situation to give advice on is where there is no apparent natural orientation for quadrats. If it is possible and practical, the best method to use under these circumstances is to map the plants because such data can be analyzed using the mapped point methods which do not have an imposed directionality, or converted secondarily to quadrat data with several different orientations to investigate the effect of direction on the spatial pattern detected. In many instances, as mentioned above, mapping will be facilitated by dividing the sample area into smaller subdivisions, which are mapped separately (cf. Gill 1975; Podani 1984).

If mapping is not a possibility, there are several alternatives. The first is to use a grid of quadrats at some particular orientation and then to use some of the methods described in Chapter 4 to determine whether the pattern has a directional component or to use methods of analysis that limit the effects of the grid's orientation (such as the random pairing method described in Chapter 5). If the anisotropy of pattern is the focus of the study, sets of transects with a range of orientations might be the most sensible approach. Interpreting the results from a set of transects at different angles is made easier by examining an example. Figure 2.5 illustrates the effect of transect angle on the scale of pattern detected when the pattern is anisotropic. When the transect is oriented along the length of the pattern's features, the scale of pattern detected is large, but when transects are oriented across the features, the detected scale is small. Comparisons of angles and scales in transects should make it possible to interpret the pattern as perceived by the set of transects.

Another question to be decided when using quadrats is their size. For most kinds of vegetation and most kinds of study, extreme quadrat sizes (20 m × 20 m or 1 mm × 1 mm) are not practical. Large quadrats make it impossible to get accurate estimates of density for the whole sampling unit. Another problem with large quadrats is that they may have too much internal variability and the information on that variability is lost by reducing it to one observation. Very small quadrats may create problems because of the inaccuracies of position and there may be practical difficulties of needing very large numbers of them to cover sufficient distance to detect pattern at larger scales. Small quadrats may also sample the same plants repeatedly, increasing the redundancy in the data. Some

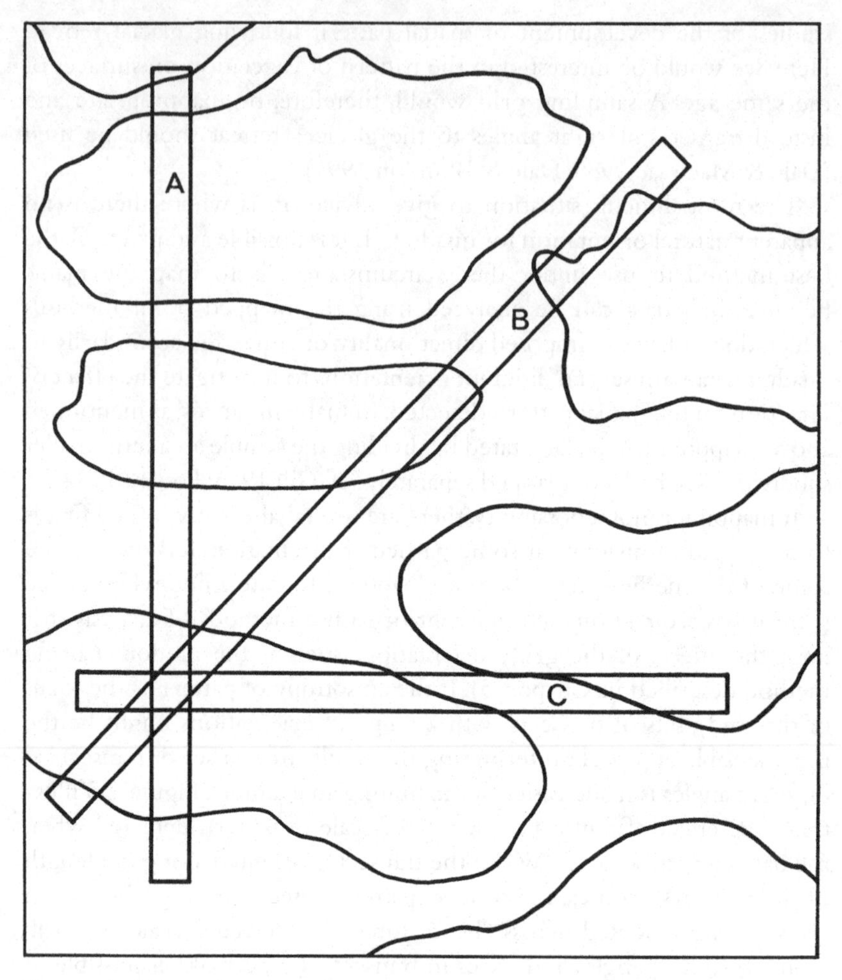

Figure 2.5 Where the pattern is anisotropic, transects at different angles detect different scales of pattern. Transect A intersects four patches and four gaps; transect C intersects two patches and one gap, thus giving a larger scale of pattern.

authors have deliberately chosen the size of their sample unit so as to have most of them contain only one rooted plant or ramet (cf. Stowe & Wade 1979). For the purpose of spatial pattern analysis, Greig-Smith (1983) advises that the safest procedure is to use the smallest size of quadrat that is practical. That choice minimizes the probability that small scales of pattern will be missed. Clearly, the best quadrat size is related both to the scales of pattern that are of interest and to the overall scale and purpose of the study.

Another decision to be made about sampling with quadrats is the kind of data that will be recorded from them. The most common approach is to use ocular estimates of percentage cover for each identified taxon (cf. Kent & Coker 1992). This method has the advantage of being quantitative, nondestructive and relatively efficient. The disadvantage is that it is subjective and the values may vary depending on the observer. One alternative is to record merely species presence or absence, which has the advantage of being fast and more repeatable. The disadvantages, that it is not fully quantitative and does not distinguish between small rare plants and large common ones, may be less of a problem when the quadrats are small. In our Mt. Robson study (Dale & Blundon 1990), we found that the patterns detected with density data and with presence/absence data were similar; but that will probably not be true in all vegetation types.

Clipping quadrats and separating the biomass by species for weighing provides high-quality data, but it is both destructive and laborious. In some circumstances, where there are identifiable units such as stems that can be counted, the density of the units of each species in a quadrat may be a useful measure. For instance, in his classic 1973 paper, Hill illustrates his discussion of methods with data from Greig-Smith & Chadwick (1965), which consists of the counts of three size classes of plants of *Acacia ehrenbergiana*. Whichever kind of data is chosen to be recorded, it is probably true to say that obtaining high-quality data requires care and effort.

Lines

Line intercept sampling may be viewed as a modification of the transect sampling using strings of small contiguous quadrats in which the quadrats are shrunk to width zero. This method is distinct from the 'line transect' method, used to estimate density of animal populations (Eberhardt 1978; Burnham *et al.* 1980). Their method involves counting the animals seen while the observer is travelling along a transect but our approach includes only the plants that actually intersect the sampling line. The appropriateness of the method will depend on the structure of the vegetation, but it is most easily used when the plants are more-or-less two dimensional, as in a moss lawn or a mosaic of saxicolous lichens, so that any particular point can be occupied by only one or very few different species. It is also recommended for use in very sparse vegetation or in shrublands (Kent & Coker 1992). The procedure is to establish a line (usually the edge of a tape measure) and then to record the segments of the line that intersect each species; for example 0 to 3.1 cm species A, 3.1 to 4.2 cm B, and so on

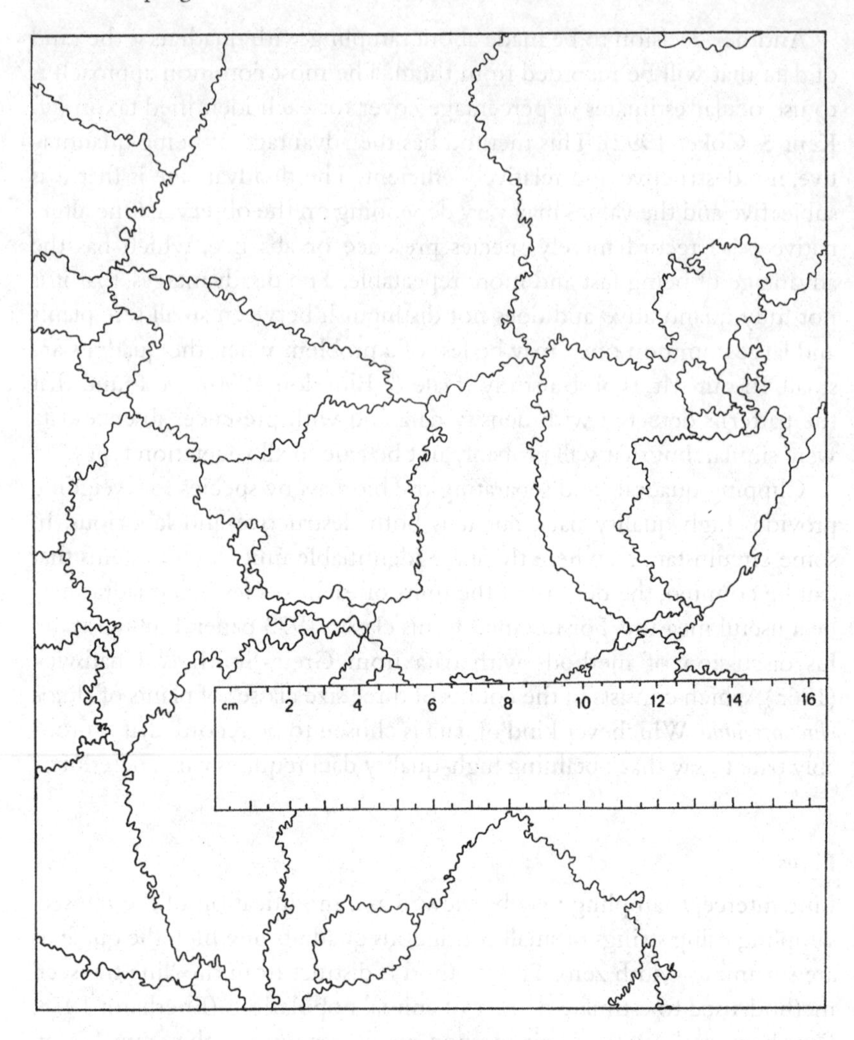

Figure 2.6 A line is established; the segments of the line that intersect each species are recorded. Along the upper edge, 0 to 3.1 cm is species A, 3.1 cm to 4.2 cm is B, 4.2 cm to 5.4 cm is C, 5.4 cm to 6.6 cm is A, 6.6 cm to 7.8 cm is D, and so on.

(see Figure 2.6). Data collected by this method have been used to study various characteristics of the arrangement of seaweed species on environmental gradients (Dale 1984, 1986, 1988a). Westman (1983) used the line intercept method to determine floristic composition in xeric shrublands in southwestern North America. He measured canopy cover along four 25-m line transects in each of a number of 25 m × 25 m plots and then used the data to examine species associations.

Mapped mosaic

If the vegetation forms a mosaic of distinct domains with clear boundaries, such as may be found in saxicolous lichen communities, it may be possible to extend the line intercept sampling into two dimensions, and to record what is essentially a map of the vegetation with the boundaries and identities of the domains. The simplest such exercise would be to map the presence of a single species, including everything else in the 'absent' category; Diggle (1981) discusses a map of the incidence of *Calluna vulgaris* (heather) in a 10m×20m plot as an example of a binary mosaic. In some situations, it may be possible to carry out this sampling using photographs which can be converted to a variety of digital forms later. The most useful digital form is probably the conversion of the photograph into a numerical map, usually by treating the vegetation mosaic as a grid of numbers or symbols. This can be done by creating a grid in which the grid units are coded for the dominant species or type in it. While the conversion to a single grid imposes a certain amount of directionality on the data, it is always possible to convert the data to several different grids with different orientations, in order to test whether that directionality is a problem.

The grid of small units is referred to as 'raster' format and there is an obvious tradeoff between many small units which preserve more of the map's detail and fewer larger units which require less room for data storage (Bailey & Gatrell 1995). Maps can also be coverted to digital form in 'vector' format, in which each feature, such as a boundary between mosaic patches, is recorded as a sequence of (x,y) coordinates. Again there is a tradeoff in that more points preserve more detail of a particular feature, but result in greater storage requirements (Bailey & Gatrell 1995).

Similar kinds of data can be acquired from satellite images which are essentially grids of the intensities of several parts of the electromagnetic spectrum. The grid units are referred to as 'pixels' and the image is usually analyzed by classifying the pixels either based on the spectral characteristics of known features in the landscape (supervised classification) or by a multivariate analysis (unsupervised classification) (cf. Lillesand & Kiefer 1987). In LANDSAT TM (Thematic Mapper) images, the pixels are about 30m×30m, which is not exactly high resolution, but may be a scale suitable for spatial pattern analysis at the landscape level. Musik & Grover (1991), for instance, used LANDSAT images to investigate textural measures of landscape pattern. The images available from the SPOT satellite system have a finer scale but are more expensive to acquire.

O'Neill *et al.* (1991) successfully used digitized maps of lower resolution in a study of spatial hierarchies by using a wagon-wheel of 32 transects radiating out from the center of the map.

The development and analysis of digital maps derived from remote sensing has been greatly enhanced in recent years by the improvement in the availablity and sophistication of hardware and software for Geographical Information Systems (GIS). While these systems were developed for application to spatial data at the geographical scale, there is no reason why they cannot be applied to smaller scale data sets. GIS software is developing rapidly and there are many different kinds available; many have the capability for some of the spatial analysis techniques described in subsequent chapters of this book, but GIS software and spatial analysis have yet to become fully integrated (Bailey & Gatrell 1995).

The combination of remote sensing and GIS manipulation has not been used extensively for spatial pattern analysis, but it is a promising approach. Goodchild (1994) describes some of the methodological issues for their application to the general areas of vegetation analysis and modelling. Turner *et al.* (1994) present TM-derived data on how fire affected the landscape in Yellowstone National Park but do not explicitly and completely investigate the scale of that heterogeneity. Mertes *et al.* (1995) used spatial statistics as part of their analysis satellite images of the Amazon basin, but found that the pixel size (30 m) was the most obvious feature of that analysis. These examples show that remote sensing and GIS techniques have the potential to be very useful for spatial pattern analysis and we expect that they will be used extensively in the future.

At a different scale, a study of spatial pattern of leaves in a forest canopy could be based on (black and white) photographs taken directly upward from the forest floor. These images can then be scanned directly into computer form for analysis (Lee 1993). Dietz & Steinlein (1996) used color photographs and digital image analysis to determine plant species cover and while the usefulness of that approach depends on the type of vegetation, it holds great promise for the rapid acquisition of detailed data for pattern analysis.

Sampling for pattern on gradients

One obvious feature of the environment that affects the arrangement of plants in space is the existence of environmental gradients. In this book, we will be considering an environmental gradient to be a directional change in the intensity of an environmental factor that is monotonic with distance. This kind is the class referred to as 'spatially continuous' by Keddy (1991) and includes cases that may give rise to obvious zonation of plants in the community. Examples include rocky tidal shores, salt marshes, ponds and the sides of mountains. A gradient affects the spatial arrangement of plants, but the kind of pattern we expect is different from the repeating features we would expect in a patterned fen or temperate forest.

Many of the sampling methods described in the previous section can be used to examine the patterns of species arrangement on environmental gradients. Not all the methods of analysis described in Chapter 8 can be applied to all kinds of data that might be collected. As always, therefore, we must consider the questions we want the data to answer before deciding on a sampling technique.

Several different methods of sampling have been used for the study of pattern on a gradient, including continuous line intercept sampling (Dale 1984), contiguous quadrats (Wulff & Webb 1969; Pielou & Routledge 1976), evenly spaced quadrats (Mandossian & MacIntosh 1960) and a combination of floristic studies on an altitudinal gradient and data from herbarium sheets (Auerbach & Shmida 1993). Where contiguous quadrats are used, rectangular rather than square quadrats may be preferable in order to adjust the ratio of the spacing between the centers of adjacent quadrats and and the area sampled by each one; for instance, Pielou & Routledge (1976) used transects of quadrats 0.5 m wide but 1 m long in a study of saltmarsh zonation. On the other hand, Johnson *et al.* (1987) sampled the zones around prairie marshes using contiguous quadrats 1 m wide and 0.5 m long. A general comment on the choice of quadrat size and shape is that if there are several species boundaries in a single quadrat, the quadrats are probably too long.

The problems with noncontiguous quadrats, as described above for ordinary pattern analysis, may not be so pronounced in the study of some aspects of pattern on gradients. The loss of detail and refinement, however, may be a critical problem for some studies. Just as in sampling for pattern analysis in the absence of a clear gradient, unbroken data sets such as those from line intercept sampling or from contiguous quadrats that are small relative to the rate of change along the transect are the safest choices.

Sampling for pattern relative to other plants

In looking at spatial pattern relative to the positions of other plants, we are usually investigating the association of different species or of different growth stages of the same species. Association, like many other phenomena discussed in this book, is scale dependent! Here, somewhat different considerations apply than when sampling for spatial pattern relative to a fixed frame of reference. For instance, in placing quadrats, we recommended the use of contiguous quadrats for spatial pattern analysis, but in investigating association contiguous quadrats may introduce an undesirably large amount of spatial autocorrelation into the data. Spatial autocorrelation makes statistical tests too liberal, producing more apparently significant results than the data actually justify. Clearly, the choice of the size of the sample unit is also affected; for instance, it is clear that samples that would usually contain only one rooted plant might not be useful. While contiguous quadrats introduce autocorrelation problems, they do permit the examination of association at a range of scales; using one size of quadrat not in a contiguous array permits the evaluation of association at only one scale. It is possible, under some circumstances, to quantify the spatial autocorrelation and remove its effects from statistical testing (Dale *et al.* 1991; Borcard *et al.* 1992).

Points

Having just said that we would not want to use sampling units that were so small that they usually contained only one plant, it may seem odd to discuss the use of points as a sampling technique, since clearly a point is even smaller than the smallest quadrat. Points can be used in two ways, however, either as representations of plants' positions in a map or as a sampling method, usually as an objective way of choosing plants for investigation. We shall discuss the use of mapped plant positions at length in Chapter 7, but briefly, to look at species associations, we tally the number of plants of each species at a range of distances from a plant of the target species and then compare those numbers to the expected distribution based on a null hypothesis of independent arrangement of the species.

Point sampling involves the use of regularly or randomly placed points to designate certain plants in the vegetation, which can be thought of as the primary or initial plants of the sample. The procedure is then to identify the plants that are neighbors of the initial plant, usually choosing only the one that is closest to the original point on the initial plant, or the one that is touching the initial plant closest to the original point. That kind of sampling is often referred to as 'point-contact' sampling and is widely used in studies of vegetation that can be treated as essentially two dimensional. It was pioneered by Yarranton (1966) in a study of bryophyte communities on Dartmoor and has been used extensively in studies of cryptogamic communities (eg., John 1989; John & Dale 1995). It has also been used in studies of more structurally complex communities such as pastures (Turkington & Harper 1979a,b; de Jong *et al.* 1980) and the successional vegetation on subalpine moraines (Dale *et al.* 1993). The analysis and interpretation of this kind of data require some care; we will not discuss the details of analysis here, but see Dale *et al.* (1993) for a summary.

Point-contact sampling is promoted as being superior to quadrat sampling because it looks at the vegetation 'from the plant's-eye view' (Turkington & Harper 1979a,b), whereas quadrats by the choice of size look at the vegetation from a more human view-point. It may not be true, however, that a plant's contact is indicative of the most important of its neighbors. For example, in some arid tropical communities, cacti may be strongly associated with the shade of tall perennials such as trees and shrubs, sometimes referred to as nurse plants (Franco & Nobel 1989; Arriaga *et al.* 1993). The relationship does not require physical contact and point-contact sampling might not detect it.

A variant of the point-contact method was used by Stowe & Wade (1979) and Whittaker (1991). The technique examines circles of radius r, centered on n points arranged in a line with a distance of $2r$ between successive points. The data are records of the species of plant rooted closest to the sample point within distance r; these are the primary data. Whittaker (1991) also recorded all other species rooted within radius r as the secondary data. Because the circular samples are touching, this method is related also to the contiguous quadrat technique.

A similar concept is to use the stems of the plants themselves as the centers of sampling circles and record the closest plant or all the other plants within a given distance. For example, Mahdi & Law (1987) used circles of radius 3 cm centered on ramets of seven target species in a limestone grassland. Their choice of 3 cm was based on the concept of a ramet's neighborhood, citing Silander & Pacala (1985). This kind of plant-centered sampling is very useful for looking at very specific kinds of spatial pattern such as evidence of asymmetric association, whether positive or negative. This approach would not be practical for general community analysis, because it requires a set of samples for each species. On the other hand, only such intensive sampling provides the means to examine asymmetric association.

One form of asymmetric positive association is 'nucleation', the phenomenon of plants of one species increasing, in their immediate vicinity, the successful establishment by plants of another species (Yarranton & Morrison 1974). For instance, we tested for nucleation during primary succession on moraines by using circular quadrats centered on plants of the genus *Hedysarum*; it is a nitrogen-fixing plant and therefore a potential enhancer of colonization by other species (Blundon *et al.* 1993). The size of the quadrat was the average size of 100 randomly chosen *Hedysarum* plants on each moraine. Using this technique, we found that recruitment of later successional species was significantly greater in the *Hedysarum* patches than elsewhere on the moraine.

Location of sampling

One of the most difficult issues for plant ecologists studying spatial pattern or studying vegetation in general is the location of the sampling area; especially if we wish to be as objective as possible in our approach. The reason it is a difficult issue is that it is not always possible to explain the logic behind decisions to sample at one location rather than another. One of the most frequently cited criteria is that vegetation that looks homogeneous is chosen for study to avoid any obvious discontinuities. Often, we will not want to study vegetation where the spatial pattern is too obvious. We are tempted to choose areas where the pattern may be sufficiently subtle that there is some challenge in elucidating it, but where we suspect that there is some underlying pattern with an important relationship to community processes. Perhaps we should pursue more studies of the obvious for comparing spatial pattern. The subtle underlying pattern we search for may actually be the result of stochastic factors acting locally to give the appearance of order, but which produce disorder at larger scales.

The search for subtlety is not always true of studies of communities on spatially continuous gradients, in which sampling location will be chosen, often, where there are obvious responses to the environmental factors. If we are looking to study the response of vegetation to a gradient, we will choose a location where there is clear pattern of zonation, because we then know that there is a response to be studied.

In some cases the search for homogeneity has a sound basis, particularly if we want to look at population processes. To study self-thinning in tree populations, it is obvious that pure stands of the species of interest will greatly reduce the variability compared to a stand with many species of many different morphologies and phenologies.

Concluding remarks

At the beginning of this chapter, we said that the sampling method used to investigate spatial pattern will depend on a number of factors, based on both theoretical and practical considerations. The subsequent discussion and breadth of examples will not have changed the truth of that statement, but will have amplified and supported it. We can offer the usual advice that we have probably all ignored at some time, and then regretted, which is that the hypotheses to be tested and the methods of analysis should be decided upon before sampling begins.

Recommendations

1. Because the choice of sampling method depends, in part, on the method of analysis that will be used to answer the question of interest, read the appropriate material on analysis in later chapters of this book before making that decision.
2. In using the point pattern approach, complete mapping is the best sampling method.
3. In general, continuously sampled data from line intercept sampling or from strings or grids of contiguous quadrats are preferable. For some kinds of data, this kind of sampling may be impractical or impossible.
4. The quadrats should be as small as is practical to avoid missing small scales of pattern. More is better than too few.
5. The shape of quadrats will depend on the desired relationship between the distance between the centers of quadrats and the area included in each one.
6. If practical, record the most quantitative form of data; it can always be converted to categories later.
7. For mapping vegetation, 'high-tech' methods such as remote sensing and GIS manipulation should be considered.
8. Most of the recommendations for sampling apply to studies of patterns on spatially continuous gradients.

3 · *Basic methods for one dimension and one species*

Introduction

Several methods have been proposed to detect the scale of pattern in vegetation; most of them analyze density data in strings or rectangular arrays of contiguous quadrats by examining how variance depends on the size of blocks of quadrats which are lumped together in the analysis (e.g., Greig-Smith 1952; Hill 1973; Usher 1975). In this chapter, we will review and illustrate the basic methods for studying the spatial pattern of a single species in one dimension along which there is no environmental gradient. The kind of data under consideration are therefore density or presence/absence data collected in a string of contiguous quadrats (see Chapter 2).

Data

We will begin by considering a standard pattern consisting of a regular square wave and let the scale of the pattern be B quadrat units. Throughout the transect, gaps of B quadrats, each of density 0, alternate regularly with patches of B quadrats, each with density d. There are several ways in which this basic pattern can be modified to be made less regular:

1. The pattern is 'unbalanced' with the patch:gap ratio different from 1:1 but the patch size (p quadrats) and the gap size (g quadrats) are both constant for the entire length of the transect. For a given value of d, unbalanced patterns have a lower intensity than balanced patterns. In forest communities, the spatial pattern of the canopy will often be unbalanced in this way with the patches of canopy being considerably larger than the gaps between them. For example, in the mature phase of *Fagus japonica-Fagus crenata* forest in Japan, the trees in the tallest stratum have 90–95% canopy cover so that gaps in the canopy are small (Peters & Ohkubo 1990,

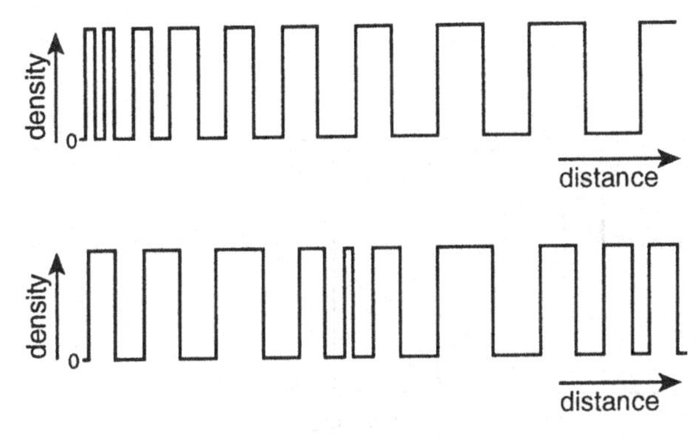

Figure 3.1 The upper part shows a nonstationary spatial pattern which exhibits a simple trend in scale. The lower part shows more complex nonstationarity which may be difficult to evaluate.

Figure 2). On the other hand, in arid systems, competition for moisture may result in the patches of vegetation being so well spaced that along a transect, gaps are considerably larger than patches (Montaña 1992).

2. The patch: gap ratio is variable along the transect, but the scale is constant with the lengths of adjacent patch-gap pairs adding consistently to around $2B$. This kind of variability could arise from differences in the ages and sizes of individual plants or clones that form the patches or from differences in the number of individuals in the patches. The overall scale of pattern might then be controlled by an underlying environmental factor such as moisture in a hummock-hollow system.

3. The scale of pattern varies along the transect; for example, exhibiting a trend from a scale of 6 quadrats at one end of the transect to 18 at the other. This characteristic is sometimes termed lack of stationarity, because the rate of turnover from patch to gap and from gap to patch is not stationary along the length of the transect. If the spatial structure of the vegetation is modelled by a Markov process, the trend in scale can be interpreted as a trend in the transition probabilities from one phase to the other. It is possible to have nonstationarity that is more complex than a simple trend, but usually the more complex the behavior becomes, the more difficult it is to detect (Figure 3.1). A simple trend in density may follow a systematic change in an environmental factor such as elevation or soil moisture.

4. The pattern is 'interrupted' in that sections of the transect, some multiple of B quadrats long (say β in length), have a regular pattern of

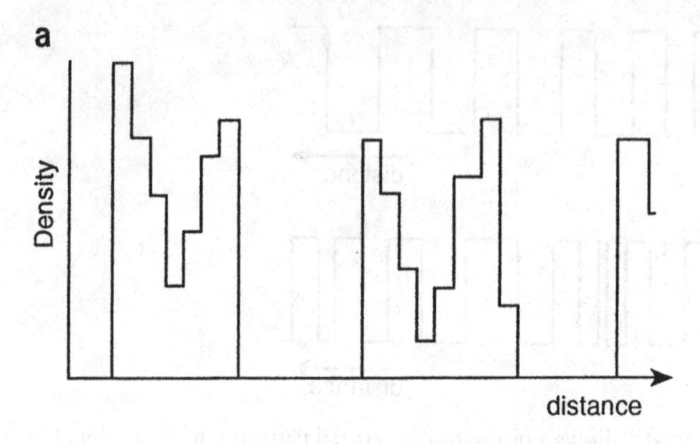

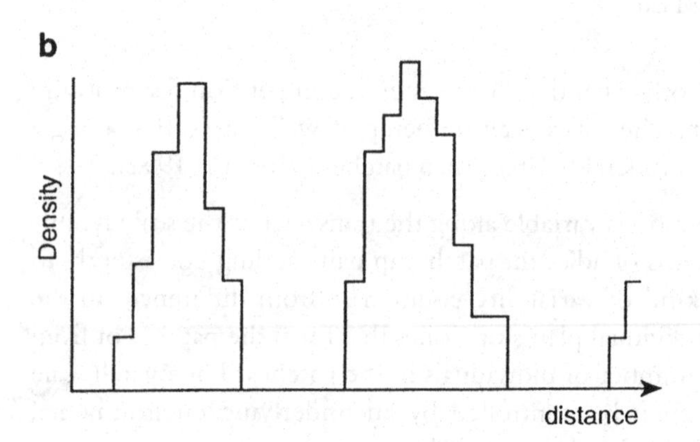

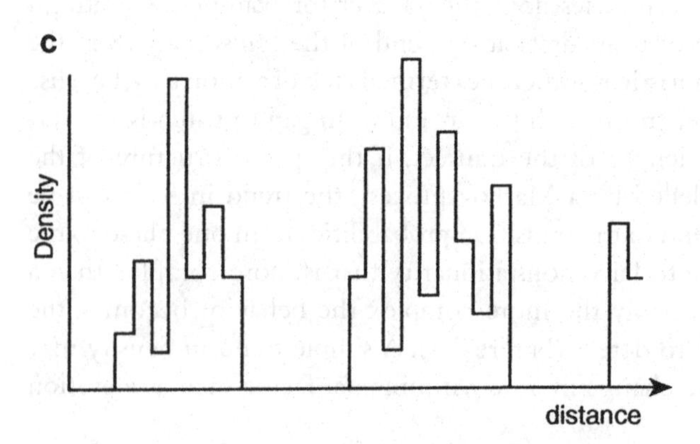

Figure 3.2 Intrapatch variability with *a* concave, *b* convex, and *c* irregularly varying densities within the patches.

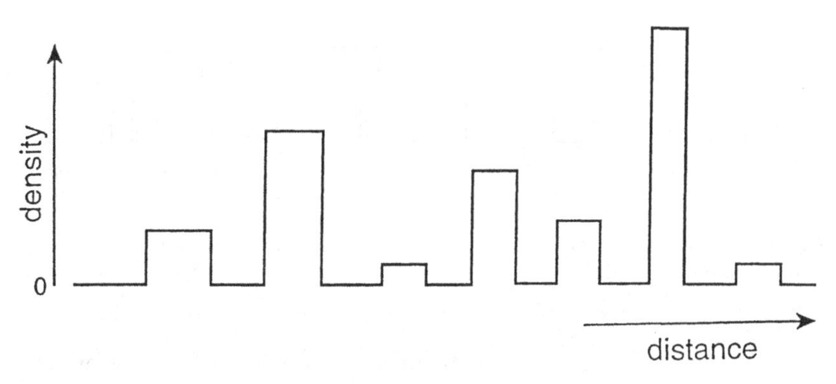

Figure 3.3 Interpatch variability resulting from the occurrence of patches with densities higher or lower than average.

scale *B* but are separated by large gaps of β empty quadrats, thus introducing a second scale of pattern, here β. For example, in an open forest, the smaller scale of pattern may arise from the canopies of individual trees and the small spaces between them, but a larger scale of pattern may arise from local disturbances of tree-fall or from permanent features that lack trees, such as wetlands or rock outcrops (Qinghong & Hytteborn 1991).

5. There may be intrapatch variability with consistently higher density in the middle of the patch (convex), consistently lower density in the center (concave), or irregular variation of density within the patch (Figure 3.2). (If the patches are consistently convex and close together, it might be possible to model the density with a sine or cosine function, rather than the square wave used in other cases.) Patches with a convex density profile may be the result of active growth and outward expansion; whereas concave density profiles may result from senescence and degeneration of the patch centers.

6. Interpatch variability may exist in which some whole patches have lower than average density and others have higher than average density (Figure 3.3). This variability may arise from differences in the age and density of the plants that form the patch or from underlying controls such as patchiness of nutrients.

7. There may be 'error' quadrats such as isolated quadrats within gaps that have nonzero density or isolated quadrats of zero density within patches (Figure 3.4). Isolated nonzero observations in gaps may represent post-disturbance residuals or advanced colonizers, whereas zeros in patches may represent very local disturbance such as broken branches or

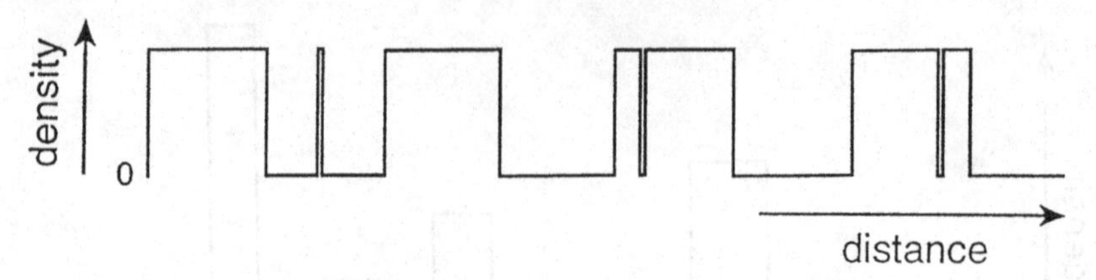

Figure 3.4 Error quadrats: isolated quadrats either of zero density within patches or of nonzero density within gaps.

the effects of pathogens or herbivores. They may also just represent incomplete development in the patch due to the plant's architecture or other feature.

The introduction of irregularity into the basic model of spatial pattern, as just described, relied in several cases on differences in density. There are, however, two main ways of looking at the plants that occur in the sampling units: density on the one hand and presence/absence on the other. In our analysis of the Mt. Robson moraine data which will be used to illustrate some of the methods described in this chapter, we found that the analysis of density data as presence/absence data gave essentially the same results (Dale & Blundon 1991). That will not always be the case as will be illustrated below.

It is possible for a single data set to combine more than one scale of pattern and there is more than one way in which scales can be combined. The combination can be additive, so that the density found in a quadrat is the sum of the densities attributable to each scale of pattern (Figure 3.5). Additive combination would be expected if density is controlled by several nutrients in the soil, each of which exhibits spatial variation in concentration at different scales. Of course, in presence/absence data, truly additive combination is not possible. The combination of scales can be multiplicative so that a quadrat has nonzero density only if it occurs in a patch of each of the component scales of pattern (Figure 3.5). Multiplicative combination might occur when several factors, which vary at different scales, completely prevent plant growth beyond certain threshold intensities. Presence/absence data lend themselves well to the multiplicative combination of component patterns. Interrupted patterns, described above, can result from the multiplicative combination of patterns.

The quadrat variance methods that will be described below detect the scale and intensity of patterns in vegetation, but most do not give directly the sizes of the patches and gaps that make up the pattern. In some

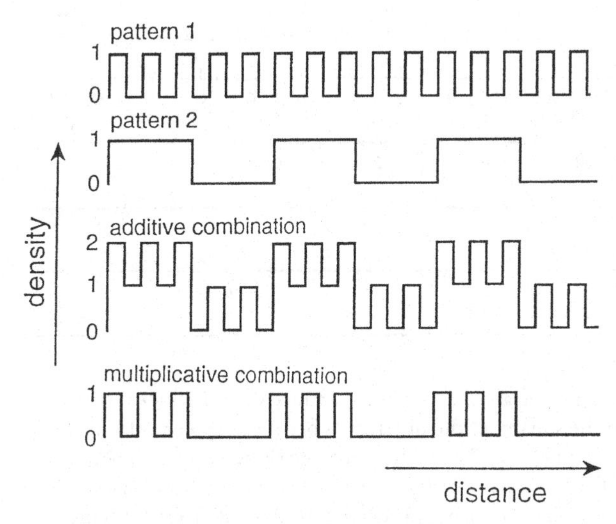

Figure 3.5 The first two lines show the two scales of pattern to be combined. The third line shows the two scales of pattern combined additively, so that the density in any quadrat equals the sum of its densities attributable to each scale of pattern. The fourth line shows the scales of pattern combined multiplicatively, where a quadrat has a nonzero density only when it corresponds to a patch in each of the component scales of pattern.

applications of spatial pattern analysis, the relationship between gap size and patch size may be as important as the pattern's scale. For example, during primary succession, the vegetation may consist of aggregations of patches. At first, as the patches grow, the scale of pattern related to the initial spacing of the individuals remains constant with patch size increasing and gap size decreasing. Finally the patches begin to coalesce and the smaller scale of pattern is lost, so that only the pattern due to the aggregations of patches remains. This was illustrated in Chapter 1. In a study of primary succession then, knowing the contribution of patch and gap size to pattern will be important.

There are several methods available for spatial pattern analysis that are based on the calculation of variances over a range of spatial scales and these come in two basic types. The first type includes blocked quadrat techniques in which the quadrats are combined into groups of particular sizes (blocks) and in which scale is related to the size of the blocks and to the spacing between the centers of adjacent blocks. The second category of quadrat variance methods can be referred to as spaced quadrat methods, because the quadrats remain as individual units rather than being combined into blocks, and the scale is determined only by the spacing or distance between quadrats.

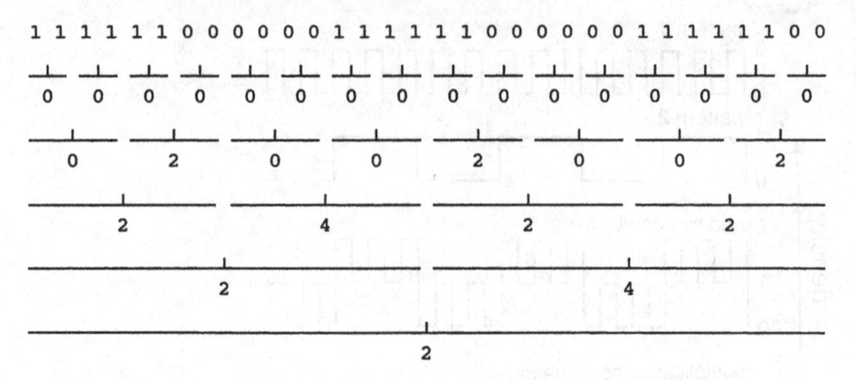

Figure 3.6 Blocked quadrat variance (BQV): the variance of each block size is calculated as the sum of the squares of the differences between adjacent blocks divided by the number of quadrats. Those differences are shown below the lines representing the pairs of blocks. For $b=1$ the variance is 0; for $b=2$ the variance is $3\times 2^2/32=0.375$; for $b=4$ the variance is $(3\times 2^2+4^2)/32=0.875$; for $b=8$ the variance is $(2^2+4^2)/32=0.625$; and for $b=16$, the variance is $2^2/32=0.125$.

Blocked quadrat variance

The original method for analyzing contiguous quadrat data is due to Greig-Smith (1952) as modified by Kershaw (1957), and is generally referred to as blocked quadrat variance (BQV). Ver Hoef *et al.* (1993) refer to this method as the nested analysis of variance (ANOVA) approach and Carpenter and Chaney (1983) call it hierarchical ANOVA. The method combines the quadrats into nonoverlapping blocks by powers of two and calculates a variance for each block size 1, 2, 4, 8 . . .

For block size one the variance is:

$$V_B(1)=2/n\sum_{i=1}^{n/2}\left(x_{2i-1}-x_{2i}\right)^2/2, \tag{3.1}$$

where x_i is the density in the ith quadrat and n is the number of quadrats in the transect; n is itself a power of 2.

For block size two the variance is:

$$V_B(2)=4/n\sum_{i=1}^{n/4}\left(x_{4i-3}+x_{4i-2}-x_{4i-1}-x_{4i}\right)^2/4. \tag{3.2}$$

In general for block size b, $b=2^k$

$$V_B(2^k)=\sum_{i=1}^{n/2b}\left(\sum_{j=2b(i-1)+1}^{i2b-b} x_j-\sum_{j=i2b-b+1}^{i2b} x_j\right)^2/n$$

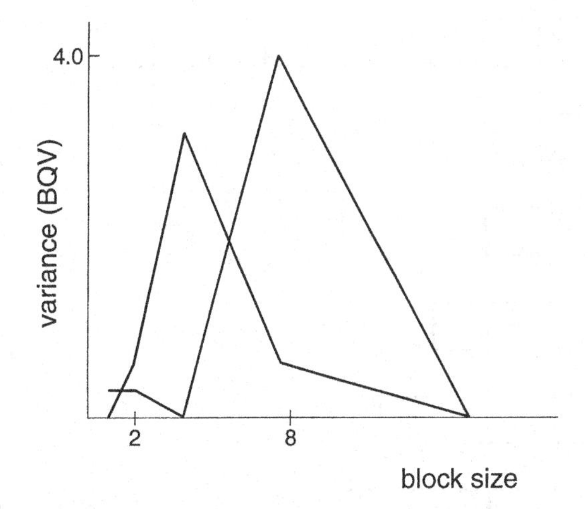

Figure 3.7 The results of BQV depend in part on starting position. The true scale is 8 and if the pattern is in phase with the quadrats (that is, it starts 1111111100000000 ...) there is a peak at block size 8. If the pattern is not in phase with the positions of the quadrats (that is, it starts 1111000000001111 ...), there is a peak at block size 4 (cf. Errington 1973, Figure 1).

$$= 2b\sum_{i=1}^{n/2b}\left(\sum_{j=2b(i-1)+1}^{i2b-b}[x_j - x_{j+b}]\right)^2/n. \qquad (3.3)$$

The variance is then plotted as a function of block size and peaks in the plot are interpreted as indicative of scales of pattern.

This method can only examine block sizes that are powers of two and clearly relies on the transect length being chosen appropriately as a number that is itself a power of 2. This property becomes a greater liability as the length of the transect or the largest block size examined grows. For example, suppose we have a transect that is $2^{10} = 1024$ quadrats long and variance is equally great at block sizes 256 and 512; all that we could conclude would be that there was pattern with a scale somewhere between 250 and 500 quadrats. The method has a number of other limitations that have been pointed out by several authors. For instance, the results depend to some extent on the starting position of the blocking and the variance peak can occur at a block size smaller than the actual scale of pattern (Errington 1973). For example, if $B = 8$, with $p = 6$ and $g = 10$, when the pattern begins with the 6 dense quadrats, the peak occurs at 8, but if it begins with 4 empty quadrats, the peak is at 4 (see Figure 3.7). Another drawback is that the size of the blocks is confounded with the distance between their centers (Goodall 1963). These concerns gave

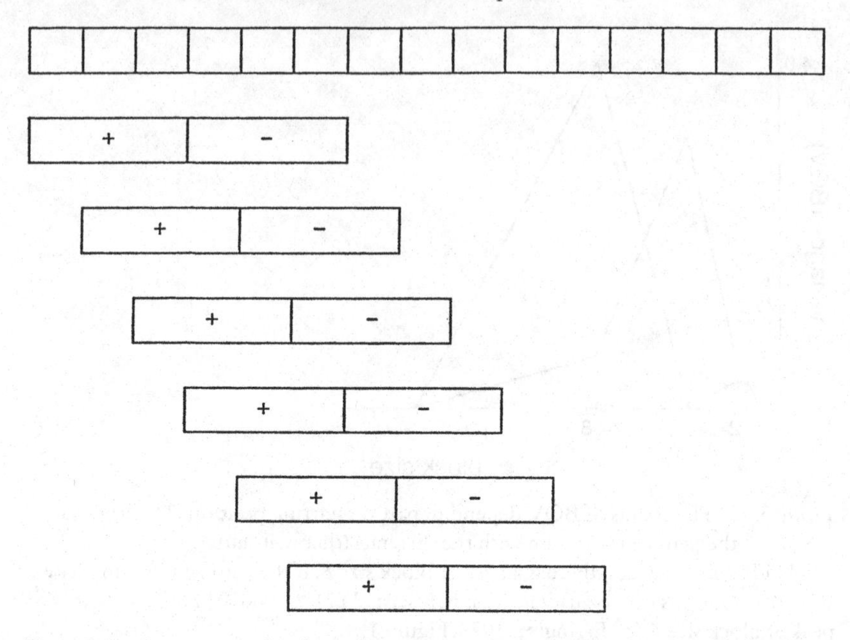

Figure 3.8 In two-term local quadrat variance (TTLQV), variance is calculated as the average of the square of the difference between the block totals of all possible adjacent pairs of block size *b*. The values in the quadrats of the '+' block are added together and the values in the '−' block are then subtracted. The differences are squared and then averaged over all positions.

rise to a number of new methods that avoided some of the problems of BQV.

Local quadrat variances

Two new methods were published by Hill in 1973, both of which use a complete range of block sizes and which, in essence, average over all possible starting positions for the blocking. These two methods are two-term local quadrat variance (TTLQV) and its three-term variant (3TLQV). The variance in TTLQV is calculated as:

$$V_2(b) = \sum_{i=1}^{n+1-2b} \left(\sum_{j=1}^{i+b-1} x_j - \sum_{j=i+b}^{i+2b-1} x_j \right)^2 / 2b(n+1-2b). \tag{3.4}$$

Essentially, what it is doing is calculating the average of the square of the difference between the block totals of adjacent pairs of blocks size *b* (see Figure 3.8). All possible contiguous adjacent pairs of blocks are examined.

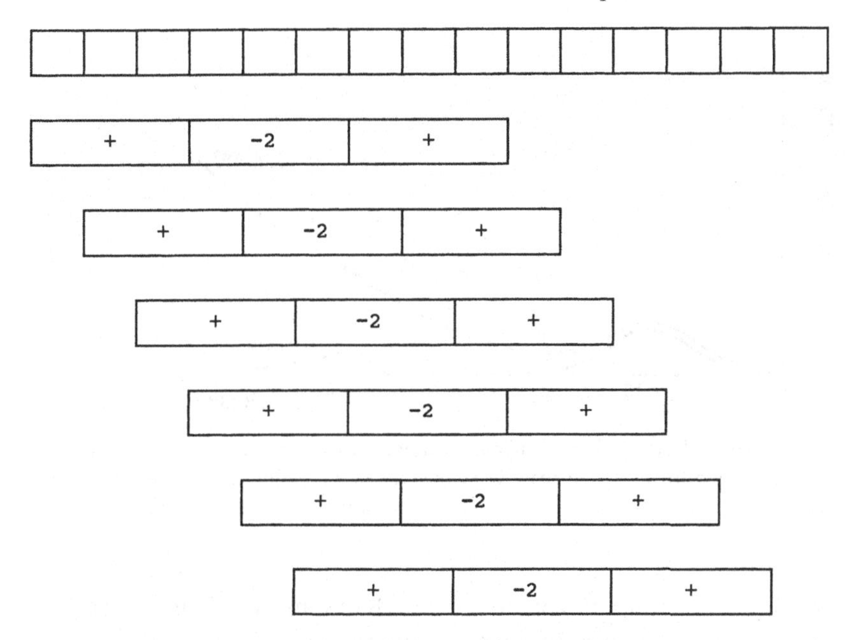

Figure 3.9 Three-term local quadrat variance (3TLQV) examines the squared differences among trios of adjacent blocks size b, by subtracting twice the total of the middle block from the sum of the two surrounding blocks. The differences are squared and averaged over all positions.

The three-term variance 3TLQV is similar:

$$V_3(b) = \sum_{i=1}^{n+1-3b} \left(\sum_{j=1}^{i+b-1} x_j - 2 \sum_{j=i+b}^{i+2b-1} x_j + \sum_{j=i+2b}^{i+3b-1} x_j \right)^2 / 8b(n+1-3b). \qquad (3.5)$$

It examines the average of squared differences among trios of adjacent blocks of size b, by subtracting twice the total of the middle block from the sum of the two that surround it (Figure 3.9). The principal difference between TTLQV and 3TLQV is that the latter is less sensitive to trends in the data (see Figure 3.10).

In Hill's (1973) original formulation, the divisor of 3TLQV contained '6' where we have '8'; the modification makes the variances calculated by the two methods directly comparable, as is demonstrated below. In both TTLQV and 3TLQV, the variance is plotted as a function of block size and peaks in that plot are interpreted as being indicative of pattern at that scale. These methods are illustrated in Figure 3.11.

It is generally agreed that TTLQV and 3TLQV are the recommended techniques for pattern analysis (Ludwig 1979; Lepš 1990b) and so we will

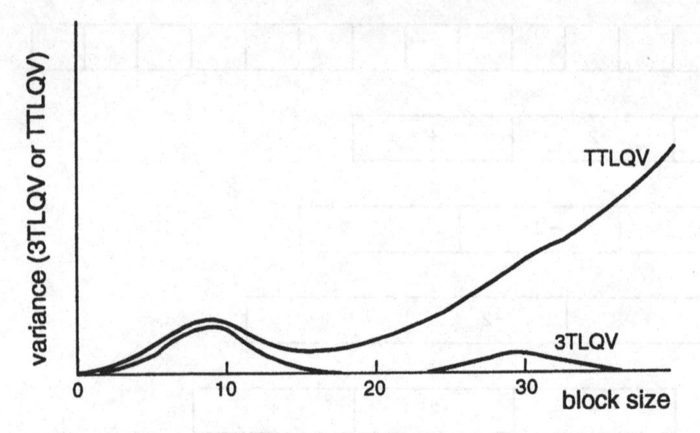

Figure 3.10 3TLQV is less sensitive to trends in data than is TTLQV. The data were derived from a simple square wave of scale 10 with a linear trend in the density of the patches and the gaps. A trend in patch density only does not produce this effect.

investigate the properties of those methods in some detail. In doing so, we will investigate the relationship between scale and variance peak position, the combination of several scales of pattern in a data set and suggested measures of the consistency and regularity of pattern based on those calculations.

Let us first consider data with only one scale of pattern, consisting of strings of g empty quadrats alternating with strings of p quadrats, all with

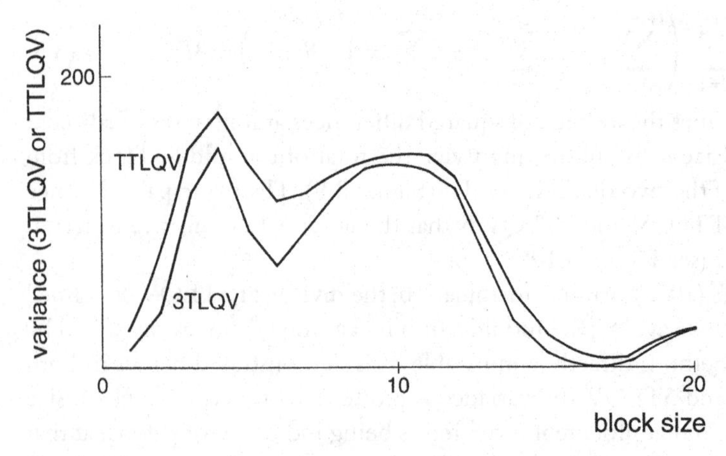

Figure 3.11 TTLQV and 3TLQV analysis of data with scales of 4 and 10. The positions of the variance peaks reflect those scales closely.

density d. Let q be the smaller of p and g. By examining Figure 3.12, it is clear that in a perfect pattern, there are $B+1-q$ pairs of blocks in each cycle of B pairs that contribute q^2d^2 to the sum that is the numerator of the TTLQV variance. The remaining pairs of blocks contribute $[(q-2)d]^2$, $[(q-4)d)^2 \ldots$, $0d^2$ or $1d^2, \ldots [(q-4)d]^2$, $[(q-2)d]^2$, with the middle term depending on whether q is even or odd.

Putting these kinds of terms into the equation for $V(B)$, we get:

$$V_2(B) = \left\{ (qd)^2(B+1-q) + 2\sum_{j=1}^{[q/2]} ((q-2j)d)^2 \right\} / 2B^2$$

$$= d^2 \left\{ q^2(B+1-q) + 2\sum_{j=1}^{[q/2]} (q^2 - 4jq + 4j^2) \right\} / 2B^2$$

$$= d^2 q \{ q^2(B+1-q)/2 + q^3/2 - q^2(q+2)/2 + q(q+1)(q+2)/6 \} / B^2$$

$$= d^2 q \{ 3q(B+1-q) + (q-1)(q-2) \} / 6B^2. \quad (3.6)$$

where the square brackets indicate the integer part of the division. For example, if $g=7$ and $p=5$ and $d=20$, then:

$$V_2(6) = \frac{20^2 \times 5(15 \times 2 + 4 \times 3)}{6 \times 36} = 389. \quad (3.7)$$

In many applications we may not know q, and so to derive an estimate of d from $V_2(B)$ we will either have to rely on some complementary method to find p and g and thus q, or assume that g and p are approximately equal. If we follow the second option and assume that $p=g$ so that $q=B$, the equation for $V_2(B)$ simplifies to:

$$V_2(B) = d^2(B^2+2)/6B. \quad (3.8)$$

Thus, when a TTLQV or 3TLQV analysis gives a single peak at block size B, the estimated intensity of the pattern that gives rise to the peak is:

$$I = \sqrt{6BV(B)/(B^2+2)}. \quad (3.9)$$

When there are two or more scales of pattern, their intensities cannot be estimated directly in this way, because the variance calculated at block size b consists of two components: variance due to pattern at scale B_1, $v(b,1)$, and variance due to pattern at scale B_2, $v(b,2)$ where the lower-case v is used to indicate an additive component of variance. If there are two scales of pattern that are combined additively, i.e., $x_i = x_i(1) + x_i(2)$ then the corrected sums of squares (and here the variances as well) due to the

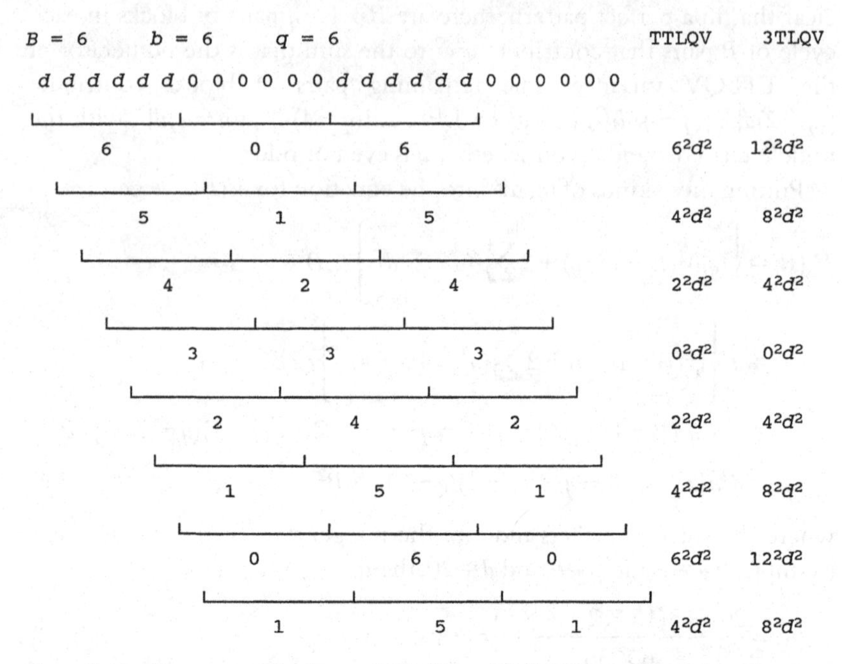

Figure 3.12 Terms contributing to TTLQV and 3TLQV in artificial data. The TTLQV values are based on the left pair of blocks indicated. When the pattern is perfectly regular and the block size, b, is the same as the scale of the pattern, B, each term for 3TLQV is exactly 4 times the TTLQV term. *a* Size of smaller phase, q, is the same as scale, $B=6$. *b* $q=5$. *c* $q=4$.

two patterns are also additive (cf. Sokal & Rohlf, 1981 p. 198). Where b is any block size and B_1 is the scale of the first pattern intensity d_1, we can calculate $v(b,1)$ by considering a perfect pattern of B_1 0's alternating with $B_1 d_1$'s such as is illustrated in Figure 3.13. Let h be the smallest difference between b and an even multiple of B_1: $h=\min(b \bmod 2B_1, 2b-b \bmod 2B_1)$, where min means minimum and mod means modulo.

If $h<B_1/2$, then:

$$v(b,1)=d_1^2\left(h^2+2\sum_{j=0}^{h-1}j^2\right)/2bB_1=d_1^2\left(h^2+\frac{(h-1)h(2h-1)}{3}\right)/2bB_1$$

$$=\frac{d_1^2 h}{6bB_1}(3h+2h^2-3h+1)=\frac{d_1^2 h}{6bB_1}(2h^2+1). \tag{3.10}$$

If $h\geq B_1/2$, then:

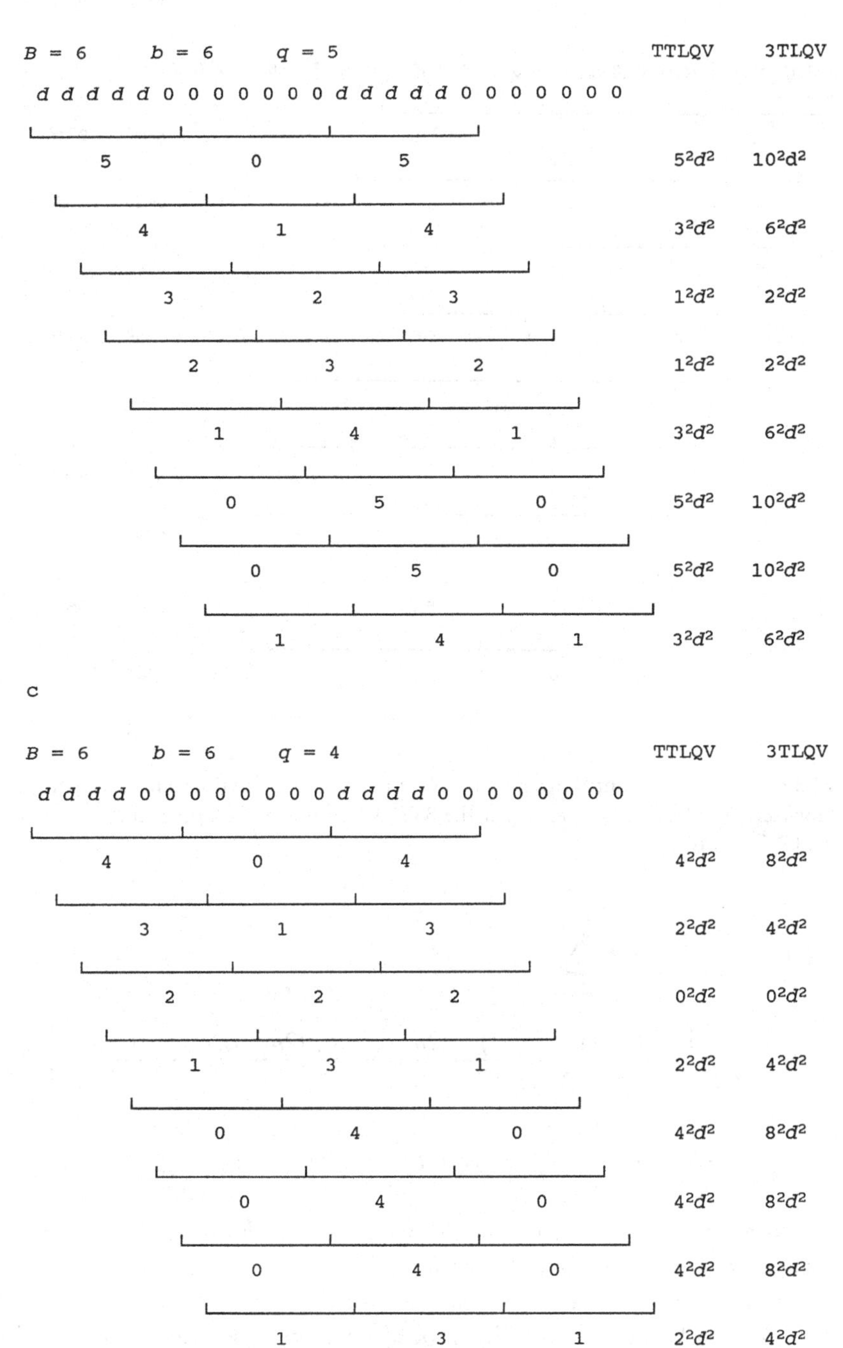
b
B = 6 b = 6 q = 5
TTLQV 3TLQV
d d d d d 0 0 0 0 0 0 0 d d d d d 0 0 0 0 0 0 0
5 0 5
5²d² 10²d²
4 1 4
3²d² 6²d²
3 2 3
1²d² 2²d²
2 3 2
1²d² 2²d²
1 4 1
3²d² 6²d²
0 5 0
5²d² 10²d²
0 5 0
5²d² 10²d²
1 4 1
3²d² 6²d²
c
B = 6 b = 6 q = 4
TTLQV 3TLQV
d d d d 0 0 0 0 0 0 0 0 d d d d 0 0 0 0 0 0 0 0
4 0 4
4²d² 8²d²
3 1 3
2²d² 4²d²
2 2 2
0²d² 0²d²
1 3 1
2²d² 4²d²
0 4 0
4²d² 8²d²
0 4 0
4²d² 8²d²
0 4 0
4²d² 8²d²
1 3 1
2²d² 4²d²

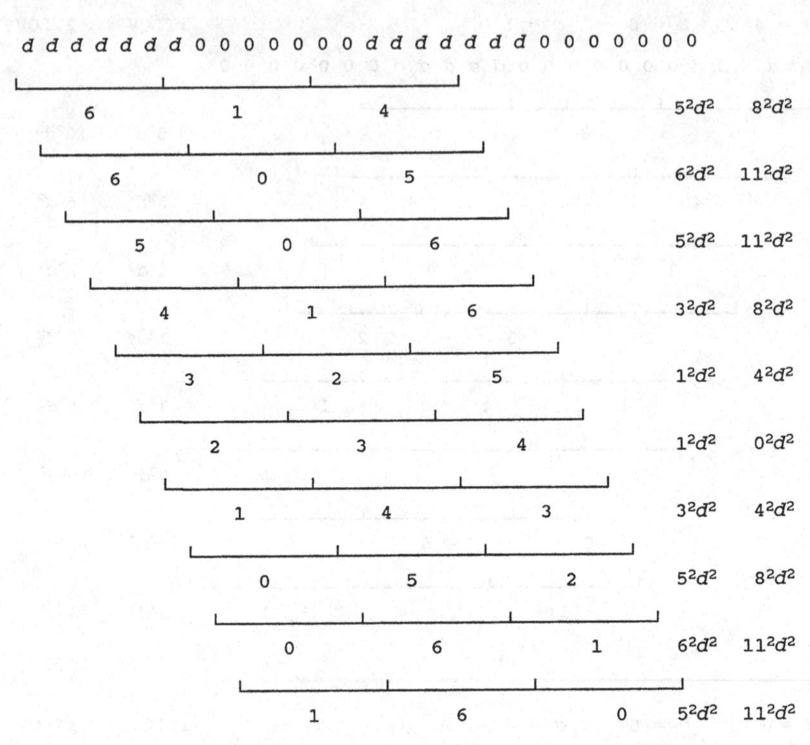

Figure 3.13 Terms contributing to TTLQV and 3TLQV in artificial data. When block size b and scale B are not equal, the 3TLQV terms are not 4 times the TTLQV terms.

$$v(b,1)=\frac{d_1^2}{2bB_1}\left[h^2+2\sum_{j=2h-B_1}^{h-1}j^2+2\sum_{j=1}^{h-1-B_1/2}(2j)^2\right]$$

$$=\frac{d_1^2}{2bB_1}\left[h^2\frac{(h-1)h(2h-1)}{3}+\frac{(2h-2-B_1)(2h-B_1)(2h-1-B_1)}{3}\right]$$

$$=\frac{d_1^2}{6bB_1}[h(2h^2+1)-(2h-B_1)(2h-B_1-1)(2h-B_1+1)]. \qquad (3.11)$$

(Note that apart from the single occurrence of b in the divisor, the variance depends only on d, B_1 and h.)

Knowing the variances, we can solve for the values of d_1 and d_2, and thus for $v(b,1)$ and $v(b,2)$. The procedure is best described using an example:

$B_1=6$; $B_2=14$. Let $h(i,j)$ be the smallest difference between i and an even multiple of j so that $h(B_1, B_2)=6$ and $h(B_2, B_1)=2$ because $14-2\times6=2$. $V_2(6)=770$; $V_2(14)=960$.

$$v(B_2,1)=d_1^2\times2\,(2\times2^2+1)/(6\times6\times14)=0.035\,d_1^2.$$
$$v(B_1,2)=d_2^2\times6(2\times6^2+1)/(6\times6\times14)=0.87\,d_2^2.$$
$$v(B_2,2)=d_2^2(14^2+2)/(6\times14)=2.36\,d_2^2.$$
$$v(B_1,1)=d_1^2\,(6^2+2)/(6\times6)=1.05\,d_1^2.$$

Therefore:

$$0.035\,d_1^2+2.36\,d_2^2=960,$$
$$1.05\,d_1^2+0.87\,d_2^2=770.$$

Multiply the upper of these two equations by 30:

$$1.05\,d_1^2+70.8\,d_2^2=28800.$$

Subtract to get:

$$69.9\,d_2^2=28030.$$

Divide by 69.9:

$$d_2^2=401.0.$$

Substituting back:

$$d_1^2=401.5.$$

Therefore the total variance of 770 at block size 6 is made up of 424 due to the pattern at scale 6 and 346 from scale 14. The total of 960 at block size 14 is mainly due to the pattern at that scale which contributes 946 (cf. Figure 3.14.)

Thus, given two peak variances associated with two scales of pattern, it is possible to partition each of them into two additive components. This is true also for three or more peaks, but of course the procedure would involving solving more equations for more unknown intensities. It is not clear under what circumstances such multiple-peak decomposition would be useful. If there are two patterns that are combined multiplicatively, it is obvious that additive decomposition will produce less satisfactory results. One approach to decomposing that kind of pattern is to use a log transform of the data and then treat them as an additive combination. We have not tested this suggestion.

There is another important consequence of the equations given above, because from them we can derive that for a perfect square wave pattern of B d's alternating with B 0's, for $B>4$, then if $b=h=B$:

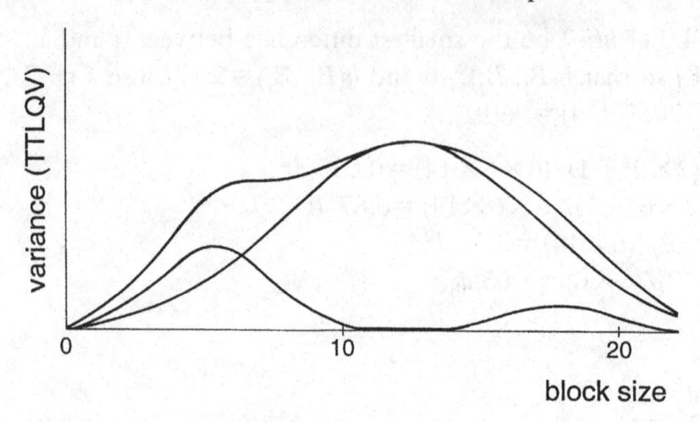

Figure 3.14 The variance plot of spatial pattern with two scales of pattern, 6 and 14, combined additively is the sum of the variance plots derived from separate patterns of the same scales.

$$V_2(B) = d^2[B(2B^2+1) - (2B-B-1)(2B-B)(2B-B+1)]/6B^2$$
$$= d^2[2B^3+B-(B-1)(B)(B+1)]/6B^2$$
$$= d^2\frac{B^3+2B}{6B^2} = d^2\frac{B^2+2}{6B}. \quad (3.12)$$

and if $b=h=B-1$, then:

$$V_2(b) = d^2\frac{(B-1)(2B^2-4B-1)-(B-3)(B-2)(B-1)}{6B(B-1)}$$
$$= d^2\frac{(2B^2-4B+3)-(B^2-5B+6)}{6B}$$
$$= d^2\frac{B^2+B-3}{6B}. \quad (3.13)$$

Therefore:

$$V_2(B) = d^2(B^2+2)/6B, \quad (3.14)$$

and

$$V_2(B-1) = d^2(B^2+B-3)/6B. \quad (3.15)$$

Comparing these, $V_2\ (B-1) > V_2(B)$ whenever $B^2+B-3 > B^2+2$; that is, when $B>5$. For example, if $B=20$, $V_2(20)=3.35d^2$ but $V_2(19)=3.475d^2$.

Following the same procedure, we can show that $V_2(B-2) > V_2(B$

-1) when $14<B<25$, $V_2(B-3)>V_2(B-2)$ when $24<B<34$, and so on. With increasing scale, the peak variance deviates more and more from the true scale. The relationship between the peak variance and the true scale of the pattern is approximately (Dale & Blundon 1990):

$$B=b+[30b/255]+1 \text{ for } b>5 \tag{3.16}$$

where the square brackets indicate the integer part of the division. That equation was formulated based on the positions of variance peaks for $B=2$ to $B=300$.

The phenomenon of the variance peak drifting from the actual scale of pattern seems like a serious drawback to the TTLQV approach, although we now know how to correct for it. The problem can be completely eliminated by changing the divisor in the formula for variance by leaving out the b in the divisor:

$$V_2(b)=\sum_{i=1}^{n+1-2b}\left(\sum_{j=i}^{i+b-1}x_j-\sum_{j=i+b}^{i+2b-1}x_j\right)^2/2b(n+1-2b) \tag{3.17}$$

becomes

$$V_c(b)=\sum_{i=1}^{n+1-2b}\left(\sum_{j=i}^{i+b-1}x_j-\sum_{j=i+b}^{i+2b-1}x_j\right)^2/2(n+1-2b). \tag{3.18}$$

Based on the equations presented above, if $b=h=B$ then:

$$V_c(B)=(B^2+2)/6; \tag{3.19}$$

if $b=B-1$ then $h=B-1$ and

$$V_c(B-1)=(B^2+B-3)(B-1)/6B. \tag{3.20}$$

which is less than $V_c(B)$ provided $B>0$. $V_c(B+1)=V_c(B-1)$ and so the peak in the variance will always occur at block size B (Figure 3.15). The problem with removing the term b from the divisor is that the resonance peaks do not diminish as in TTLQV (Figure 3.16). There is a clear tradeoff between the disadvantages of peak drift and nondiminishing resonance peaks, which we will encounter in other related techniques. In analyzing real ecological data, the precise position of the variance peak and the precise scale of pattern may not be as important as the ease of interpreting the analysis. Because resonance peaks can be very confusing, the drift of the variance peak may be the lesser of the two problems, and the original formulation of TTLQV is to be preferred.

For a pattern that is a perfect square wave of B empty quadrats alternating with B quadrats of density d, the intensity of the pattern, I, is equal

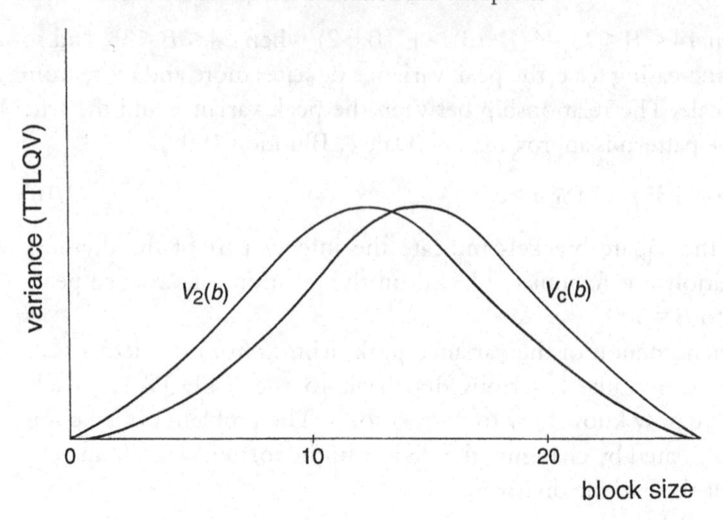

Figure 3.15 A comparison of the original TTLQV with its modified form, V_c, which omits b from the divisor: V_c does not have the peak drift problem of TTLQV. The variance peak is exactly at the scale of the pattern. Here the scale of pattern is 15; V_2 has a peak at 13, V_c at 15. The curves are scaled to be the same height.

to the patch density, d. I and d can be derived directly from the known variance, $V_2(B)$, as described above:

$$I = \sqrt{6B\, V_2(B)/(B^2+2)}. \tag{3.21}$$

For a pattern that is of unknown regularity and unknown patch:gap ratio but which is known to have a single scale of pattern, that function of the variance can be used as a measure of the pattern's intensity. The value of I calculated from the variance can be compared with d_{avg}, the average density of quadrats in patches. Where the two values are similar, any irregularity in the pattern can be attributed to density differences. On the other hand, if the intensity is low compared to the average patch density, the pattern's irregularity can be attributed to differences in patch and gap sizes.

Where there are several scales of pattern, only that portion of the variance at block size B that is attributable to the pattern of scale B should be used to calculate I.

On examining Figure 3.12, it is obvious that, in this instance, the contributions of each element to the 3TLQV variance is exactly four times the contribution of an equivalent element to TTLQV. Dividing the 3TLQV variance by 8 (compared to the 2 for TTLQV), the two tech-

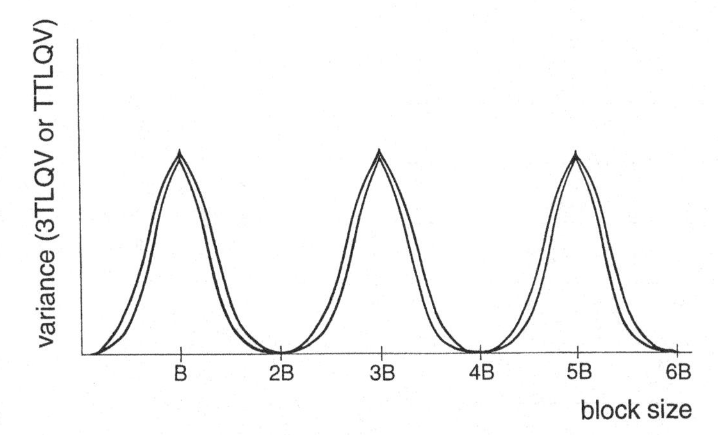

Figure 3.16 The modified local quadrat variances, in both two-term and three-term form (lower curve), produce resonance peaks at block sizes 3B, 5B, and so on, that do not diminish.

niques will give the same values. 3TLQV will always give the same variance as TTLQV for $b=B$, but that is not the case when $b \neq B$.

By examining figures like Figures 3.13 and 3.17, we can show that if $b=B-1$, 3TLQV will be made up of terms such as:

$$S_3=(2B-3)^2d^2+\sum_{i=1}^{(B-3)/2}(4i)^2d^2, \tag{3.22}$$

when B is odd, and:

$$S_3=(2B-3)^2d^2+\sum_{i=1}^{(B-2)/2}(4i-2)^2d^2, \tag{3.23}$$

when B is even.

Both these terms reduce to:

$$S_3=(2B^3-14B+15)d^2/3, \tag{3.24}$$

giving, for a half cycle of $B/2$ terms

$$V_3(B-1)=d^2\frac{2B^3-14B+15}{12B(B-1)}.$$

This means that $V_3(B-1)>V_3(B)$ when $B>7$. For example, for $B=20$, $V_3(20)=3.35d^2$ as for V_2 and $V_3(19)=3.451d^2$ which is greater than $V_3(20)$ but less than $V_2(19)$.

By a similar process, we find that if $b=B-2$:

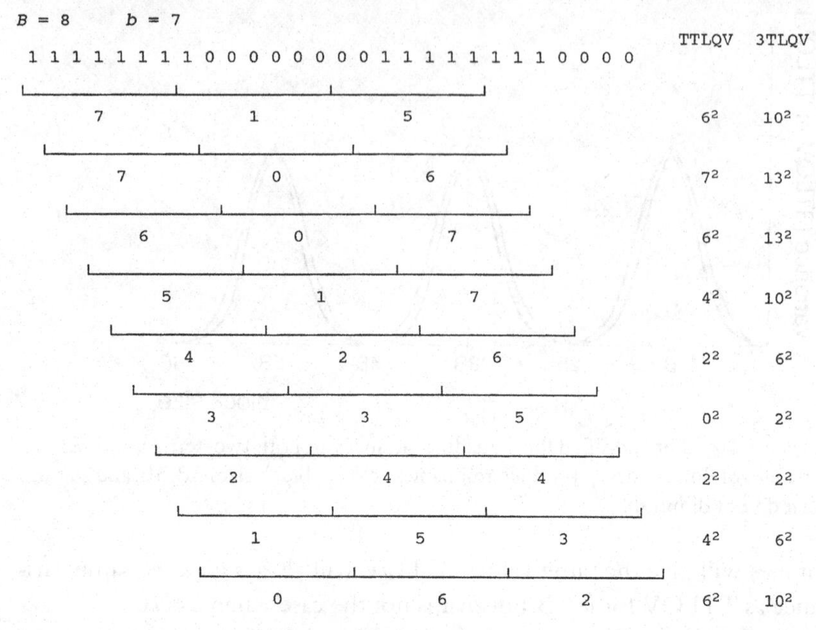

Figure 3.17 The calculation of $V_3(B-1)$ when B is even. If the patch density is d rather than 1, all terms are multiplied by d^2.

$$2S_3 = 3(2B-6)^2d^2 + 2(2B-9)^2d^2 + 2\sum_{i=1}^{(B-6)/2}(4i)^2d^2$$

$$S_3 = (2B^3 - 68B + 165)d^2. \tag{3.25}$$

$$V_3(B-2) = d^2\frac{2B^3 - 68B + 165}{12B(B-2)}. \tag{3.26}$$

This means that $V_3(B-2) > V_3(B-1)$ when $B>22$. The relationship between b and B for 3TLQV is (Dale & Blundon 1991):

$$B = b + [(b+9)/13.5] \text{ when } b>7. \tag{3.27}$$

The result indicates that, as B increases, the difference between B and the block size that gives the peak variance increases more slowly with 3TLQV than with TTLQV (Figure 3.18). Another advantage of 3TLQV compared with TTLQV is that it is less sensitive to trends in the data (Lepš 1990b), as is shown in Figure 3.10. The only apparent disadvantage of 3TLQV may be that the maximum block size that can be examined may be smaller than that which can be examined using TTLQV. Ludwig and Reynolds (1988) suggest that variances such as TTLQV (or PQV

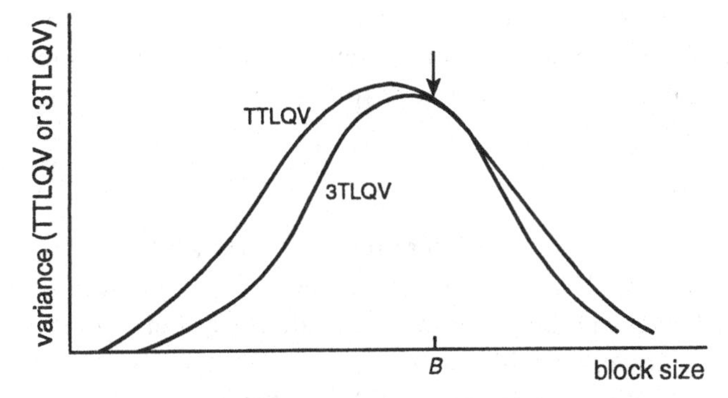

Figure 3.18 The drift of the peak variance is less in 3TLQV (lower curve) than in TTLQV (upper curve). The real scale of pattern is $B = 30$.

which follows) should not be calculated for block sizes or distances greater than 10% of the transect length. The reason they cite for this advice is that at sizes greater than 10%, the estimates will lack precision. The restriction to 10% of the total length requires further investigation.

Paired quadrat variance

To avoid the confounding of block size with distance, pairs of single quadrats at specified distances can be examined. If the pairs are chosen at random, the method is referred to as random paired quadrat variance, RPQV, or if all possible pairs are used then it is just paired quadrat variance, PQV (Ludwig & Goodall 1978). The variance in PQV at distance b is calculated as:

$$V_P(b) = \sum_{i=1}^{n-b} (x_i - x_{i+b})^2 / 2(n-b). \tag{3.28}$$

Again, peaks in the plot of variance as a function of distance are interpreted as scales of pattern. Ver Hoef *et al.* (1993) have demonstrated the close relationship between PQV and TTLQV, showing that PQV can be used to approximate TTLQV.

There is a close relationship between PQV and one of the basic methods from the field of geostatistics called the variogram. Several methods that were originally developed in the field of geostatistics, for analyzing the spatial dependence of geological phenomena (David 1977), are now being transferred to applications in ecology. The variogram is one of them (Rossi *et al.* 1992). Suppose that position in a multidimensional space is measured by vector $\mathbf{z}$ and that there is a variable, Y, representing a spatial process, the value of which depends on position and thus is a function of $\mathbf{z}$. Where $\mathbf{h}$ is a vector of the separation of two points in the multidimensional space, the variogram, $2\gamma(\mathbf{h})$, is a function of separation $\mathbf{h}$:

$$2\gamma(\mathbf{h}) = \mathrm{Var}\left[Y(\mathbf{z}+\mathbf{h}) - Y(\mathbf{z})\right]. \tag{3.29}$$

Rossi *et al.* (1992) point out the interesting fact that while mining geologists are credited with showing the usefulness of variograms, the approach was used earlier by biomathematicians such as Matérn (1947) who applied it to a study of Swedish forests.

To investigate the relationship between PQV and the variogram, consider the variation in density in one dimension. Where Y is a spatial density process and ζ measures position in one dimension, then the variogram is:

$$2\gamma(h) = \mathrm{Var}\left[Y(\zeta+h) - Y(\zeta)\right]. \tag{3.30}$$

Where h is the distance of separation. The estimate of the variogram, $2g$ or $2\hat{\gamma}$, is calculated as:

$$2g(h) = \sum_{i=1}^{n-h}\left[Y(i+h) - Y(i)\right]^2/(n-h). \tag{3.31}$$

Substituting x_i, the density in the ith quadrat, for $Y(\zeta)$ and b for h will give an equation similar to the one for PQV (3.28). The term 'semivariogram' also appears in geostatistical methodology; it is just half the variogram, that is $\gamma(h)$. In fact PQV (or RPQV) in a one-dimensional transect estimates the semivariogram (see also Ver Hoef *et al.* 1993).

The PQV method is closely related also to the calculation of the autocorrelation in the data. Let $C(h)$ be the covariance for quadrats at spacing h:

$$C(h) = \mathrm{Cov}\left[Y(\zeta), Y(\zeta+h)\right]. \tag{3.32}$$

Then $\gamma(h) = C(0) - C(h)$ (Ver Hoef *et al.* 1993); Figure 3.19 illustrates this dependence. Rossi *et al.* (1992) point out that the correlogram $\rho(h)$, which is the correlation coefficient as a function of distance h, is also related by simple equations:

$$\rho(h) = C(h)/C(0)$$

or

$$\rho(h) = 1 - \gamma(h)/C(0). \tag{3.33}$$

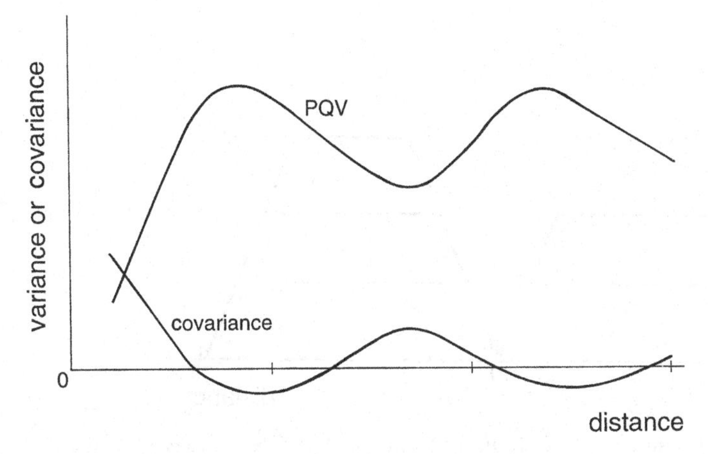

Figure 3.19 The close relationship between PQV (upper curve) which estimates the variogram, and the covariance (lower curve). The input scale of pattern is 8.

The above discussion has described PQV which examines all pairs of quadrats separated by a given distance. There is a statistical advantage to randomly selecting pairs of quadrats at a given distance *without replacement*, because the estimates of variance at different block sizes are then independent (Goodall 1974). This method is called the random paired quadrat variance (RPQV) for obvious reasons. The main disadvantage of the method is that, because the pairs must be chosen without replacement to ensure independence, the proportion of possible pairs at any

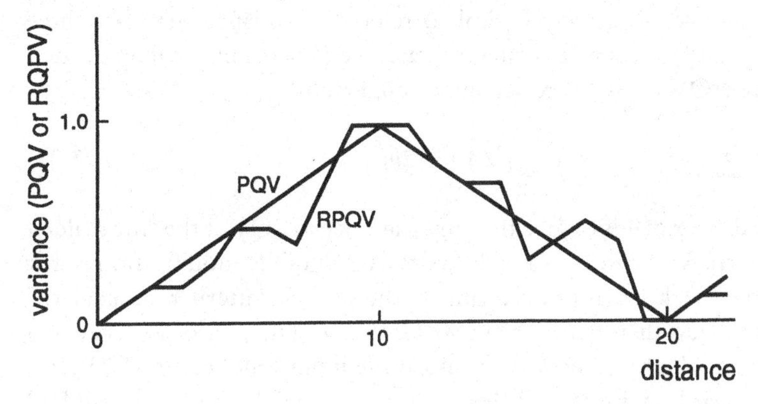

Figure 3.20 The inaccuracy of random paired quadrat variance (RPQV) compared to paired quadrat variance (PQV) (big triangle) caused by lower sampling intensity because sampling must be without replacement. Artificial data from a square wave with scale 10.

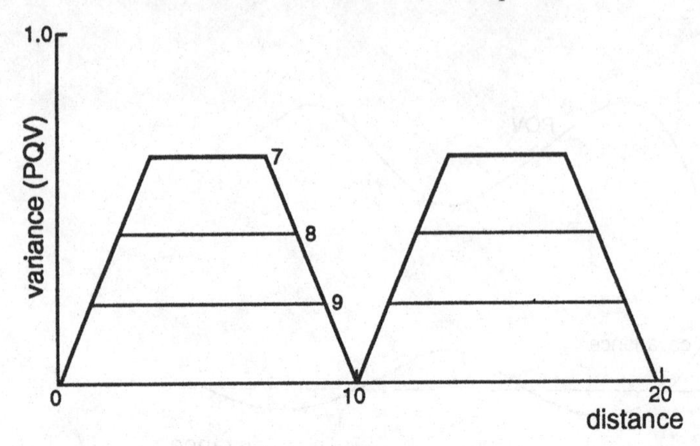

Figure 3.21 For simple artificial data, PQV does not give sharp peaks when the size of the larger phase is greater than the scale of pattern ($Q > B$). Here the scale is 5, with $Q = 7, 8$, and 9 as indicated. This will occur for any unbalanced pattern.

given distance that are used is small (and decreases with the range of distances examined) resulting in estimates of low precision (Figure 3.20).

In analyzing artificial data, a weakness of the PQV method became obvious in a way that would not appear clearly in an analysis of field data. Given the usual square wave underlying model, when patch size equals gap size, PQV gives variance peaks, but when they are unequal the result is flat-topped trapezoids, as in Figure 3.21. They do not give a clear indication of scale. A suggestion for improving the sensitivity of PQV for unequal patch and gap size is to use a three-term spaced-quadrat method. It does not seem entirely logical to refer to a variance that uses three terms as 'three-term paired-quadrat variance'; a name like 'triplet quadrat variance', tQV, seems more accurate, designated V_t:

$$V_t(b) = \sum_{i=1}^{n-2b} (x_i - 2x_{i+b} + x_{i+2b})^2 / 4(n-2b). \tag{3.34}$$

The divisor of 4 is to give the same variance as PQV at the true scale of the pattern. As Figure 3.22a–c shows, the result of the modification is that when the block size is not the same as the scale of pattern, the calculated variance is less than that in the two-term case. The plot of variance as a function of block size now has peaks at the input scale (Figure 3.23). It is not yet clear how important this modification will be when the method is applied to the analysis of field data.

In comparing PQV methods with TTLQV and 3TLQV, the paired quadrat methods have the disadvantage that the resonance peaks do not

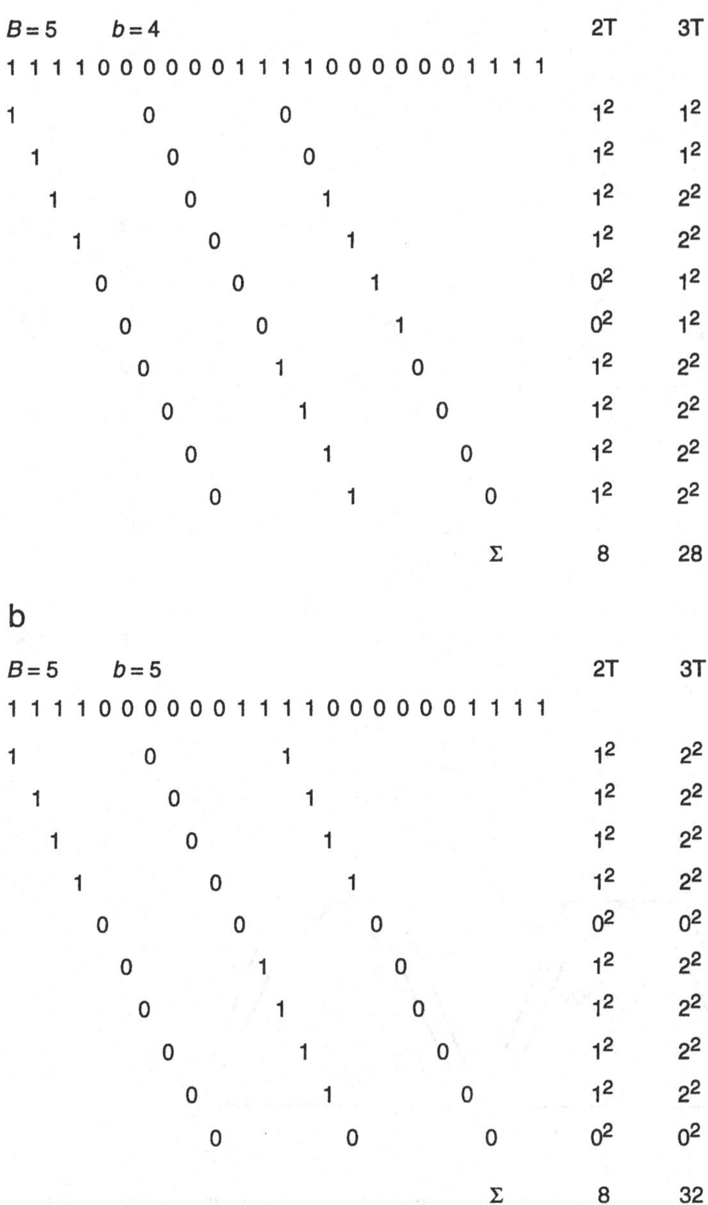

Figure 3.22 The calculation of spaced quadrat variance in its two-term form, PQV, and in the three-term form, tQV. The three-term method is divided by 4 giving a variance sum of 7 when $b = B - 1$ (*a*), or when $b = B + 1$ (*c*), but it gives 8 when $b = B$. The two-term version gives 8 in all three cases.

C

$B = 5$ $b = 6$

1 1 1 1 0 0 0 0 0 0 1 1 1 1 0 0 0 0 0 0 1 1 1 1

			2T	3T
1	0	1	1^2	2^2
1	0	1	1^2	2^2
1	0	0	1^2	1^2
1	0	0	1^2	1^2
0	1	0	1^2	2^2
0	1	0	1^2	2^2
0	1	0	1^2	2^2
0	1	0	1^2	2^2
0	0	1	0^2	1^2
0	0	1	0^2	1^2
		Σ	8	28

Figure 3.22 (cont.)

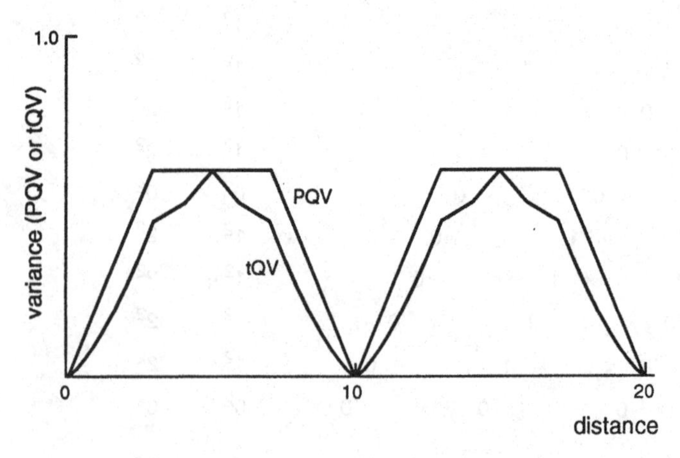

Figure 3.23 The two- and three-term versions of spaced quadrat variance using artificial data with a scale of 5 and gap size 7. The three-term version, tQV, gives a peak at the scale, not just a plateau.

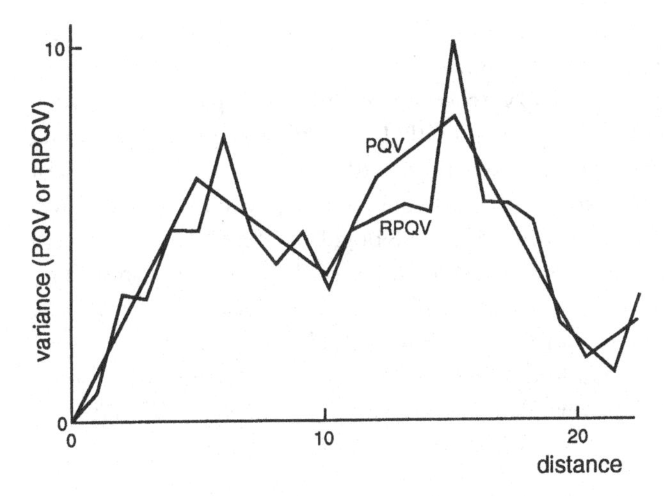

Figure 3.24 PQV and RPQV on artificial data with scales of 5 and 12. In PQV, the scale of 12 appears indistinctly as a shoulder; in RPQV it is lost.

diminish in height. Thus, if there are several scales of pattern in the data the variance plot can be very confusing. For instance, in Figure 3.24 the PQV plot has only a shoulder in the variance plot corresponding to the scale of pattern at 12, because of the resonance peak at 15 due to the pattern at scale 5. Carter and O'Connor (1991) tried to solve the problem of choosing between PQV and TTLQV by using PQV for small block sizes and TTLQV for larger block sizes. It is certainly a good idea to use more than one method, but their solution does not avoid the TTLQV's drawback of increasing peak drift at larger block sizes.

In that study, they examined the spatial pattern of a two-phase mosaic grassland in southern Africa, the phases being dominated by *Setaria incrassata* and *Themeda triandra* (Carter & O'Connor 1991). They used one-dimensional transects but placed some across the 2° slope of the site and some up and down the slope. PQV detected similar scales of pattern in both dominant species, but the pattern was anisotropic, with a scale of about 9m upslope and 4m along the contour. Interestingly, while the TTLQV analysis seems to show evidence of spatial pattern at larger scales (Carter & O'Connor 1991, Figure 3), the authors conclude that no significant contagion was found for either species at larger scales. The significance of peaks was assessed using Mead's '2-within-4' randomization test (Mead 1974) and it is not clear whether that approach is appropriate for TTLQV. There may, therefore, be real spatial pattern at larger scales. The authors conclude that the patchy spatial structure of the mosaic may be related to tree canopy, clonal growth, grazing and competitive exclusion (Carter & O'Connor 1991).

Schaefer (1993) used PQV to compare the spatial pattern of several species in burned and in old-growth boreal plant communities. The hypothesis being tested was whether, with succession, small scales of spatial pattern are lost so that the scale of pattern is greater in older stands. The communities sampled were mature bog, Jack Pine (*Pinus banksiana*) sand plain, mixed coniferous forest and mixed deciduous forest. The species used for single-species analysis included *Vaccinium vitis-idea, Arctostaphylos uva-ursi*, and *Cornus canadensis*. There was no general tendency for the older stands to have larger scales of pattern; in fact, for some species, the scale of pattern detected was smaller (Schaefer 1993, Table 3).

New local variance

The last of the related variance methods was suggested by Galiano (1982a), and is called new local variance, NLV; it was also proposed in both a two-term and three-term version. In the two-term form, the variance is calculated as:

$$V_N(b) = \sum_{i=1}^{n-2b} \left| \left(\sum_{j=i}^{i+b-1} x_j - \sum_{j=i+1}^{i+2b-1} x_j \right)^2 - \left(\sum_{j=i+1}^{i+b} x_j - \sum_{j=i+b+1}^{i+2b} x_j \right)^2 \right| / 2b(n-2b). \tag{3.35}$$

The NLV was proposed as a method for detecting the size of patches rather than as a method for detecting the scale of pattern. What it actually does is to detect the average size of the smaller phase of the pattern, whether it is the gaps or the patches. For example, a regular pattern with $p=7$ and $g=3$ produces a maximum value of V_N at $b=3$, and a pattern with $p=3$ and $g=7$ produces a V_N graph that is identical, also with a peak at $b=3$ (Figure 3.25a,b).

Galiano's NLV consists of the differences between adjacent TTLQV terms. Let $T(b,i)$ be the TTLQV term for block size b beginning at quadrat i:

$$T(b,\mathrm{i}) = \left(\sum_{j=1}^{i+b-1} x_j - \sum_{j=i+b}^{i+2b-1} x_j \right)^2 \tag{3.36}$$

$$V_N(b) = \sum_{i=1}^{n-2b} |\ T(b,i) - T(b,i+1)\ | /\ 2b(n-2b). \tag{3.37}$$

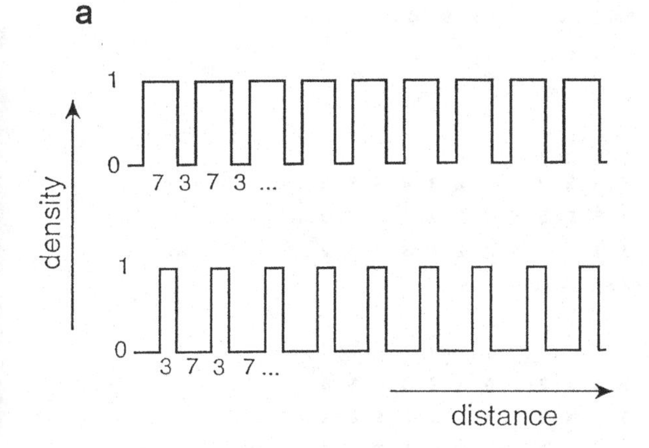

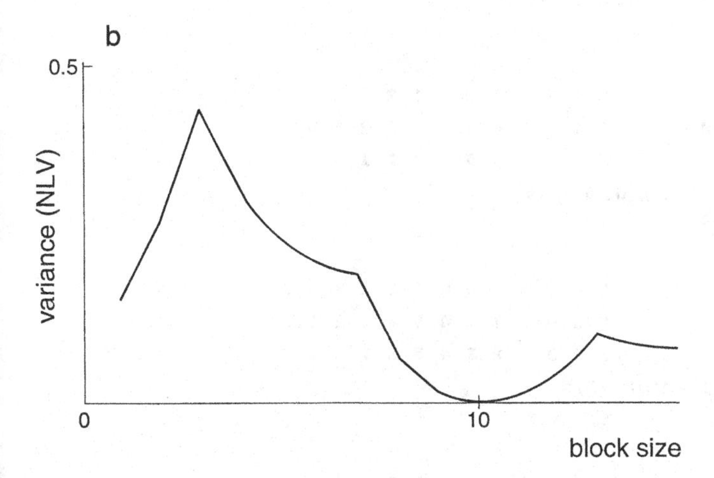

Figure 3.25 New local variance (NLV) detects the average size of the smaller phase of the pattern, whether it is the patches or the gaps. Part *b* is NLV as a function of block size for either pattern shown in *a*.

For the purposes of this discussion, it is easiest to deal with the model of a regular square wave made up of strings of g quadrats of density 0 alternating with p quadrats of density 1. Our findings can be modified for other average densities, if necessary. The scale of the pattern is B, and q, h, and b are as defined above for Equations 3.6 and 3.10. For example if $B = b = p = q = 7$, NLV is calculated from terms such as:

$$|7^2 - 5^2| + |5^2 - 3^2| + |3^2 - 1^2| + |1^2 - 1^2| + |1^2 - 3^2| + |3^2 - 5^2| + |5^2 - 7^2| + \ldots$$
$$= 49 - 25 + 25 - 9 + 9 - 1 + 1 - 1 + 9 - 1 + 25 - 9 + 49 - 25 + \ldots$$

11111000000111110000001111...

$B = 6$; $q = 5$.

$b = 8$; $h = 4$.

$S_i(b)$	5	4	3	2	1	1	2	3	4	5	5	5...
$S_{i+b}(b)$	4	5	5	5	5	4	3	2	1	1	2	3...
$T^{0.5}$	1	1	2	3	4	3	1	1	3	4	3	2...

$V_N = (4^2 - 1^2)/(6 \times 8) = 0.31$

$b = 7$; $h = 5$.

$S_i(b)$	5	4	3	2	1	0	1	2	3	4	5	5...
$S_{i+b}(b)$	2	3	4	5	5	5	4	3	2	1	0	1...
$T^{0.5}$	3	1	1	3	4	5	3	1	1	3	5	4...

$V_N = (5^2 - 1^2)/(6 \times 7) = 0.57$

$b = 6$; $h = 6$.

$S_i(b)$	5	4	3	2	1	0	0	1	2	3	4	5...
$S_{i+b}(b)$	0	1	2	3	4	5	5	4	3	2	1	0...
$T^{0.5}$	5	3	1	1	3	5	5	3	1	1	3	5...

$V_N = (5^2 - 1^2)/(6 \times 6) = 0.67$

$b = 5$; $h = 5$.

$S_i(b)$	5	4	3	2	1	0	0	0	1	2	3	4...
$S_{i+b}(b)$	0	0	0	1	2	3	4	5	4	3	2	1...
$T^{0.5}$	5	4	3	1	1	3	4	5	3	1	1	3...

$V_N = (5^2 - 1^2)/(6 \times 5) = 0.8$

$b = 4$; $h = 4$.

$S_i(b)$	4	4	3	2	1	0	0	0	0	1	2	3...
$S_{i+b}(b)$	1	0	0	0	0	1	2	3	4	4	3	2...
$T^{0.5}$	3	4	3	2	1	1	2	3	4	3	1	1...

$V_N = (4^2 - 1^2)/(6 \times 4) = 0.625$

Figure 3.26 The calculation of NLV when the scale of the pattern is 6 but the patch size is 5. NLV has its maximum value at $b = 5$. B is the scale of pattern, b is the block size, and q is the smaller of patch size and gap size. h is the smallest difference between b and an even multiple of B. $S_i(b)$ is a sum of b terms starting at position i. $T^{0.5}$ is the difference between sums and V_N is the new local variance.

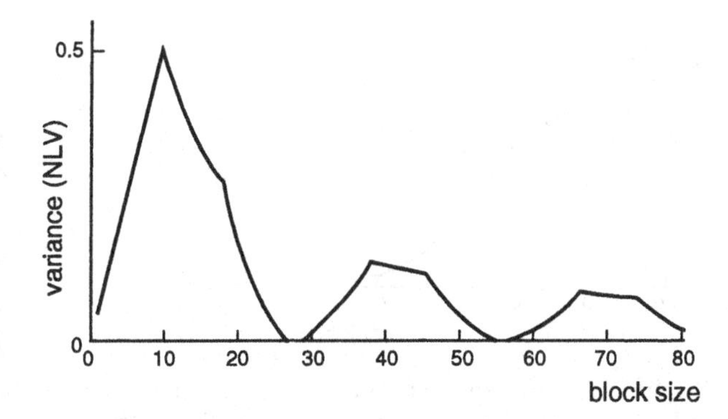

Figure 3.27 Where B is the scale of pattern and q is the size of the smaller phase, resonance peaks in a plot of NLV occur at block sizes $2B+q$, $4B+q$, and so on. Here $B=14$ and $q=10$, and so the resonance peaks are at 38 ($2\times14+10$) and 66 ($4\times14+10$).

Notice that the only terms that actually add to NLV are the 'pivotal' terms where the T's stop increasing or stop decreasing. Therefore, only the maximum and minimum values of T in the cycle of B values are of importance, as reference to Figure 3.26 will confirm. Based on this fact, we can show that if q is even:

$$V_N(b)=\min(h^2,q^2)/bB, \tag{3.38}$$

and if q is odd, then:

$$\begin{aligned} V_N(b) &= \min\,(h^2,q^2)/bB \text{ for } 1\leq h\leq(q-1)/2 \\ &= \min\,(h^2-0.5,\ q^2-0.5)/bB \text{ for } (q-1)/2<h<B-(q-1)/2 \\ &= \min\,(h^2-1,\ q^2-1)/bB \text{ for } B-(q-1)/2\leq h\leq B. \end{aligned} \tag{3.39}$$

Clearly, if neither q nor h is small, the approximate equation $V_N=\min(h^2,q^2)/bB$ will be adequate. For $B=20$, $q=17$ and $b=h=19$, the exact formula gives $V_N=0.7579$, and the approximate formula gives $V_N=0.7605$.

The equations have several results. First, they explain why V_N peaks at $b=q$, the smaller of the patch and gap sizes. For simplicity, let us examine a case where q is even and less than the overall scale, B. For $b=q-1, h\leq q$ and $V_N(q-1)=(q-1)/B$ which is less than $V_N(q)=q/B$, but when $b=q+1$, $h=q+1$ and $q<h$, so $V_N(q+1)=q^2/(q+1)B$ which is also less than $V_N(q)$.

The second feature that follows from the equations is that the peak variance does not diverge from q as B increases. The third fact that we can derive is that resonance peaks of the peak at q occur at $2B+q$, $4B+q$, $6B+q$, . . . rather than $3q$, $5q$ and so on (Figure 3.27). The equation also

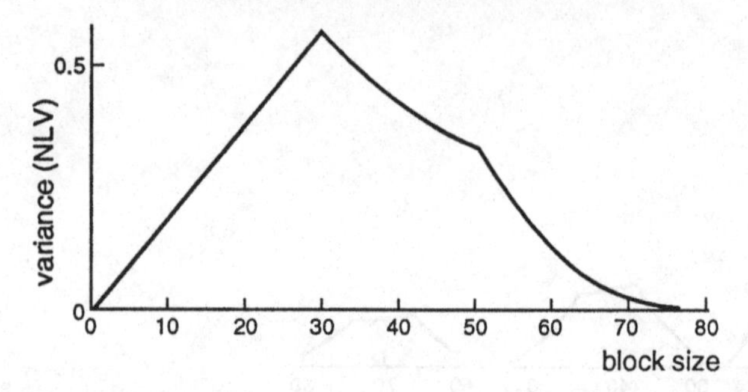

Figure 3.28 Where B is the scale of pattern and the size of the smaller phase, q, is less than B, the unimodal peak of NLV clearly consists of three parts. Here $B=40$ and $q=30$; the three parts are $0<b<30$; $30<b<50$ and $50<b<80$.

explains why, when $q<B$, the resulting unimodal graph is clearly in three segments (Figure 3.28). Using the approximate equation, when $b<q$ then $V_N(b)=h^2/bB=b/B$; when $q<b<2B-q$, $V_N(b)=q^2/bB$; and when $2B-q<b<2B$, $V_N(b)=h^2/bB=(2B-b)^2/bB$.

As stated above, we have presented equations based on a density of 1 in the patches. A modification to the equations for another density, d, is just to multiply V_N by d^2.

Our discussion of the NLV has, so far, centered on the two-term version, but Galiano (1982a) also proposed a second version based on 3TLQV, call it V_G. Let $T_3(b,i)$ be the 3TLQV term for block size b, beginning at quadrat i:

$$T_3(b,i)=\left(\sum_{j=1}^{i+b-1} x_j - 2\sum_{j=i+b}^{i+2b-1} x_j + \sum_{j=i+2b}^{i+3b-1} x_j\right)^2 \tag{3.40}$$

$$V_G(b)=\sum_{i=1}^{n-3b} | T_3(b,i)-T_3(b,i+1) | / 8b(n-3b). \tag{3.41}$$

As can be seen in Figure 3.29, the response of the three-term version to simple pattern is much more complex and, until its properties are understood better, we should use the two-term version when this approach is thought desirable.

We have stated that the two-term version detects the average size of the smaller phase, whether it is the patch or the gap. This statement refers to the phase that is locally smaller. A transect that has gaps of 20 and patches of 60 in the first half and gaps of 60 and patches of 20 in the second half will give a NQV peak at 20, not at 40 which is the average

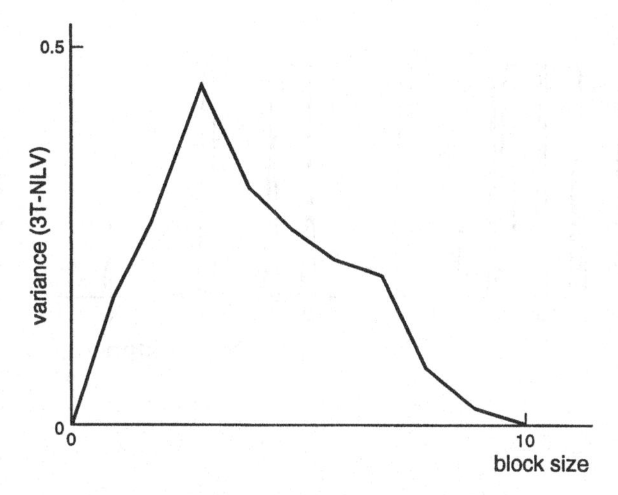

Figure 3.29 The three-term version of Galiano's NLV produces a multipartite curve for simple artificial data. As in Figure 3.28, $B=40$ and $q=30$.

size of either phase. This is an important feature for the interpretation of the results if there is nonstationarity in the pattern.

Combined analysis

Given that there are many methods that can be used to analyze spatial pattern (and there are more to follow), it is already clear that no single analysis will recover all the characteristics of spatial pattern. Before discussing more methods, we should consider the quadrat variance methods as a set and see whether there are combinations of methods that will give better information than when the methods are used separately.

Of the methods designed to detect the scale of pattern, 3TLQV has many advantages, including insensitivity to trends, less peak drift than TTLQV and decreasing resonance peaks. One current disadvantage is that a general equation for $V_3(b \mid B)$ is not known, but it is certainly possible to derive it based on the TTLQV model. A disadvantage of both TTLQV and 3TLQV, when used in isolation, is that pattern intensity measurement is based on the assumption that $p=g$. Our suggestion for a combined analysis is to use 3TLQV in combination with the two-term version of NQV which detects q, the average length of the locally smaller phase. Knowing q, and $V(B)$, we can solve for d [because TTLQV and 3TLQV give the same $V(B)$], the average density in patches. This process

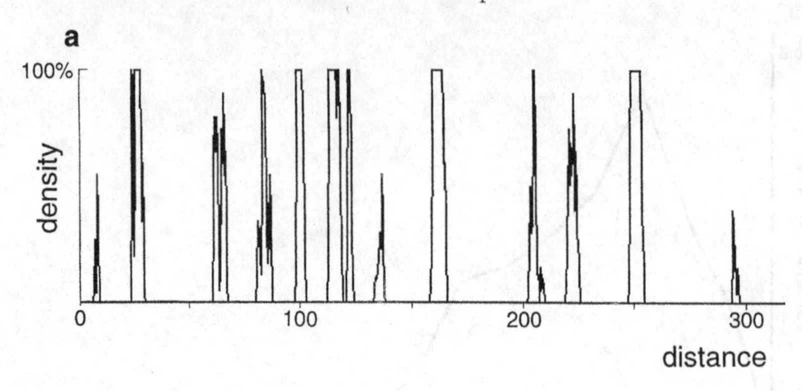

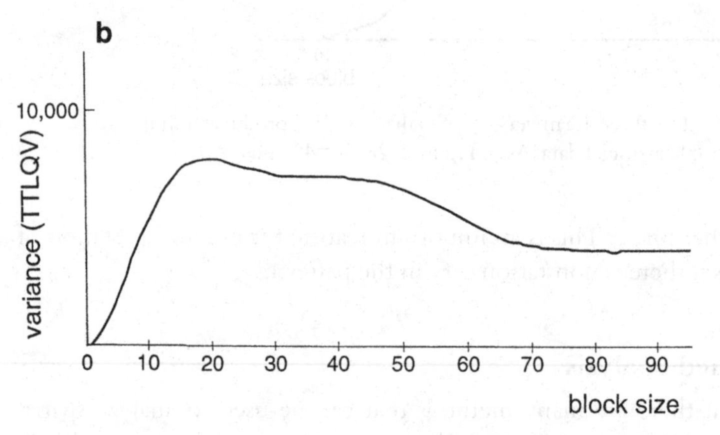

Figure 3.30 *Picea engelmannii* at Robson moraine 3. *a* Density data, *b* TTLQV, *c* 3TLQV, *d* NLV.

will enable us to distinguish between decreases in intensity caused by the pattern being unbalanced and decreases due to changes in density.

As an illustration, we will use data from moraines near Mt. Robson, British Columbia (53°N, 110°W), in the Canadian Rockies (Dale & Blundon 1990). The moraines are at an elevation of about 1650 m and lie between 1.49 and 0.66 km from the Robson Glacier. They were formed in approximately 1801, 1891, 1912, 1933, and 1939 and are referred to as moraines 1, 3, 5, 7 and 8 (Heusser 1956).

The moraines are sampled using transects of 600 contiguous 10 cm × 10 cm quadrats. The transects were placed along previously surveyed lines running along the ridge of each moraine. The estimated percentage cover of all species was estimated, but the examples cited in this and following chapters concentrate on *Hedysarum boreale*, *Dryas drummondii*, *Picea engelmannii*, and the three species of *Salix*.

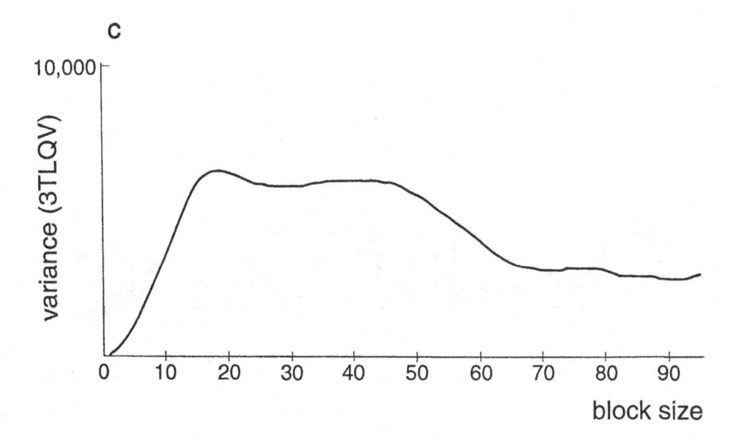

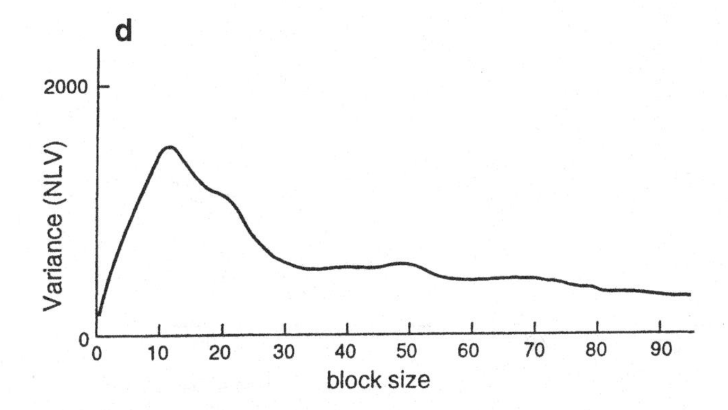

In this example, we will look at three major species on one of the moraines: *Picea*, *Hedysarum*, and *Dryas drummondii* on moraine 3. The *Picea* density plot (Figure 3.30a) shows a clear alternation of 13 patches and 14 gaps and so we will be very surprised if our analyses do not show a clear pattern of scale about 20 quadrats. The TTLQV and 3TLQV peaks around block size 18 correspond to that scale of pattern (Figure 3.30b, & c). The second peak in the 3TLQV plot, around 40, reflects the uneven distribution of the major patches along the transect. The sharp peak in the NLV graph at 11 quadrats (Figure 3.30d) suggests the average patch size which is confirmed by an examination of the density plot (Figure 3.30a)

On the other hand, the *Hedysarum* density plot is more complex and we might suspect that there were two scales of pattern in the data, six or seven major clumps with smaller scale pattern within them (Figure 3.31a).

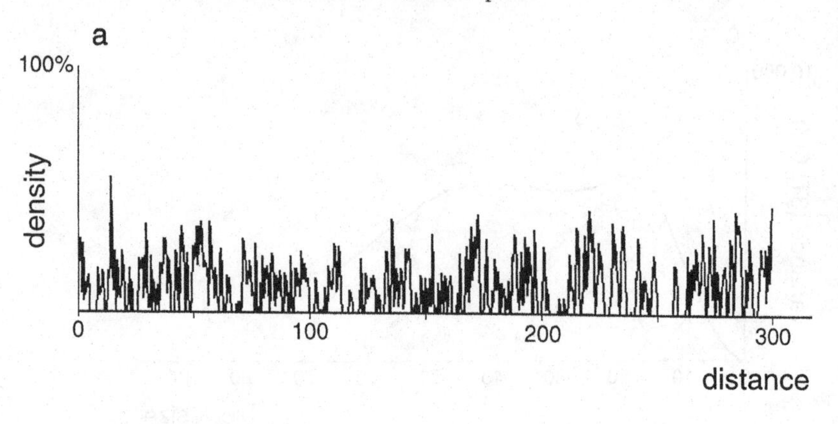

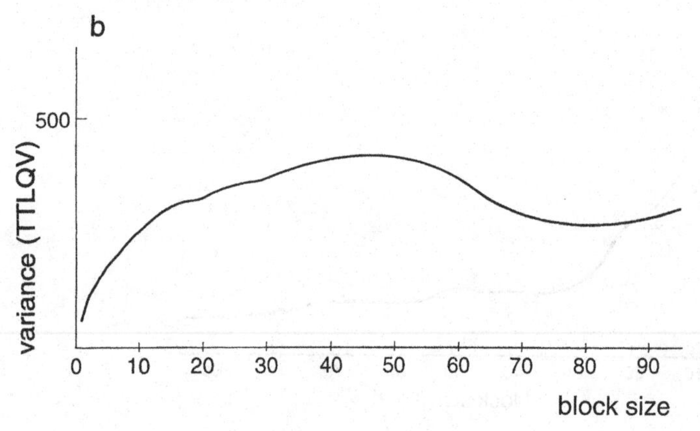

Figure 3.31 *Hedysarum mackenzii* at Robson moraine 3. *a* Density data, *b* TTLQV, *c* 3TLQV, *d* NLV.

The TTLQV and 3TLQV graphs reflect this perception, with major peaks around block size 45 and evidence of smaller scales of pattern around 18 and 30 (Figure 3.31b,c). The NLV plot is very interesting, giving a sharp peak at block size 3, indicating that the densest patches were small, but decreasing very slowly with many small peaks indicating a range of patch sizes (Figure 3.31d). Figure 3.32 shows the PQV and tQV (paired and triplet quadrat variances) analysis of the *Hedysarum* data.

Dryas drummondii occurred as patches of patches, as can be seen from the density plot (Figure 3.33a). The 3TLQV graph depicts this structure more clearly than TTLQV, with a plateau around block size 5 and a peak around 45 (Figure 3.33b,c). The NLV plot picks out the clusters

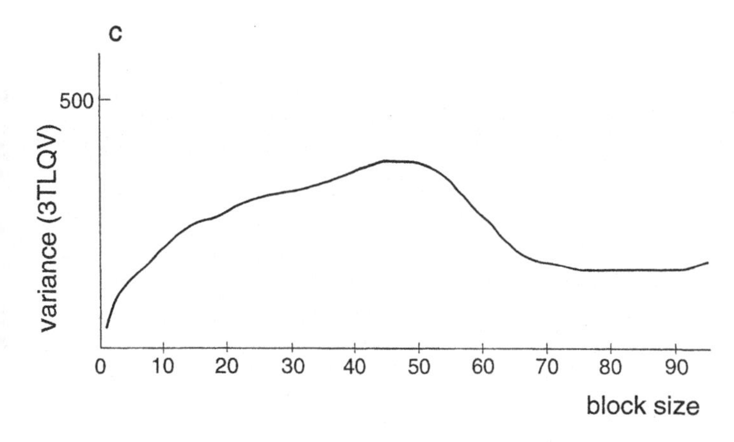

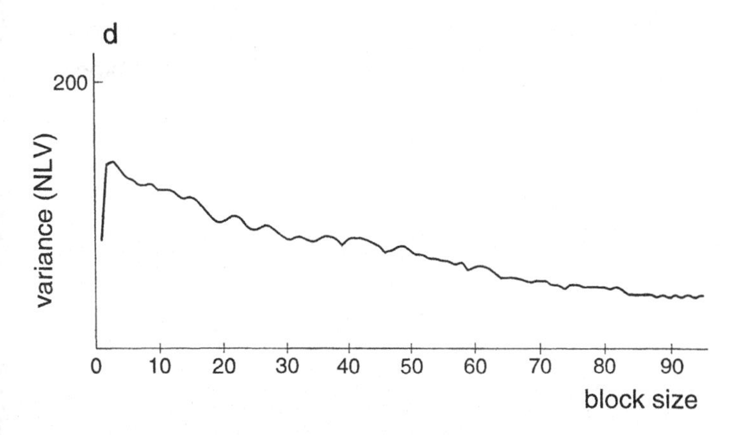

of patches structure with a peak around 6 and a second around 22, which matches the situation seen in the density plot very well (Figure 3.33d).

Most of the species on most of the moraines had evidence of more than one scale of pattern and there was no apparent trend towards a reduction in the number of scales of pattern on the older moraines nor in their regularity. What is interesting about the three species discussed in greater detail above is that while they have very different growth forms (tree, herb, dwarf shrub) and those differences are reflected in some features of their spatial pattern, they all seem to be responding to larger scale factors that produce scale in the range of 40 to 45 quadrats (about 13m).

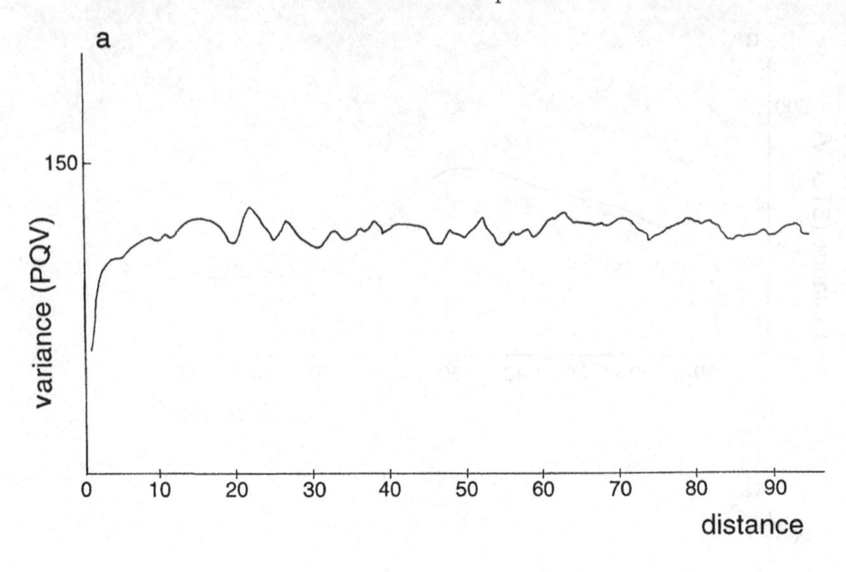

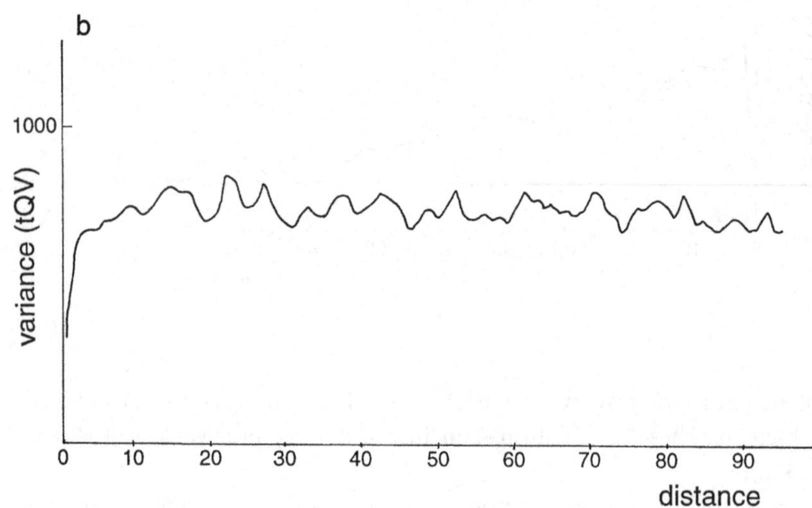

Figure 3.32 *a* PQV and *b* tQV applied to the same *Hedysarum* data as in Figure 3.31.

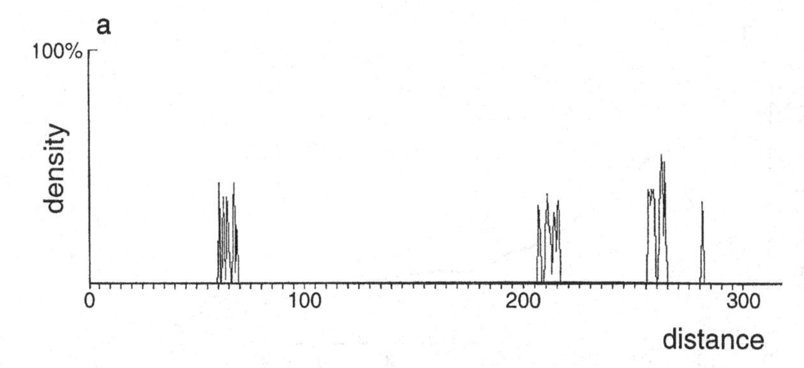

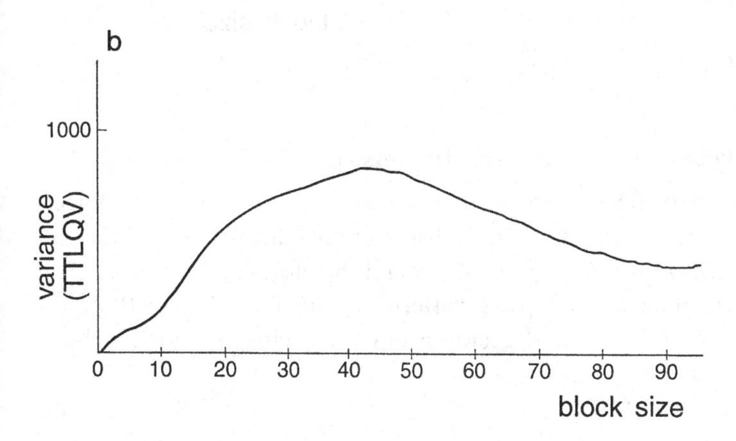

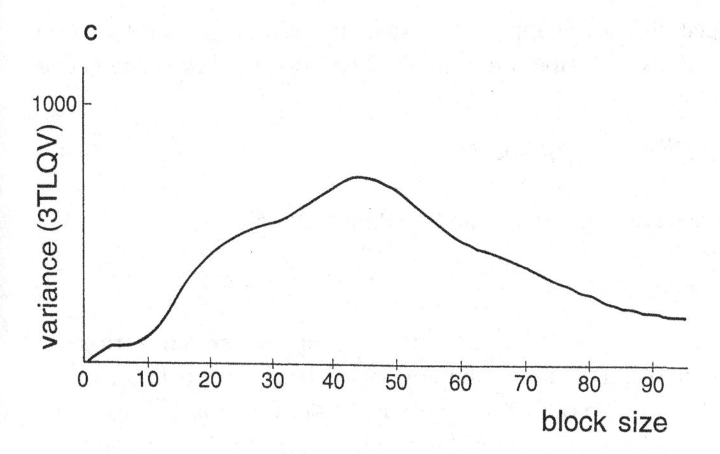

Figure 3.33 *Dryas drummondii* at Robson moraine 3. *a* density data, *b* TTLQV, *c* 3TLQV, *d* NLV.

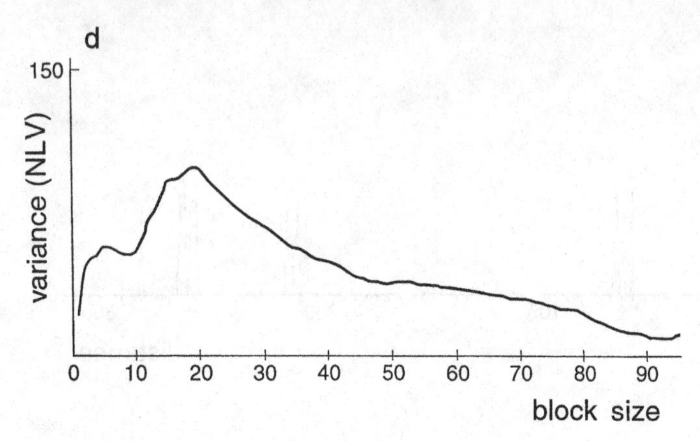

Figure 3.33 (cont.)

Semivariogram and fractal dimension

The concept of fractals was introduced in Chapter 1: they are objects with fractional spatial dimension. That fractional dimension is a measure of the spatial complexity of the object and therefore holds promise as an approach to the study of spatial pattern. In our discussion of PQV, we introduced the concept of the semivariogram, γ, which is estimated by

$$g(b) = \sum_{i=1}^{n-b} (x_i - x_{i+b})^2 / 2(n-b). \tag{3.42}$$

The fractal dimension approach to spatial pattern analysis is based on the slope of the semivariogram estimate. The slope is often calculated as (Palmer 1988):

$$m(b) = [\log g(2b) - \log g(b)] / \log 2. \tag{3.43}$$

Then the fractal dimension at scale b is (Phillips 1985):

$$\mathcal{D}(b) = [4 - m(b)]/2. \tag{3.44}$$

If there is no real pattern in the data, the slope of the semivariogram will be close to 0 and the fractal dimension will be about 2.0. Large values of $\mathcal{D}(b)$ indicate a scale of pattern of b because densities that are dissimilar at distance b (high variance and g) but similar at distance $2b$ (low variance and g) give a large negative slope and thus high values of $\mathcal{D}$ (Palmer 1988). Dissimilarity at separation b and similarity at separation $2b$ fits well with our definition of the scale of pattern, as half the distance between similar phases of a mosaic. Note that similarity at distance B and dissimilarity at $2B$, for a regular pattern, result in similarity at $3B$ and dissimilarity at $6B$. Because the divisor in the formula for $m(b)$ does not change, there will be nondiminishing resonance peaks in $\mathfrak{D}$ at $3B$, $5B$, and so on, resulting from pattern at scale B.

Palmer (1988) demonstrates the response of this method to the characteristics of artificial data and found that it could detect more than one scale of patchiness. He applied the method to field data in a multispecies form, to which we will return in Chapter 5, but the main finding was that $\mathfrak{D}$ seldom had peaks much greater than 2.0 indicating that the spatial pattern was weak.

As an illustration, we turn to data from sedge meadows on Ellesmere Island, consisting of three transects from Sverdrup Pass, (79°N, 80°W), designated BMS, BRS, and CRT, and one from Alexandra Fiord (79°N, 76°W), OWT (Young 1994). Each transect was 100 m long and was sampled with 1000 10 cm × 5 cm quadrats; presence/absence data were recorded. The most frequent species were *Eriophorum triste*, *Carex aquatilis*, *Salix arctica*, *Arctagrostis latifolia*, and *Eriophorum scheurchzeri*.

Figure 3.34 illustrates the fractal dimension analysis of *Dryas integrifolia* at the OWT transect in the Ellesmere Island data, together with the 3TLQV analysis for comparison. The species was present in 442 of the 1000 quadrats and therefore has the potential to display strong pattern. Both analyses show spatial pattern at scales 5 and 23; the fact that the values of $\mathfrak{D}$ are close to 2.0 at their peaks indicates that the pattern is weak. This conclusion is confirmed by the intensities associated with the two variance peaks: 0.23 and 0.14.

Spectral analysis

The last major method that we will describe in this chapter is spectral analysis, which also has some relationship with the variogram approach. The basic concept is to fit sine waves to the density data with a range of frequencies and starting positions and then see which fit the data best (Figure 3.35). The sine wave may not be an appropriate underlying model in many applications because it has equal lengths of high and low density, whereas in real vegetation patches and gaps are often very unequal (Figures 3.30a and 3.33a). A second difference between the sine model and real data is that the model has smooth transitions between the phases but the transitions in real data are often abrupt (Figure 3.30a).

Spectral analysis is made easier by the fact that the weighted sum of a

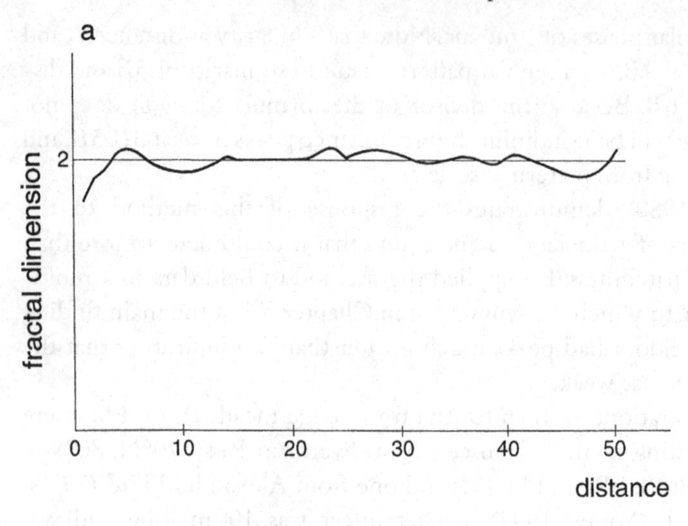

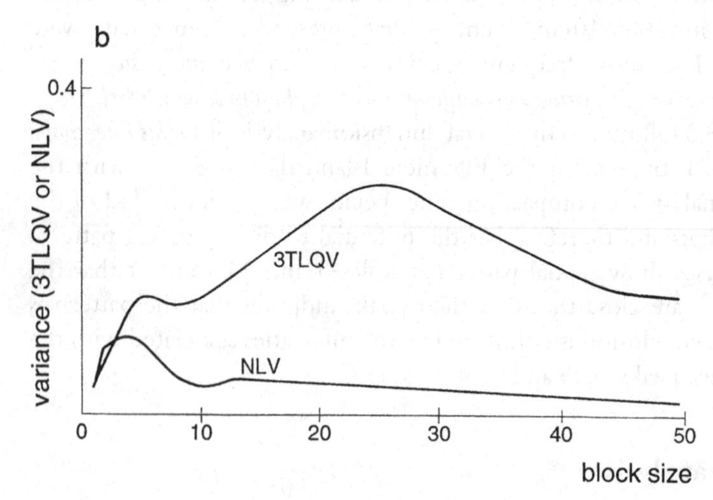

Figure 3.34 Analysis of *Dryas integrifolia* data from the OWT transect in the Ellesmere Island data. *a* Fractal dimension, *b* 3TLQV and NLV. Both approaches show pattern at scales around 5 and 23.

sine function and a cosine function of the same frequency or period gives a sine wave the position of which depends on the weights assigned to the two functions. Therefore, we express the density in the *i*th quadrat, x_i, as the mean density plus a weighted sum of sine and cosine functions:

$$x_i = \bar{x} + \sum_{p=1}^{n/2-1} c_p \cos(2\pi ip/n) + s_p \sin(2\pi ip/n). \tag{3.45}$$

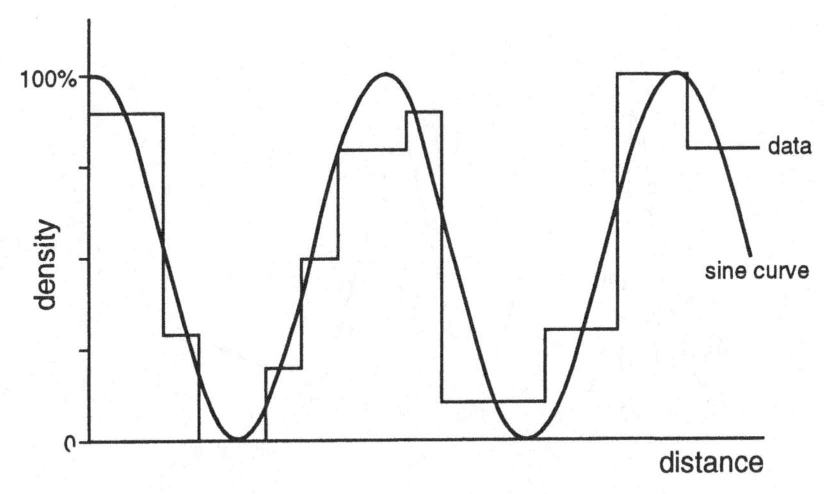

Figure 3.35 A sine wave fit to density data.

The weights for the cosine and sine functions are c_p and s_p:

$$c_p = (2/n)\sum_{i=1}^{n} x_i \cos(2\pi i p/n) \qquad (3.46)$$

and

$$s_p = (2/n)\sum_{i=1}^{n} x_i \sin(2\pi i p/n). \qquad (3.47)$$

We evaluate the reduction in the sum of squares due to fitting this sine/cosine wave of period p/n using what is referred to as the periodogram, I_p:

$$I_p = n(c_p^2 + s_p^2). \qquad (3.48)$$

I_p is proportional to that reduction in the sum of squares (Ripley 1978).

The periodogram values are also estimates of what is called spectral density, which is related to pattern in the data. Plots of I_p as a function of p often exhibit a great deal of fluctuation and are usually smoothed for interpretation. This can be accomplished by using a moving average, either with or without weighting. For example:

$$I_p' = (I_{p-1} + I_p + I_{p+1})/3 \qquad (3.49)$$

or

$$I_p'' = (I_{p-2} + 2I_{p-1} + 4I_p + 2I_{p+1} + I_{p+2})/10. \qquad (3.50)$$

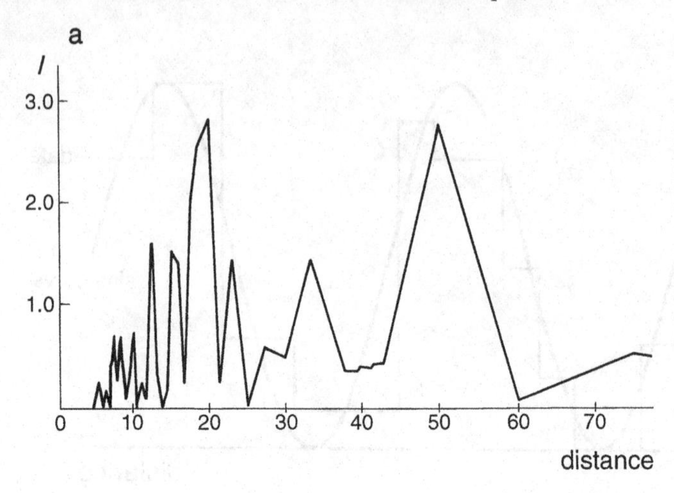

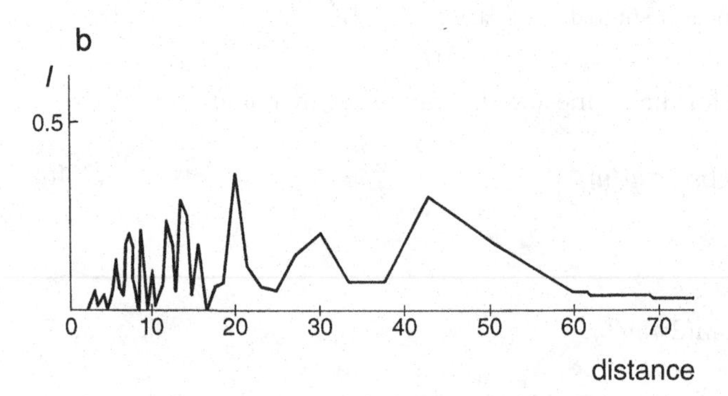

Figure 3.36 Spectral analysis of field data at Robson Moraine 3. *a Picea engelmannii*, *b Hedysarum mackenzii*.

The periodogram plots amplitude as a function of wavelength, but in many applications it is more common to plot amplitude as a function of frequency. In such a diagram, a peak at frequency f gives resonance peaks at $2f$, $3f$, $4f$, and so on. From the point of view of spatial pattern analysis, the periodogram may be easier to read, since it is more directly comparable to the results of other analysis techniques that have a distance measure on the x-axis. Because wavelength is the distance between two density peaks or two density troughs in the spatial pattern, wavelength is twice the scale of the pattern. Given the same data, spectral analysis should produce large amplitudes at wavelengths corresponding to twice the scale of the pattern as detected by quadrat variance methods.

A major disadvantage of the spectral analysis method is that the position and size of peaks in the periodogram depend on the smoothing used

(Ripley 1978). Usher (1975) concluded that it is difficult to interpret amplitude-frequency graphs, because of the difficulty in distinguishing which peaks reflect real pattern in the data from spurious and resonance peaks. A further disadvantage is that, although the significance of the whole periodogram can be tested, the significance of individual peaks cannot (Ripley 1978). Our own experience would lead us to echo some of Usher's misgivings, since we encountered apparently spurious peaks in analyzing artificial data. Figure 3.36a shows the spectral analysis of the *Picea* data from Figure 3.30a; the major peaks at 20 and 50 match those in the 3TLQV analysis. Figure 3.36b shows the spectral analysis of the *Hedysarum* data from Figure 3.31a; there is no clear match between its results and the 3TLQV analysis.

Other methods

A variety of other methods have been proposed but have not risen to prominence in application. For instance, in their book *Statistical Ecology*, Ludwig and Reynolds (1988) comment that measures of dispersion such as Morisita's index, I_δ, can be used to examine spatial pattern. Where x_i is the number of individuals of a particular kind in the ith quadrat, with mean $\bar{x}$, sample variance s^2, and total n, the index I_δ is defined as:

$$I_\delta = n\left\{1 + \frac{s^2 - \bar{x}}{\bar{x}^2}\right\}/(n-1). \qquad (3.51)$$

This index obviously has a close relationship with the variance:mean ratio, to which we will return in Chapter 7.

For pattern analysis, the idea is that if I_δ is calculated from individual blocks for a range of block sizes, its value will remain more or less constant until the mean clump size is reached and then it increases. As Figure 3.37 shows, for artificial square wave data, the method produces a peak at a block size that is twice the scale. When there is more than one scale of pattern in the data, the method does not produce clear results.

A number of studies have used this approach to look at the scale of the clumping of trees, including Veblen (1979), Lamont and Fox (1981) and Taylor and Halpern (1991). In the last study, the authors were examining the structure and dynamics of forests dominated by *Abies magnifica* in the Cascade Range, California. They recorded the number of trees in each 5 m × 5 m quadrat of a 100 m × 100 m plot. The quadrats were combined into 10 m × 10 m, 15 m × 15 m, . . . squares and Morisita's index calculated at each size. In the two plots examined, the strongest clumping was in the

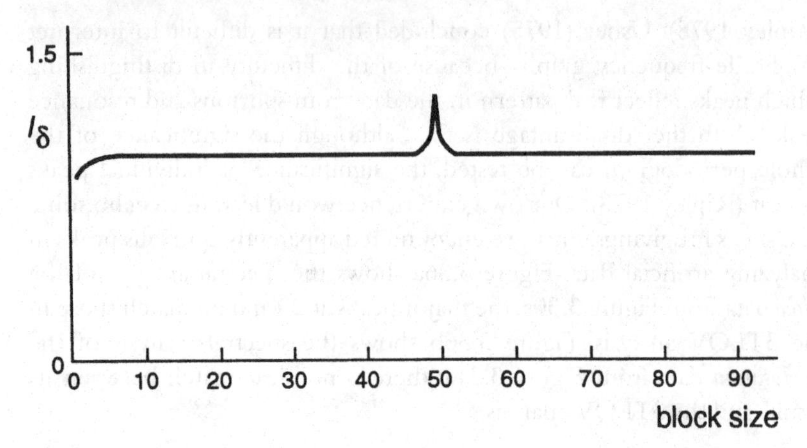

Figure 3.37 Morisita's I_δ, a measure of dispersion, as a function of block size for artificial data with a single scale of pattern at 25.

20 m × 20 m squares, a scale that corresponded to the sizes of canopy gaps (Taylor & Halpern 1991).

Dale and MacIsaac (1989) introduced a method based on combinatorics to detect the sizes of patches and gaps. (Combinatorics is the branch of discrete mathematics that deals with finite problems of counting, selection, and arrangement of mathematical objects.) The method is based on converting the density data to ranks, which produces what is essentially a permutation of the integers $1, 2, 3, \ldots n$. For each block size b, the number of runs of b quadrats that are all with ranks smaller than the two quadrats which immediately precede and immediately follow the run (b-dips) are counted by the variable m_b. Figure 3.38 illustrates the concept of b-dips. The expected value and the variance of m_b can be calculated using the formulae of Dale and Moon (1988):

$$E(m_b) = \frac{2n}{(b+1)(b+2)}. \tag{3.52}$$

$$\mathrm{Var}(m_b) = \frac{2nb(4b^2+5b-3)}{(b+1)(b+2)^2(2b+1)(2b+3)}. \tag{3.53}$$

The standardized value of b-ups is then plotted as a function of b:

$$w_b = [m_b - E(m_b)] / \sqrt{\mathrm{Var}(m_b)}. \tag{3.54}$$

Peaks in that plot show common patch sizes. Similarly, runs of b ranks that are greater than those that flank them (b-ups, cf. Figure 3.38) are counted, u_b, and standardized:

```
                                    ┌────────────────┐                    ┌──────────────────┐
22 18 14 9 6 1 4 8 11 15 26 20 19 12 10 5 2 3 7 23 17 21 24 25 13
         └──────────┘                 └────────────┘
```

Figure 3.38 b-dips and b-ups in a ranking of quadrat densities. Two 5-dips are underlined and two 5-ups are overlined. All the ranks in a 5-dip are smaller than the ranks immediately preceding and following. All the ranks in a 5-up are larger. Numbers 9 to 8 and 10 to 7 are 5-dips; 15 to 12 and 23 to 25 are 5-ups.

$$y_b = [u_b - E(u_b)]/\sqrt{\mathrm{Var}(u_b)}. \tag{3.55}$$

Peaks in the plot of this standardized plot show common gap sizes.

This approach has not been used very frequently, perhaps because it is unfamiliar compared to a TTLQV-based method such as NLV. Another drawback is that there is no guarantee that patch and gap sizes detected in this way will have an average equal to the scale of the pattern as detected by other methods. It may also be confusing that the appearance of any long dip or up will appear statistically significant, with a standardized value greater than 2.0, merely because they are so improbable. Because the method is based on ranks, the magnitude of differences is not taken into account. The last problem for this method is that it is very sensitive to 'error' quadrats, that is single quadrats of nonzero density in the midst of an obvious gap or of zero density in the midst of an obvious gap. Galiano's NLV is less sensitive to error quadrats; for instance, for well-defined artificial pattern it can reliably find the smaller phase with error rates as high as 20%. For these reasons, the combined analysis is preferable.

To end this review of methods used for single-species spatial pattern analysis, we will describe two statistics that are more general in their range of application than just pattern analysis because they can be used for that specific purpose. They are the closely related statistics known as Moran's I and Geary's C.

$$I_{\mathrm{M}} = n\sum_{i=1}^{n}\sum_{j=1}^{n} w_{ij}\,(x_i - \bar{x})(x_j - \bar{x}) \Big/ \sum_{i=1}^{n}(x_i - \bar{x})^2 \sum_{i=1}^{n}\sum_{j=1}^{n} w_{ij} \tag{3.56}$$

$$C_{\mathrm{G}} = (n-1)\sum_{i=1}^{n}\sum_{j=1}^{n} w_{ij}(x_i - x_j)^2 \Big/ 2\sum_{i=1}^{n}(x_i - \bar{x})^2 \sum_{i=1}^{n}\sum_{j=1}^{n} w_{ij}. \tag{3.57}$$

The x_i are the data and the w_{ij} are weights taking the value 1 when the pair x_i and x_j are to be included in the calculation and 0 when they are not. In many applications, the w's are referred to as proximity indices because they are 1 when the two observations are neighbors and 0 otherwise (Bailey & Gatrell 1995). For spatial pattern analysis, we can calculate

I_M or C_G as a function of separation h, so that w_{ij} is 1 when observations i and j are separated by distance h and 0 otherwise. The close relationship of both statistics, when used in this way, to the autocorrelogram, $\rho(h)$, and to the variogram and PQV as discussed earlier is obvious.

One advantage of this approach is that the values associated with each distance can be tested individually for statistical significance. The overall result can be evaluated by determining whether at least one value exceeds the Bonferroni criterion, which is the significance level $\alpha' = \alpha/m$, where m is the number of distances tested (Legendre & Fortin 1989). In a study of the spatial pattern of tree species at a site in Québec, Legendre and Fortin (1989) found that the correlogram for the distribution of *Tsuga canadensis* was significant overall. Significantly high values of I_M for distance classes 1 and 2 (57 m and 114 m) were interpreted as patch size and the next peak at distance class 9 (485 m) indicated the distance between patches. The significantly low values at classes 4, 11, and 12 confirmed this interpretation.

Concluding remarks

Based on our studies of the methods presented in this Chapter, applied both to artificial data of known structure and to field data, we can conclude that no single method of analysis can tell us everything we want to know. In spite of the problem of peak drift, 3TLQV seems to be the best single method, but it should be complemented by looking at measures of pattern intensity and tQV. Where the pattern is weak, measures of intensity can be compared with average densities in occupied quadrats to see whether density fluctuations or patch and gap length irregularity contribute more to the pattern's weakness. NLV can be used to detect the size of the smaller phase, which in most instances will be patch size. Until we understand more clearly how the three-term version works, the two-term version of NLV is to be preferred. Because of the problems with interpretation and the often inappropriate underlying sine/cosine model, with its smooth transitions and equal phase sizes, spectral analysis is not recommended. It is yet to be proven that the fractal approach offers advantages over the other methods.

In field studies of pattern in one dimension, a general feature is that even single species display several scales of pattern. That is, there may be a hierarchy of patchiness in the plants' arrangement in space. It is also common that different species in the same community display different intensities and scales of pattern, which cannot always be explained easily by reference to plant size or growth habit. The prediction that small scales of pattern tend to be lost as succession proceeds does not seem to be supported by the studies that have tested it.

Recommendations

1. Use 3TLQV combined with tQV to detect scale.
2. Use the two-term NLV to detect the size of the smaller phase.
3. The pattern intensity calculated from the variance (TTLQV or 3TLQV) should be compared to the average density in patches to determine the relative importance of density irregularities and patch and gap size irregularities.

4 · *Spatial pattern of two species*

Introduction

Vegetation is patchy at a range of spatial and temporal scales, and so even within what might be recognized as a single plant community, the plants of different species are not really expected to be arranged homogeneously and independently. Natural groupings of species may arise from biological interactions or from shared and divergent responses to abiotic factors. In some cases, the community is viewed as a mosaic of patches, with each phase of the mosaic being characterized by a set of species' abundances. This phenomenon in plant communities has given rise to the patch dynamics approach to studies of vegetation (van der Maarel 1996).

The existence of nonrandomness in species arrangement is the context in which the multivariate analysis method of classification takes place. Classification can be used to organize samples, like quadrats, into hierarchical categories based on the similarity of species composition. The composition and strength of the associations within and between groupings is an important aspect of the plant community's structure. It is reasonable to begin to investigate this structure by examining the relationships of pairs of species because these pairwise interactions can then be amalgamated loosely or exclusively into larger groupings. We will therefore examine methods designed to evaluate the joint spatial pattern of pairs of species and thus the scales at which they are positively or negatively associated.

When we look at the spatial pattern of a single species, we are examining the arrangement in space of two mosaic phases, places where the species is present (perhaps at variable density) and places where the species is absent. In looking at the joint pattern of two species, A and B, there are four mosaic phases to be considered: both absent, both present, A absent and B present, A present and B absent, remembering that, for

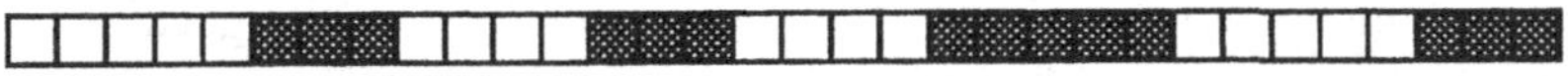

Figure 4.1 A transect in which there are only two species present and which account for all of the length, a two-phase mosaic. The scale of pattern is just the average length of segments occupied by a single species. Here the lengths are 5, 3, 4, 3, 4, 5, 5, and 3, giving a scale of 4.1.

some kinds of data, presence may include a range of densities. The analysis of the joint pattern then quantifies the spatial arrangement among these four phases.

There is a clear relationship between joint pattern analysis and examining the association of pairs of species. The spatial association of two species is positive when the plants of the two species tend to be found together, and it is negative when they tend not to be. Depending on the method used, the investigation of two-species pattern can be interpreted as a multiscale study of pairwise association, determining at what scales two species are positively associated and at what scales they are negatively associated. In this chapter, we will concentrate on one-dimensional approaches, but the extensions to two-dimensional studies will be obvious.

As always, the method of analysis used will depend on the nature of the vegetation, the kind of data recorded and the questions we want answered. The studies can be classified according to whether a single point in space can be occupied by only one or by more than one species and by whether the two species of interest are the only ones present or there are others.

At most one species per point

If the vegetation is such that each point can be occupied by only one species (as some mosaics of saxicolous lichens), we could sample it in one dimension using the line intercept method described in Chapter 2 or using strings of very small quadrats, labelling each with the species that occupies the majority of its area. If there are only two species and no joint absences (or the joint absences are collapsed), we are dealing with a mosaic of two phases and, in a one-dimensional sample, they must alternate along the transect and the (first) scale of pattern is easily calculated as the average of the lengths of line occupied by a single species (Figure 4.1). The consistency of the pattern can be evaluated by looking at the variance of the lengths of the runs of each species. A similar straightforward approach can be used when there are two obvious mosaic phases, even if one has more than one species, such as in the striped vegetation of arid regions where vegetation stripes alternate with bare ground (*cf.* Montaña 1992; White 1971).

Line intercept data can be converted to quadrat form for analysis and, because the original data are continuous, there is a choice of the size of 'quadrat' used. The value for each quadrat is just the proportion of the chosen subsection of the line transect that is occupied by a particular species. When the data are collected as quadrat data initially, with only two species, the first scale of pattern is the average length of single-species runs of quadrats. Pielou (1977a, Chapter 15) provides a discussion of the lengths of single-species runs based on several different underlying models. For example, if the sequence is the realization of a first-order Markov model with two states, A and B, corresponding to the two species, there is a simple relationship between the lengths of single-species runs and the transition probabilities. Let p_{AB} be the probability that the occurrence of A is immediately followed by that of B, and let L_A be the length of a run of A's; then the expected run length is $E(L_A)=1/p_{AB}$. Using parallel notation, the expected length of runs of species B is $E(L_B)=1/p_{BA}$. Although this relationship is simple, its usefulness may be limited because Pielou (1977a) suggests that a Markov model will seldom be a tenable hypothesis for real data. We will discuss Markov models further in Chapter 5 in the context of multispecies pattern.

In addition to being used to look at the lengths of single-species runs, contiguous transect data can also be analyzed using a variety of methods based on single-species quadrat variance methods, which will be described below.

When more than just the two species are present, the methods need to be modified somewhat. Call the two species of interest A and B, and lump all other species or bare substrate into category O. The data consist of lengths or runs of A, B, and O. Under these circumstances, three different definitions of scale might be of interest:

1. The average distance between the centers of patches of A and B when they abut (Figure 4.2).
2. The average distance between the centers of patches of A and B when they are separated by a patch of O (Figure 4.2).
3. The average distance between their centers in either case.

In a sense, the first is related to a scale of positive association, since they are in contact, and the second is related to a scale of negative association

Figure 4.2 There are three ways of defining the joint scale of species A and B: (1) the average distance between the centers of patches when the two species abut (small arrowheads); (2) the average distance between the patch centers when the species are separated by the other phase (large arrowheads); (3) the average of both. Here the scale is 3.0 based on the first definition, 6.75 on the second, and 4.25 on the third.

since they are not. The third is an average. It may, however, be difficult to interpret positive association when the species are mutually exclusive, except as an example of ecological coincidence, with the species having similar ecological requirements and capabilities.

Several species per point

In most kinds of vegetation, more than one species can occupy a single one- or two-dimensional point because of vertical structure. In this situation, there is a continuum of positive to negative association responses as a function of scale. For example, the phenomenon referred to as 'nucleation' (Yarranton & Morrison 1974) occurs where the plants of species A make their immediate neighborhood more suitable for the establishment of plants of species B. In examining presence/absence data, the two species are positively associated over the short distances of the nucleation effect, but either neutral or negatively associated at larger distances (Figure 4.3). If the amelioration of the environment does not control whether a species occurs at a particular location but affects only its growth and density, the trend in association will be similar to that produced by nucleation, but it will appear in the analysis of density data rather than in the analysis of presence/absence data.

In considering processes such as nucleation, it should be remembered that the relationship between the two species may be asymmetric; that is, species A may have a positive effect on species B, but not B on A. Many kinds of sampling used to study association, such as presence/absence or density in quadrats, will not distinguish between the two cases. If asymmetric association is the main focus of the study, the sampling regime should be designed specifically for it. For example, if nucleation by species A is the process under study, it would be best to compare the frequency of plants of other species in the neighborhood of plants of species

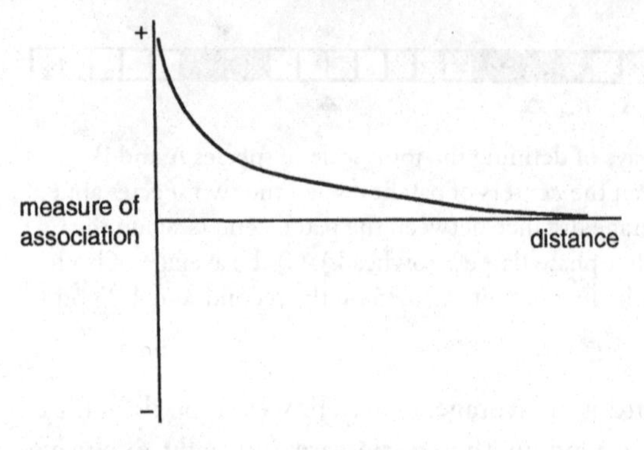

Figure 4.3 Association as a function of distance when one species has a local positive effect on the other, as in nucleation.

A with their frequency in randomly placed 'neighborhoods' of the same size.

Competition between ecologically similar species will produce a trend opposite to that of nucleation; the two species will be negatively associated at small distances but positively associated at greater distances (Figure 4.4). If the competition is strong enough to cause exclusion, this trend will be seen in the analysis of presence/absence data. If it just reduces the biomass of the competing species, the trend will show up only in density data. A comparison of the analysis of density data and of presence/absence data should distinguish between the possibilities.

As in the preceding chapter, which reviewed the range of methods available for the analysis of single-species pattern, we will now proceed to examine the range of methods available for two-species pattern analysis based on data collected in strings of contiguous quadrats.

Blocked quadrat covariance (BQC)

Many of the quadrat variance methods used to study the spatial pattern of a single species can be modified in a straightforward way to examine the spatial pattern of a pair of species. Kershaw (1960) was the first to do so, using BQV as the basic method, calculating variance as a function of block size for each of the two species, V_A and V_B, and for their combined densities V_{A+B}. Their covariance is then calculated as:

$$C_{AB} = (V_{A+B} - V_A - V_B)/2. \tag{4.1}$$

This formula is an alternative to the usual direct calculation of covariance, which for two variables x and y is of the form

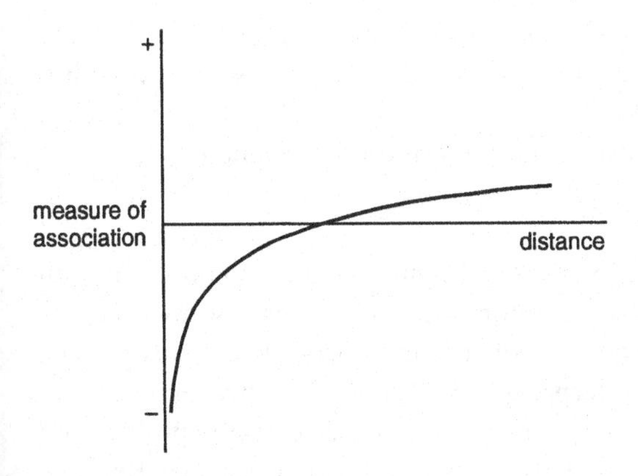

Figure 4.4 Association as a function of distance when one species has a local negative effect on the other, as in competition.

$$\mathrm{Cov}(x,y) = \left[\sum_{i=1}^{n} (x_i y_i) - \sum_{i=1}^{n} x_i \sum_{i=1}^{n} y_i / n \right] / n. \tag{4.2}$$

For example, if in 27 quadrats the two species have densities x and y as:

$x = 1\ 1\ 1\ 1\ 1\ 1\ 1\ 1\ 1\ 0\ 0\ 0\ 0\ 0\ 0\ 0\ 0\ 0\ 1\ 1\ 1\ 1\ 1\ 1\ 1\ 1\ 1$
$y = 1\ 1\ 1\ 0\ 0\ 0\ 1\ 1\ 1\ 0\ 0\ 0\ 1\ 1\ 1\ 0\ 0\ 0\ 1\ 1\ 1\ 0\ 0\ 0\ 1\ 1\ 1.$

Then:

$x + y = 2\ 2\ 2\ 1\ 1\ 1\ 2\ 2\ 2\ 0\ 0\ 0\ 1\ 1\ 1\ 0\ 0\ 0\ 2\ 2\ 2\ 1\ 1\ 1\ 2\ 2\ 2,$

and:

$xy = 1\ 1\ 1\ 0\ 0\ 0\ 1\ 1\ 1\ 0\ 0\ 0\ 0\ 0\ 0\ 0\ 0\ 0\ 1\ 1\ 1\ 0\ 0\ 0\ 1\ 1\ 1.$

At block size 1, individual quadrats:

$\mathrm{Var}(x) = (18 - 18^2/27)/27 = 6/27 = 0.2222;$
$\mathrm{Var}(y) = (15 - 15^2/27)/27 = 6.67/27 = 0.2469;$
$\mathrm{Cov}(x,y) = (12 - 15 \times 18/27)/27 = 2/27 = 0.0741;$
$\mathrm{Var}(x+y) = (57 - 33^2/27)/27 = 16.67/27 = 0.6172.$

From Kershaw's formula,

$[\mathrm{Var}(x+y) - \mathrm{Var}(x) - \mathrm{Var}(y)]/2 = (0.6172 - 0.2222 - 0.2469)/2 = 0.1482/2 = 0.0741$

as in the direct calculation. Note that the covariance 0.0741 is one-third of 0.2222 which is the variance of x. If y is modified to match x more closely, for instance $y = 111000111000000000111000111$ or $y = 111111111000111000111111111$, the covariance increases to two-thirds of that variance, 0.1481.

Kershaw suggests calculating the correlation coefficient, *r*:

$$r = C_{AB}/(V_A V_B)^{0.5}, \tag{4.3}$$

which can then be tested for statistical significance by comparison with tabulated critical values or by transformation and comparison with the t-distribution. (Of course, because of spatial autocorrelation in the data, significance tests must be interpreted with caution.) Positive and negative peaks in the graph of covariance are interpreted as reflecting positive and negative association at the scale of the block size. For instance, Kershaw (1962) studied the relationship of *Festuca rubra* and *Carex bigelowii* in a *Rhacomitrium* heath in Iceland, and interpreted a negative peak in covariance at a particular block size as resulting from competitive interaction between the species at that scale, affected by the scale of variation in microtopography.

Many of the drawbacks of BQV, described for single-species analysis in Chapter 3, remain as problems in the two-species case. The approach has therefore been superseded by other methods, which the following sections will describe.

Paired quadrat covariance (PQC) and conditional probability

Given the drawbacks of two-species pattern analysis based on BQV, an obvious step is to develop two-species versions of methods that avoid its problems in the one-species case. If the basic technique of spaced quadrats is used to calculate the variance of the two species separately and of their joint occurrence, a PQC for spacing b, $C_P(b)$, can be calculated from them using Kershaw's equation given above. An alternative formulation is as follows:

$$C_P(b) = \sum_{i=1}^{n-b} (x_i - x_{i+b})(y_i - y_{i+b})/2(n-b), \tag{4.4}$$

where x_i is the density of the first species in the ith quadrat and y_i is the density of the second.

The properties of this covariance are easily illustrated using simple artificial patterns of strings of 1's alternating with strings of 0's. Figure 4.5 shows the range of responses as the strings start matching and are sub-

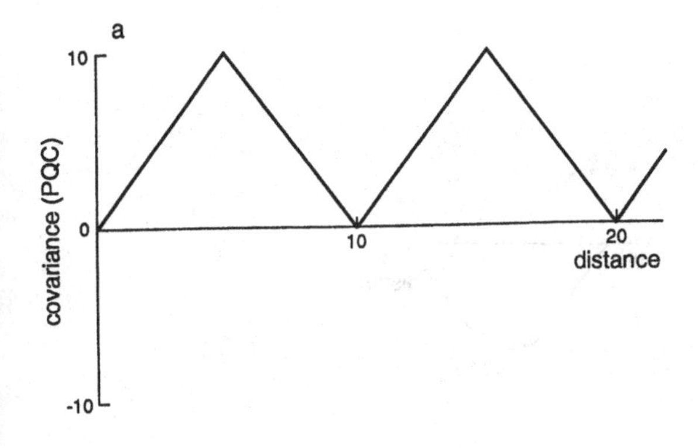

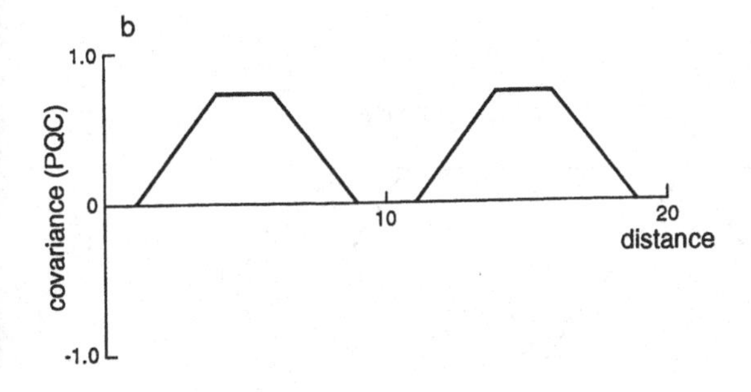

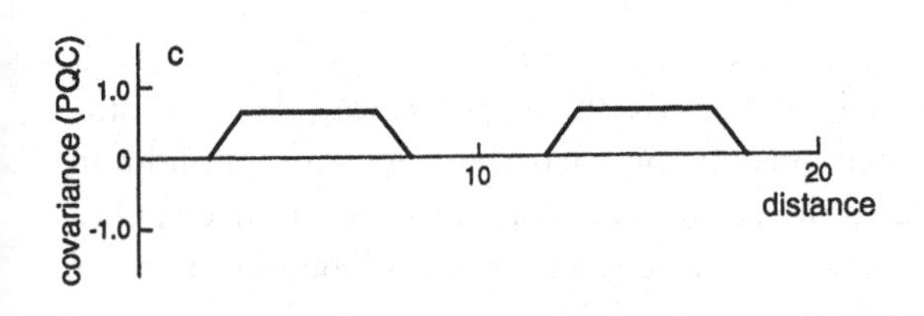

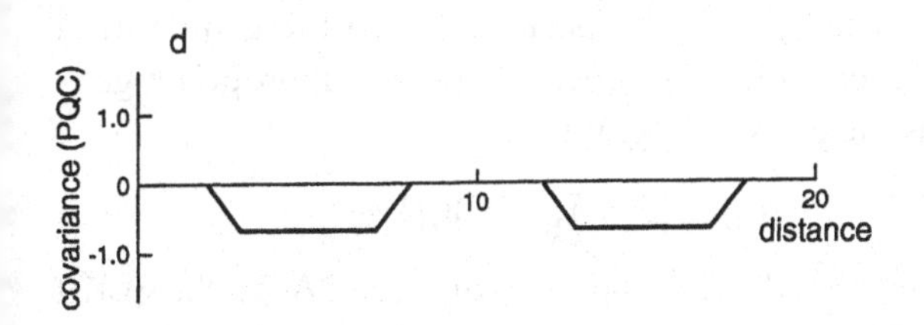

Figure 4.5 The range of response of the paired quadrat covariance as identical patterns go from perfect alignment to being completely out of phase. Here the scale of pattern is 5 and the offset of the two patterns goes from 0 in part *a* to 1 in *b*, 2 in *c*, 3 in *d*, 4 in *e*, and finally complete offset of 5 in *f*.

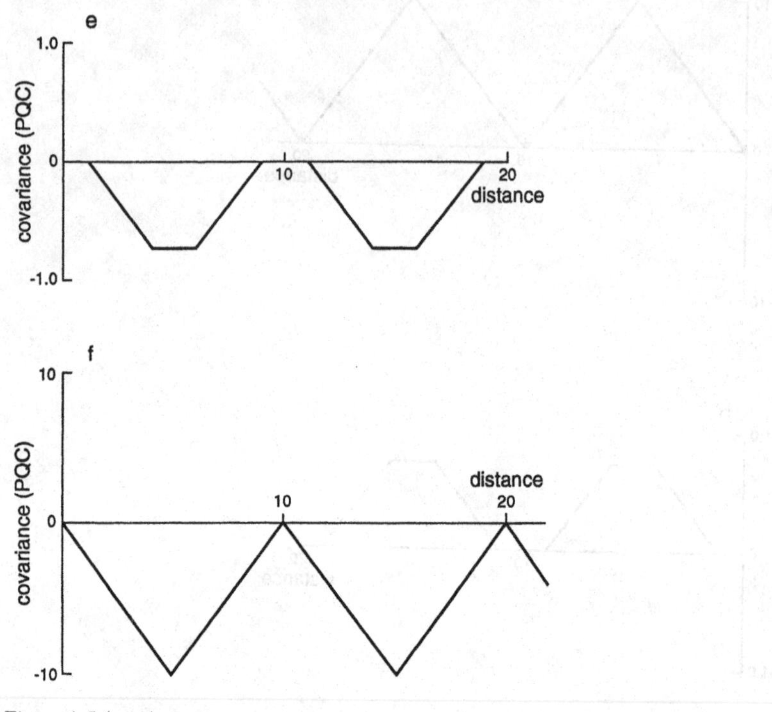

Figure 4.5 (cont.)

sequently offset until it is the mismatch that is perfect. As with other applications of the spaced quadrat approach, there are strong nondiminishing resonance peaks in the covariance plot. In the covariance application, these resonance peaks are more than a nuisance because they can lead to serious misinterpretations. We shall investigate the difficulties later in this chapter, when comparing techniques.

In Chapter 3, we related the single-species PQV to the geostatistical concept of the variogram. In the two-species case, the appropriate geostatistical equivalent is the cross-variogram:

$$2\gamma_{AB}(\mathbf{h}) = E\{[Y_A(\zeta) - Y_A(\zeta + \mathbf{h})][Y_B(\zeta) - Y_B(\zeta_i + \mathbf{h})]\} \qquad (4.5)$$

E is the expected value, Y_A is the spatial process of species A, Y_B the spatial process of species B, ζ is a measure of position, and $\mathbf{h}$ is a displacement vector. In one dimension, the displacement vector is just a distance scalar h, and the cross-variogram is estimated by (David 1977):

$$2g_{AB}(h) = \sum_{i=1}^{n-h} [Y_A(\zeta_i) - Y_A(\zeta_i + h)][Y_B(\zeta_i) - Y_B(\zeta_i + h)]/(n-h)$$

$$= \sum_{i=1}^{n-h} [x_i - x_{i+h}][y_i - y_{i+h}]/(n-h). \quad (4.6)$$

This equation is obviously the same as Equation 4.4 for PQC. Again the paired quadrat analysis is being using to estimate the (cross) variogram.

The cross-variogram can be standardized relative to the variogram values of the single species, just as the correlation coefficient relates the covariance of two variables to their individual variances. The single species variogram estimate is:

$$2g_A(h) = \sum_{i=1}^{n-h} [x_i - x_{i+h}]^2/(n-h). \quad (4.7)$$

Then the standardized variogram, $R(h)$, is:

$$R(h) = g_{AB}(h)/\sqrt{g_A(h)g_B(h)}. \quad (4.8)$$

As in the single-species case, we can formulate a three-term or triplet version of this spaced quadrat covariance:

$$C_t(b) = \sum_{i=1}^{n-2b} (x_i - 2x_{i+b} + x_{i+2b})(y_i - 2y_{i+b} + y_{i+2b})/8(n-2b). \quad (4.9)$$

The divisor of 8 is to ensure that this version gives the same value as PQC when the block size is equal to the scale. As in the single-species analysis, the three-term version, triplet quadrat covariance (tQC), produces a more angular covariance plot and is less sensitive to trends in the data. Using the notation from Chapter 3, B is the scale of pattern, p is patch size, g is gap size, and b is the block size. Figure 4.6 illustrates the calculation of PQC and tQC for $B=p=g=8$ and $b=7$ with an offset, f, of 2 between the two patterns. The relationship between the two versions is illustrated in Figure 4.7. When applied to field data in which the patterns are not distinct, the tQC curve will usually be closer than the PQC curve to the b-axis. The fact that both methods produce nondiminishing resonance peaks in the covariance is very clear in Figure 4.7, and that resonance may be a serious drawback in interpreting the covariance curves derived from data in which more than one scale of pattern is present. Figure 4.8 shows the application of PQC and tQC to the data from the OWT transect on Ellesmere Island. Most pairs of

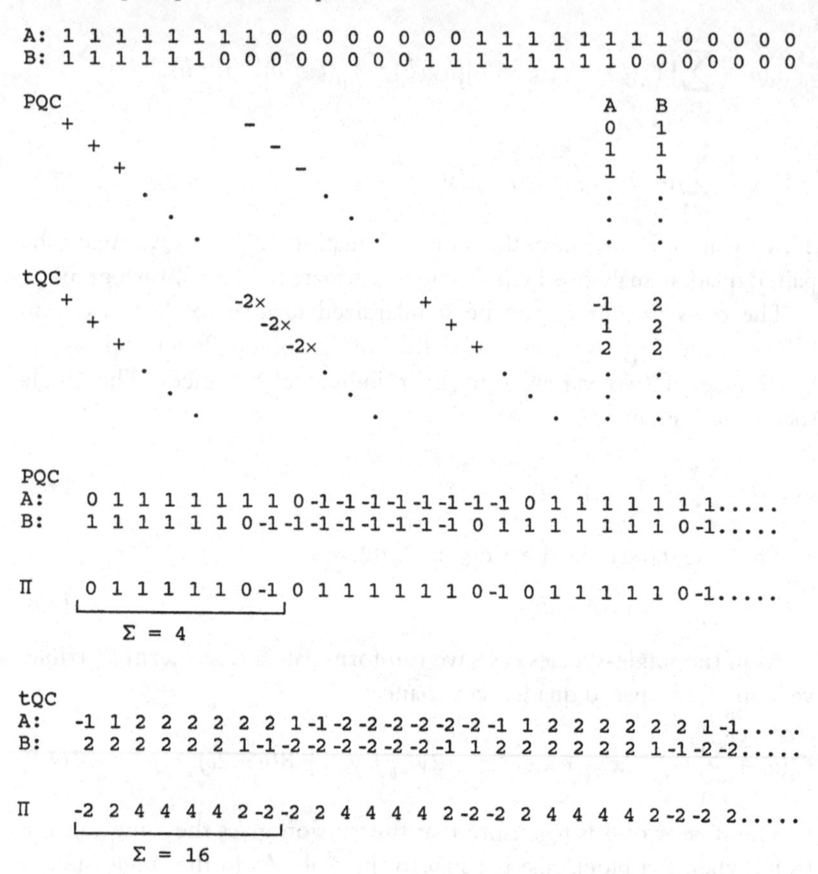

Figure 4.6 PQC and tQC for identical patterns with $p=g=8$, $b=7$ and $f=2$. The first two lines show the quadrat data for the two species A and B with 1 representing presence and 0 representing absence. The upper part of the Figure then shows how the terms $(x_i - x_{i+b})$ and $(x_i - 2x_{i+b} + x_{i+2b})$ for species A and $(y_i - y_{i+b})$ and $(y_i - 2y_{i+b} + y_{i+2b})$ for species B arise. The lower part of the Figure shows the values of those terms as a function of position, i, their product, Π, for each position and the sum of eight terms. For PQC each cycle of eight terms adds to 4 and for tQC the same sum is 16.

species show covariance, either positive or negative, at a scale close to block size 6.

Galiano (1986) describes the use of a method referred to as 'conditional probability spectra' to examine segregation between species. The method compares the probability of finding species B in a quadrat at distance b from a quadrat known to contain species A, $P(\mathrm{B}|\mathrm{A})$, with the probability that a randomly chosen quadrat contains B, $P(\mathrm{B})$. The condi-

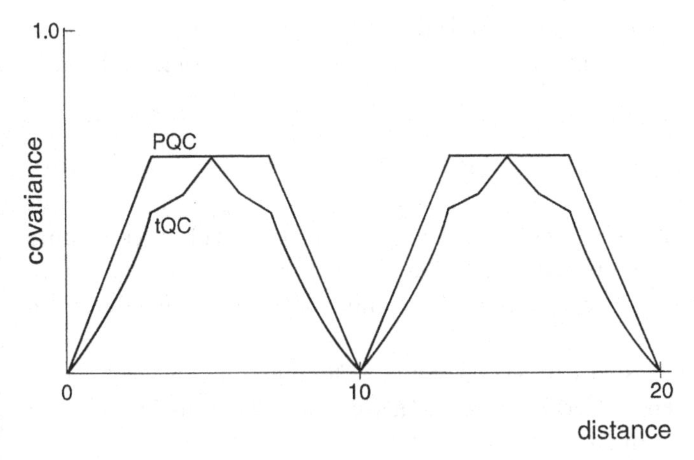

Figure 4.7 A comparison of the spaced quadrat techniques PQC and tQC. Artificial data with a scale of 5. As with the variance analyses from which they are derived, tQC produces a peak where PQC produces only a plateau.

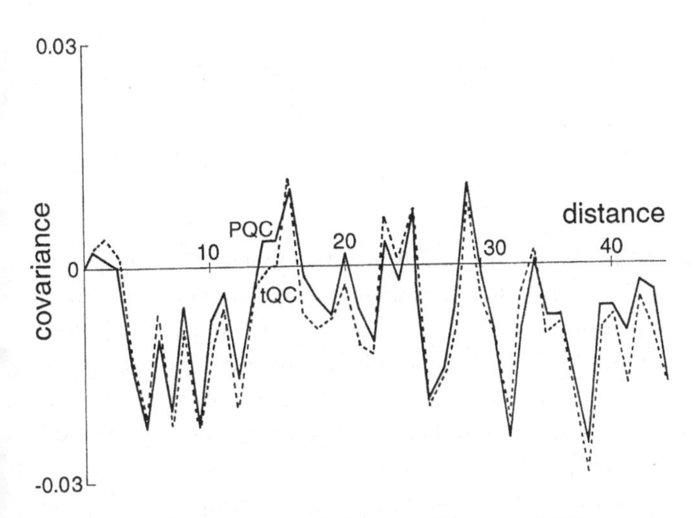

Figure 4.8 PQC and tQC analysis of data from the OWT transect on Ellesmere Island, *Carex membranacea* and *Dryas integrifolia*. After a hint of positive covariance at distance 2, there are negative peaks at 4, 6, and 8.

tional probability is estimated from the frequency with which there are two quadrats separated by distance b, one of which contains the first species and the other of which contains the second. The data are really presence/absence data and therefore the values of x_i and y_i are restricted to 0's and 1's. Based on the formula that $P(\mathrm{B}\,|\,\mathrm{A}) = P(\mathrm{B\&A})/P(\mathrm{A})$ (Gellert *et al.* 1977), and letting G(b) be the bidirectional conditional probability for distance b:

$$\mathrm{G}(b) = \sum_{i=1}^{n-b} (x_i y_{i+b} + x_{i+b} y_i) \Big/ \sum_{i=1}^{n-b} (x_i + x_{i+b}). \tag{4.10}$$

The divisor is not simply the proportion of the quadrats containing species A because quadrats that contain A and are less than b units from the end of the transect can have a B neighbor at separation b only in one direction.

The method is closely related to the PQC method described above, especially when applied to presence/absence data. Divisors ignored, the important term in G(b) is:

$$\sum_{i=1}^{n-b} (x_i y_{i+b} + x_{i+b} y_i). \tag{4.11}$$

The important term in the PQC calculation is:

$$\sum_{i=1}^{n-b} (x_i - x_{i+b})(y_i - y_{i+b})$$

which can be rewritten as

$$\sum_{i=1}^{n-b} (x_i y_i - x_i y_{i+b} - x_{i+b} y_i + x_{i+b} y_{i+b}). \tag{4.12}$$

The first and fourth terms are determined by the species compositions of the quadrats, not their spatial arrangement, so that only the central terms respond to spatial pattern. Since those two terms are the additive inverse of the important component of G(b), G(b) and PQC react similarly, but in opposite directions, to the characteristics of spatial pattern. Note that G(b) is a probability and therefore takes values between 0 and 1 whereas PQC runs from -1 to 1.

One problem with the conditional probability method is that the difference between patterns perfectly in phase and those perfectly out of phase results in what looks like a shift in the position of the probability peak. Based on the comparison with PQC, a more accurate description is that a peak has been turned into a trough, from a probability of 1 to probability 0.

In presenting the results of conditional probability analysis, Galiano (1986, Figure 4) shows 95% confidence limits. It is not clear how these were calculated but, as always in studies of spatial pattern, they should be treated with caution because of the spatial autocorrelation in the data (the quadrats, as trials, are not fully independent) and because the values for adjacent distances are not independent.

Using this approach, Galiano (1986), in a study of the herbaceous plants of an oak parkland in central Spain, showed that perennials such as *Agrostis castellana* and plants with basal rosettes such as *Plantago lanceolata* exhibited strong exclusion of other species up to distances of 10cm. Other plants such as *Trifolium* sp. and *Poa bulbosa* did not have the same negative effect on neighbors. This result suggests that the asymmetric relationship of fine-scale competitive exclusion is an important factor in the community.

Two- and three-term local quadrat covariance (TTLQC and 3TLQC)

The recommended methods for the analysis of two-species pattern are based on the single-species blocked quadrat methods TTLQV and 3TLQV, described in Chapter 3. Defining $s_b(i)$ as the sum of the densities of species A, x, in the b quadrats starting at quadrat i:

$$s_b(i) = \sum_{m=1}^{i+b-1} x_m. \tag{4.13}$$

By analogy, t_b is for sums of y and u_b is for sums of $x+y$:

$$t_b(i) = \sum_{m=1}^{i+b-1} y_m \text{ and } u_b(i) = \sum_{m=1}^{i+b-1} (x_m + y_m). \tag{4.14}$$

The two-term variance for species A is:

$$V_A(b) = \sum_{i=1}^{n+1-2b} [s_b(i) - s_b(i+b)]^2/2b(n+1-2b). \tag{4.15}$$

The three-term variance for species A is:

$$V_A(b) = \sum_{i=1}^{n+1-3b} [s_b(i) - 2s_b(i+b) + s_b(i+2b)]^2/8b(n+1-3b). \tag{4.16}$$

Similar variances can be calculated for species B using t_b , and for the combined densities using u_b. For either the two-term or three-term version, the covariance at block size b, $C_{AB}(b)$, is calculated using Kershaw's formula (Equation 4.1).

As in the single-species analysis, the three-term version of covariance calculation is preferred (cf. Greig-Smith 1983). The covariance can be calculated from the sum of products rather than the sum of squares; for example, the two-term version is:

$$C_{AB}(b) = \sum_{i=1}^{n+1-2b} [s_b(i) - s_b(i+b)][t_b(i) - t_b(i+b)]/2b(n+1-2b). \qquad (4.17)$$

In Chapter 3, we discussed a measure of the intensity of a pattern, based on the variance of the calculated for the block size of the pattern's actual scale. We can produce a similar measure of intensity for covariance:

$$I_C(B) = \sqrt{6B\,|C(B)|/(B^2+2)}, \qquad (4.18)$$

where B is the scale of pattern. This intensity can used to compare covariances at different block sizes by removing the effect of scale. We can compare the covariance with the magnitude of the single-species variances using the correlation coefficient:

$$r(b) = C_{AB}(b)/\sqrt{V_A(b)\,V_B(b)}. \qquad (4.19)$$

Some researchers have tried to use peaks in the correlation coefficient to detect scales of covariance, but our own studies have shown that it is unreliable and cannot be used in that way. (Dale & Blundon 1991; cf. Grieg-Smith 1983). You can prove this to yourself: suppose the patterns are identical for both species. If that is so then $V_A = V_B$ and $V_{A+B} = 4V_A$ for any block size; thus $C_{AB} = V_A$ and the correlation is always 1.0. Under these circumstances correlation takes its maximum value at block sizes unrelated to the scale of pattern.

The value of r can be used to evaluate the strength of the covariance by comparing it with critical values of the correlation coefficient for the appropriate numbers of degrees of freedom. For example, if the transect is 600 quadrats long and you examine block sizes up to 100, there seem to be between 300 and 600 degrees of freedom (df) for the range of block sizes studied. The 95% critical value for 500 df is 0.088, and for 300 is 0.113, and therefore a value of 0.1 can be chosen as a guideline, considering only those covariance peaks for which $|r| > 0.1$. As usual, in our studies of spatial pattern, while the argument is appealing, the term 'guideline' must be emphasized because of lack of independence. First, there is spatial autocorrelation in the data; second, the variances calculated at different block sizes are not independent. These evaluations are not strict tests of statistical significance.

In a graph of covariance as a function of block size, the important features are the positions of peaks, whether positive or negative, and their intensities. For example, Figure 4.9 shows the covariance analysis of

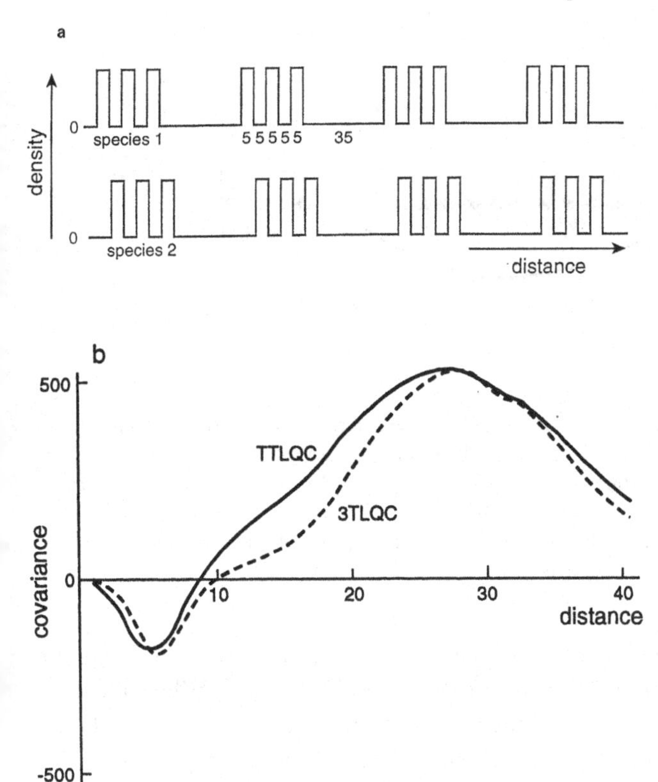

Figure 4.9 *a* Artificial data for two species that are positively associated at large distances ($b = 30$) and negatively associated at short distances ($b = 5$). *b* TTLQC and 3TLQC analysis of the artificial data in *a* showing negative covariance around $b = 5$ and positive covariance around $b = 30$.

artificial data in which the two species are negatively associated at a scale of 5 and negatively associated at a scale of 30. As in TTLQV and 3TLQV, there is a small discrepancy between the pattern scale and the block size at which a covariance peak occurs (as discussed in Chapter 3) and that complicates the interpretation of peaks. Therefore, decomposing covariance plots to look for hidden peaks or trying to partition the covariance among scales, while theoretically possible, are probably too unreliable to be practical. Two simple summary variables are N, the number of negative covariance values in the range of block sizes examined and A, the net area between the covariance curve and the block size axis (see Figure 4.10).

Artificial patterns of density and presence/absence data with various

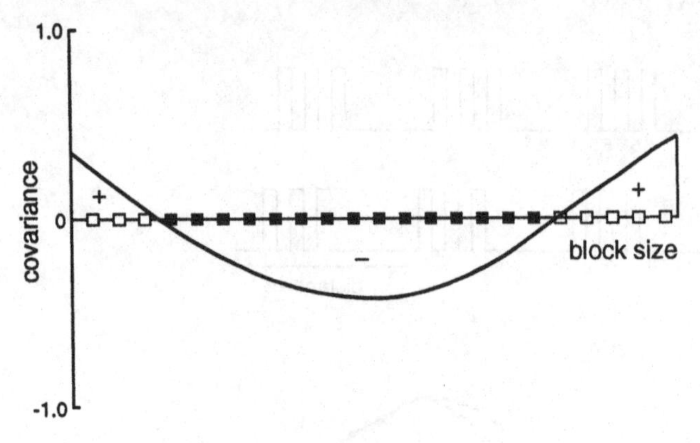

Figure 4.10 Illustration of two useful summary statistics of a covariance plot: (1) N is the number of block sizes at which the covariance was negative, here 15 out of 23; (2) A is the net area, the difference between the area above the axis labelled '+' and the area below the axis labelled '−'. In this case, the net area is negative. The two different symbols on the block size axis indicate the sign of the covariance, white for positive and black for negative.

numbers and combinations of scales of pattern can be used to explore the response of covariance analysis to features of the data. When identical patterns become more and more offset from each other, the correlation coefficient declines from 1.0 when they are perfectly in phase to −1.0 when the patches in one pattern exactly match the gaps in the other. The position of the covariance peak does not change but declines to about half of its original value and then returns to it. For instance, for a pattern consisting of patch = gap = 10 quadrats and density in the patches of 1.00, a covariance peak is found at 9 with $I_C(10) = 1.00$ when the patterns match (offset 0); $I_C(10)$ declines to 0.54 when the offset is 4 or 6 and then increases back to 1.00 when the patterns are completely opposite, with offset 10 (Figure 4.11).

When patch:gap (or gap:patch) ratios are changed from 1:1, the value of I_C also declines. Trends in patch density and in pattern scale produce measured intensities and positions of covariance peaks close to the average of that variable in the data.

Comparison of methods

So far we have looked at two classes of covariance method, PQC and tQC on the one hand and TTLQC and 3TLQC on the other. The two approaches have somewhat different properties. These differences can be

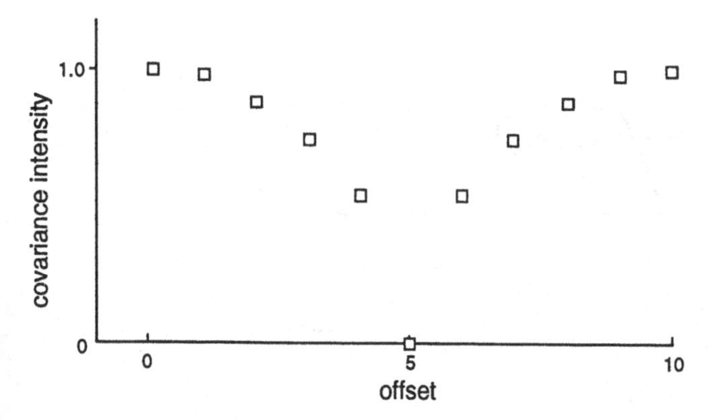

Figure 4.11 The behavior of I_c which is a measure of covariance intensity as two identical patterns of scale 10 and density 1 are offset by different amounts, f. When $f=0$, the patterns match perfectly and when $f=10$, they are completely out of phase; in both those cases I_c takes its maximum value. When $f=5$, the covariance is 0.

illustrated either with artificial or with field data. Figure 4.8 shows PQC and tQC analysis of *Carex membranacea* and *Dryas integrifolia* from the OWT transect on Ellesmere Island and Figure 4.12 shows TTLQC and 3TLQC analysis. It is obvious that the PQV-type analysis produces a variance plot that is much less smoothed because individual quadrats are used rather than the sums of blocks of quadrats. As we have commented before, spaced quadrat methods have resonance peaks that do not diminish; in covariance analysis, the result of resonance may be that important features of the data are obscured. The OWT data themselves and the TTLQC analysis show a clear positive association of the two species at a scale around 20 quadrats. In the PQC analysis this fact is obscured because of resonance from the negative peaks at 5, 7, 9 and 12. Figure 4.13 illustrates the same feature with artificial data that negatively covary at scale 6 but are positively associated at scale 30; there is no peak in the PQC plot at 30 but it is obvious in the TTLQC and 3TLQC analyses.

We used the covariance approach to two-species pattern analysis of data collected on the recessional moraines at Mt. Robson, British Columbia. We concentrated on six species: the legume *Hedysarum mackenzii*, a dwarf shrub *Dryas drummondii*, the tree *Picea engelmannii*, and three willows *Salix vestita*, *Salix glauca*, and *Salix barclayi*. Only covariance peaks for which $|r|$ exceeded 0.1 were considered. The data were analyzed twice: in the original density form and also converted into presence/absence data. We found that in some instances the covariance peaks

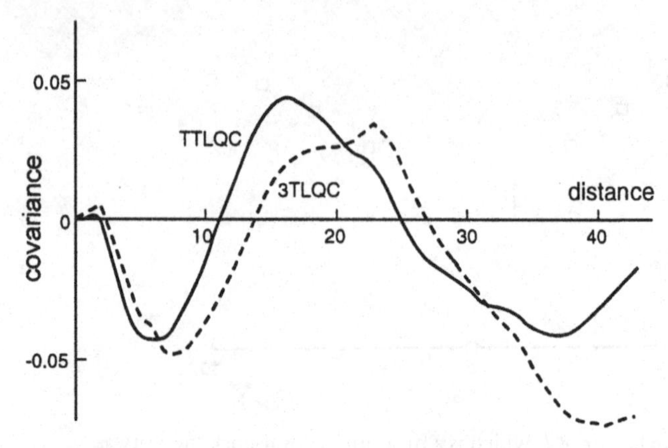

Figure 4.12 Covariance analysis of *Carex membranacea* and *Dryas integrifolia* at the OWT site on Ellesmere Island. PQC and tQC are in Figure 4.8. Here are TTLQC and 3TLQC (dashed). The blocked methods produce a more smoothed response to the spatial pattern.

were closely related to scales of pattern of the individual species, but in other cases there seemed to be little relationship. For instance, on moraine 7, *Picea* has variance peaks at 60 (density) and 63 (presence), *Dryas* has peaks at 57 and 61, and their covariance has peaks at 58 and 60. These scales match well, suggesting the existence of a natural grouping of at least two species which covary spatially. *Hedysarum* also has a variance peak in the same range, 64 (density), but there is no sign of pattern at that scale in the *Picea-Hedysarum* covariance analysis. This result shows that while a third species seems to be responding to the environment at a similar scale, it is not part of that grouping of species.

The covariance plots were more complicated than would be predicted from simple explanations of how the plants of different species interact with each other. Competitive exclusion in a pair of ecologically similar species should give a clear negative peak at small block sizes but the covariance should be positive for large block sizes, as in Figure 4.4. On the other hand, local positive association, caused by a positive influence such as nucleation, should produce a clear positive peak at small block sizes with the covariance declining to zero, as in Figure 4.3. Dale and Blundon (1990) found that the main species on the Robson moraines exhibited about four scales of pattern each; the covariance analysis of pairs of these species detected three to four scales of pattern (Dale & Blundon 1991).

Carter and O'Connor (1991) studied a two-phase mosaic in a wood-

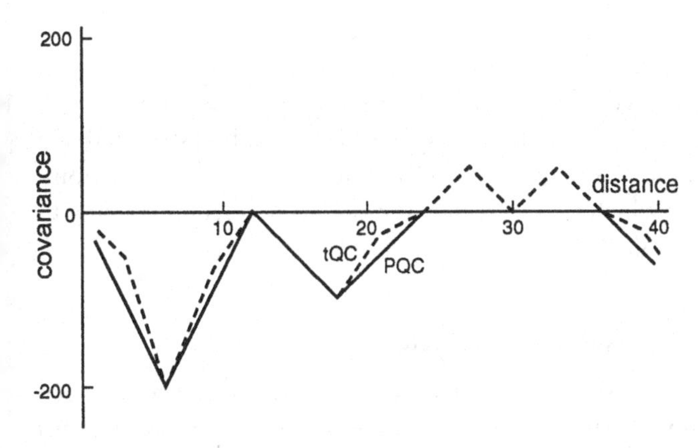

Figure 4.13 PQC and tQC (dashed) analysis of artificial data in which the two species are negatively associated at scale 5 but positively associated at scale 30, much as in Figure 4.9*a*. The TTLQC analysis resembles 4.9*b* but in the spaced quadrat method, the 5*B* resonance of the negative peak at block size 6 flattens the positive peak that should occur at $b=30$.

land savanna in South Africa; many species were present but the two phases are dominated by *Setaria incrassata* and by *Themeda triandra*. The authors analyzed the spatial pattern of the two species separately using PQV and TTLQV, but they did not use the covariance equivalents. They found that the two species that dominate the two phases were strongly negatively associated at the scale of a single quadrat, and a glance at the sample of the data presented suggests that strong negative covariance would be found at the matching scales of pattern found in the two species when analyzed separately (see Carter & O'Connor 1991, Figure 1).

The sort of covariance analysis that we have been considering would not be appropriate for data derived from a mosaic that had only two phases. If the two phases are A and B, the combined density of the two would be 100% for each quadrat, so that V_{A+B} would be constant with changing block size. Since A and B are complementary, V_A and V_B will be identical and, recalling Kershaw's formula, the covariance will therefore be an additive inverse of those curves. In this situation, then, covariance analysis can contribute no new information.

Veblen (1979) used a somewhat different approach to study the effect of scale on the association of two species of *Nothofagus* in Chile. He used plots divided into 3 m × 3 m subplots and examined the correlation of the densities of *Nothofagus pumilio* and *Nothofagus betuloides* as the subunits were combined into larger blocks. The correlation was most negative at the smallest block size (−0.8 at 3 m × 3 m) and although it increased with block size, it remained negative and decreased again at the largest block size (−0.7 at 24 m × 24 m). The two species are patchily distributed and the author attributes the negative correlation to interspecific interactions, with *N. betuloides*, which is evergreen and more shade tolerant, tending to replace the deciduous *N. pumilio*.

Extensions of covariance analysis

So far in this chapter, the analysis of association and covariance has been related to examining the relationship between a pair of species. An obvious variant of this approach is to use the same kind of analysis to examine the relationship between a species and an environmental factor. Kershaw (1964) discusses the use of BQV to investigate, in a *Rhacomitrium* heath in Iceland, the relationships of two species, *Carex bigelowii* and *Festuca rubra*, with a measure of microtopography, the height of the quadrats' centers. *Carex* was positively associated with microtopography at most scales whereas *Festuca* was negatively associated with it at most scales, both most intensely at block size 32. The negative small-scale covariance of the two species (block size 4) is attributed to the biotic interaction of competition, not to different responses to abiotic factors (Kershaw 1964).

In our Ellesmere Island data sets, we looked at the covariance of the most common species with microtopographic height. We found that species such as *Carex aquatilis* and *Carex membranacea* had negative covariance with height at a scale around 5 quadrats, but species associated with drier microsites, such as *Carex misandra*, *Dryas integrifolia*, *Polygonum viviparum* and *Saxifraga oppositifolia*, had positive covariance at the same scale. What is interesting here is that the covariances between pairs of species do not seem to reflect the response to moisture, indicating that biological interactions may be complicating the simpler patterns arising from the plants' response to an environmental factor.

One feature that has emerged from the analysis of the Mt. Robson and Ellesmere Island data sets is that the scales of covariance often do not match the scales of the individual species. This characteristic seems at first very puzzling and an interesting question is whether it can be duplicated using artificial data. The answer is yes. There are several ways in which sets of artificial data can be combined to produce covariance peaks that do not match the the original scales. One way to approach the problem is to consider what happens when two identical patterns are used; the result will depend on how the patterns are aligned, the amount of offset. Given two identical patterns of scale 20 with $g=10$ and $p=30$, if they are offset from each other by any amount from 0 to 9, there is a variance peak around 20. (Figure 4.14a shows the TTLQC and 3TLQC analysis.) If, however, the offset is 10 or 11, there is a strong negative peak at block size 10 and no evidence of pattern at scale 20 (Figure 4.14b). Equations 3.10 and 3.11 given in Chapter 3 can be used to explore why the covariance behaves this way. It is not clear how this result should affect our ecological interpretations of blocked quadrat covariance.

Other approaches

Other approaches that have not been pursued much include the possibility of a cross-covariance method and cross-spectral analysis. Cross-covariance is closely related to the geostatistical approach to estimating the cross-variogram described in 'Paired Quadrat Covariance (PQC) and Conditional Probability'. Cross-covariance functions are of the form:

$$E\{[x(i)-\mu_x][y(i+h)-\mu_y]\} \tag{4.20}$$

and

$$E\{[x(i+h)-\mu_x][y(i)-\mu_y]\} \tag{4.21}$$

where μ_x and μ_y are the means of the two spatially dependent variables x and y (Jenkins & Watts 1969).

Cross-spectral analysis compares the characteristics of two spatial series using the techniques of spectral analysis. It produces a cross-amplitude spectrum that shows whether frequency components in one variable are matched by large or small amplitudes at the same frequency in the other variable. It also gives the phase spectrum, which shows whether the frequency components of one series are ahead of or behind components of the same frequency in the other (Jenkins & Watts 1969). Kenkel (1988b) used cross-spectral analysis to compare the spectra of elevation and vegetation derived from data collected in a mire hummock-hollow system. In that instance, the vegetation spectrum was based on a multispecies analysis (see Chapter 5), rather than making strictly pairwise comparisons. The availability of this technique in software packages such as S-Plus (MathSoft Inc., Seattle, Wash., U.S.A.) will make it more accessible to researchers.

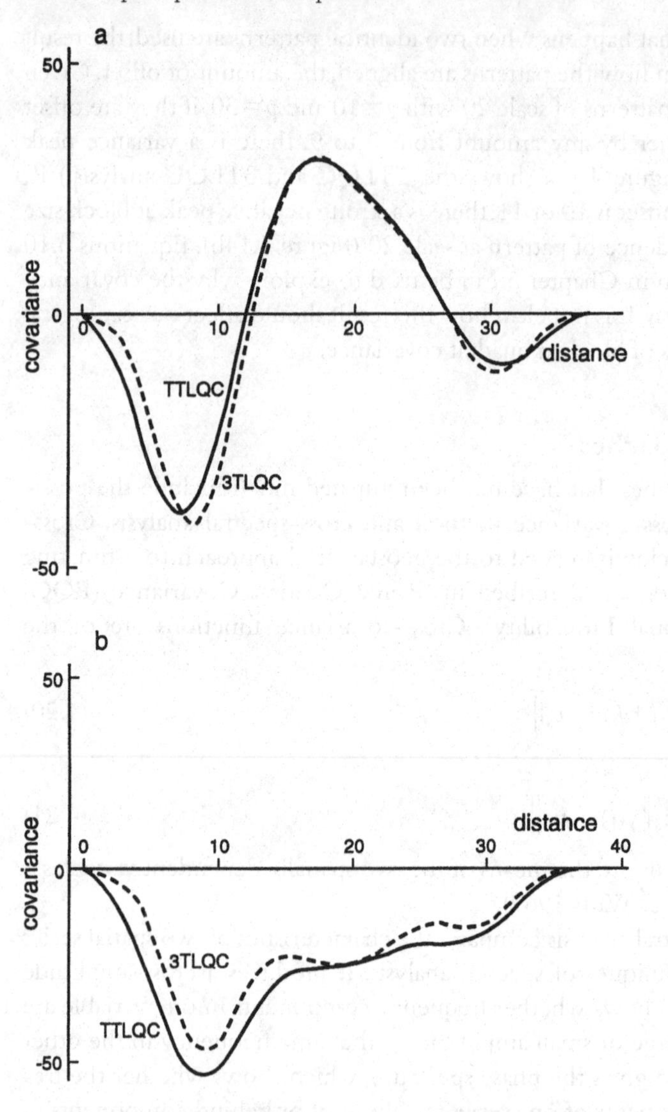

Figure 4.14 TTLQC and 3TLQC analysis of artificial data consisting of two species with scale 20, patch size 30 and gap 10. *a* When the patterns are offset by 9, the analysis correctly shows that they are negatively asssociated at that scale but positively associated at scale 20. *b* When the offset is 10 or 11, the positive peak at 20 disappears.

Relative pattern: species association

All the methods described thus far in this chapter have examined pattern defined by a fixed spatial reference, usually determined by a transect of contiguous quadrats. To finish the discussion, we should examine the situation in which the absolute positions of the plants of two species are not important but only their positions relative to each other. This involves an examination of the association of two species at only the smallest scale, but is clearly related to the detection of natural groupings of species within the community.

Using presence/absence data from quadrats, the usual method of analysis is a 2×2 contingency table which counts the number of quadrats that fall into each of four categories, where a is the number of quadrats in which both species are present, and so on:

species A		present	absent	
species B	present	a	b	
	absent	c	d	
	total			n

A goodness of fit test is carried out using the X^2 or G statistic and a significant result is interpreted as indicating positive association if $ad > bc$ and as negative association if $ad < bc$.

As has already been discussed, where there are actually more than two species in the study, not all pairwise tests are independent. In addition, if the quadrat data are collected in transects of contiguous or evenly spaced quadrats, the data from adjacent quadrats are not independent, giving spatial autocorrelation. In Dale *et al.* (1991), we describe a method for dealing with the effects of autocorrelation, by deflating the test statistic calculated, based on an underlying first-order Markov model and a Monte Carlo procedure to determine the amount of deflation needed to reduce the proportion of 1000 trials that are found to be significant to the correct level. Use of this procedure will depend on the appropriateness of a Markov model. For instance, in the Ellesmere Island sedge meadow data, while statistical testing provided no reason to reject low-order Markov models of multispecies spatial dependence, first-order models were rejected for single-species and two-species data. We would therefore not be able to use that approach in the analysis of those data. The topic of species association will be dealt with at greater length in Chapter 5, which deals with multispecies data.

Concluding remarks

The study of the joint spatial pattern of two species using covariance is closely related to the study of the scales of pairwise species associations and to the detection of natural groupings of species in a plant community. This kind of investigation should be able to find evidence for important community processes such as nucleation and competition. Paired quadrat techniques are closely related to the geostatistical approach of using cross-variograms, but the potential for confusion resulting from the existence of resonance peaks (both positive and negative) leads to preference being given to the bocked-quadrat methods of TTLQC and 3TLQC. Of the two, the three-term method is probably a better choice because it is less sensitive to trends in the data. In our own investigations, we found that density data and presence/absence data often gave similar results where both were available. We also found that results from field studies were much more complex than the simple hypothetical nucleation or competition results we had first considered.

Recommendations

1. For the analysis of two-species pattern in data from transects of contiguous quadrats, 3TLQC is the recommended method, although it is not perfect. In paired quadrat methods (PQC and tQC), the resonance peaks can cause problems for interpretation.
2. The correlation coefficient can be used to evaluate the strength of joint pattern compared to single-species patterns, but it should not be used by itself to detect scale.
3. Two-species pattern can be interpreted in terms of scales of positive and negative association. In the 2×2 contingency table approach to detecting species association, the tests of all pairs of several species are not independent. The potential effects of spatial autocorrelation on those tests must also be considered.
4. Analysis of the spatial pattern of covariance between a plant species and an environmental factor may provide different insights into the structuring of vegetation.

5 · *Multispecies pattern*

Introduction

In this chapter, we will present and discuss methods designed to examine the spatial pattern of groups of species or of whole plant communities. While it is true that plant communities are made up of individual species, we do not expect to be able to capture the essential features of the spatial structure of the whole community by compiling information on the spatial patterns of single species. Similarly, while we tend to think of species interactions as being pairwise, we know that the relationship between two species, A and B, can be modified by the presence and absence of other species (Dale *et al.* 1991). We cannot, therefore, in studies of plant communities, restrict our examination of species interactions only to pairs. Instead, we must find ways to look at the spatial structure and pattern of vegetation more holistically, by looking at many species simultaneously.

In Chapter 3, we described how the spatial pattern of a single species can be studied using methods that examine the effects of distance or block size on a calculated variance, with low variance indicating similarity and high variance indicating dissimilarity. In analyzing the spatial pattern of a single species using the data from a string of contiguous quadrats, the information for each quadrat is a single value, either some measure of the species' density, or simply 0 for absence and 1 for presence. A technique like two-term local quadrat variance (TTLQV) combines the quadrats into blocks of a range of sizes to determine which block size maximizes the difference between adjacent blocks of quadrats. The high variance is caused by strings of quadrats with low density alternating with strings of quadrats with high density. The scale of the pattern can be defined equivalently as half the average distance between the centers of successive patches or as the average distance between the centers of patches and the centers of their neighboring gaps. The two definitions give the same value because only two phases are considered (Figure 5.1).

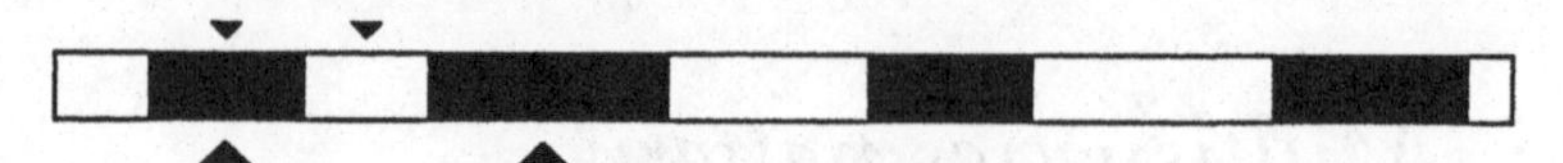

Figure 5.1 When there are only two phases to the mosaic, the average distance between the centers of adjacent patches (small triangles) is the same as half the average distance between the centers of patches of the same kind (large triangles).

In Chapter 4, we examined an obvious extension of single species pattern analysis to the analysis of the pattern of pairs of species by looking at the effect of scale on covariance (Greig-Smith, 1983; Dale & Blundon 1991). A further extension would be to study the spatial pattern of a whole community by looking at the covariance of all possible pairs, but the covariance analysis of many species pairs might be more confusing than helpful, depending on how subtle or consistent the overall pattern was. The desire to study and to quantify the spatial pattern of sets of species, including whole plant communities, has prompted researchers to develop a variety of methods for examining multispecies pattern.

One way in which the concept of multispecies pattern may have arisen is a perception that, in two dimensions, vegetation can be treated as a mosaic made up of patches of distinguishable vegetation types defined by combinations of species densities or occurrence. Each phase may not be completely homogeneous, but there is greater similarity within a patch and less between patches. As in the cycle-mosaic view of vegetation, there may be a continuum of phases that grade into each other through time, or there may be only two phases that are distinct and obvious.

There are many examples of natural systems that consist of two phases, including the vegetation stripes of some arid regions where lines of vegetation alternate with patches of bare ground (cf. Chapter 1 of White 1971, Montaña 1992). Whittaker and Naveh (1979) examined the multispecies pattern in three communities characterized by patches of shrubs in grassland: a mallee in Australia, a mesquite grassland in Texas, and *Pistacia lentiscus* woodland in Israel. All three were clearly two-phase mosaics. In a savanna grassland in South Africa, Carter and O'Connor (1991) also found a two-phase mosaic, with each phase dominated by a different perennial grass, *Setaria incrassata* in one and *Themeda triandra* in the other. In these cases in which there are only two phases and the demarcation between them is unambiguous, the definition of spatial scale is easy since half the distance between the centers of similar phases is the same as the distance between the centers of adjoining different phases.

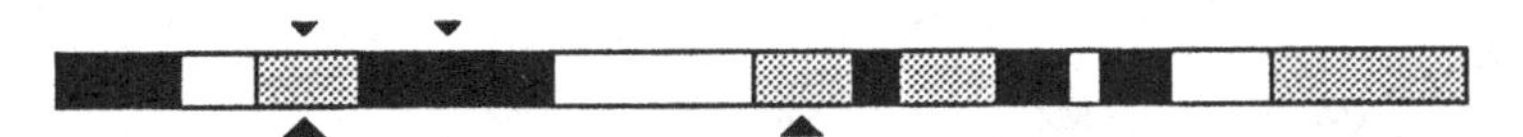

Figure 5.2 When there are more than two phases, the average distance between the centers of adjacent patches (small triangles) is less than half the average distance between the centers of patches of the same kind (large triangles).

In many other communities, however, there may be more than two phases in the mosaic, and then the 'scale' of the pattern will be different depending on whether it is defined in terms of distances between the centers of similar phases or distances between the centers of different phases (Figure 5.2). It is best to define the scale of multispecies pattern as half the distance that maximizes the probability of finding the most similar combination of species' densities (Dale and Zbigniewicz 1995). The average distance between the centers of adjoining phases that are different is the same as the average patch width, which is itself an important property of the vegetation pattern. In many natural communities, the demarcation between different phases may not be clear and the vegetation types may grade into one another, rather than having abrupt transitions. This feature may require multivariate analysis to be used in the analysis of spatial pattern.

One consequence of our definition of scale is that it is possible for different phases to have different scales of pattern (Figure 5.2). Allowing different phases to have different scales may, at first glance, seem to be a difficulty for the concept of multispecies pattern, but it probably reflects the nature of spatial structure of vegetation better than a definition that would force all phases to exhibit the same scale. The example used in Dale and Zbigniewicz (1995) to illustrate this point is a hummock-hollow system in which species that specialize in hummock tops or hollow bottoms will have a scale of pattern that is twice the scale of species that inhabit the sides of the hummocks. For instance, in bogs in Western Canada, *Sphagnum fuscum* is most abundant on the tops of the hummocks, *Sphagnum megellanicum* on the sides and *Sphagnum angustifolium* in the pools between the hummocks (Gignac & Vitt 1990). The scale of pattern of *S. megellanicum* is half the scale of the other two species (Figure 5.3). In this example, there is a single controlling environmental variable but there is more than one scale of pattern. If there are several important environmental variables, the situation may be much more complicated.

It is possible that various factors in the environment such as light, soil nutrients, moisture, and grazing all impose patchiness of different scales;

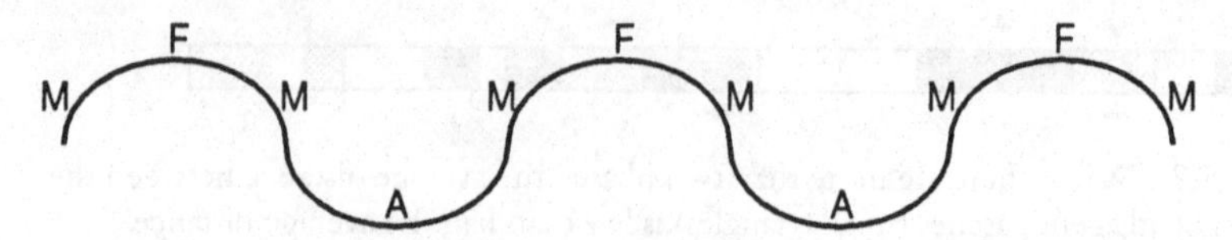

Figure 5.3 In a hummock-hollow system, species that specialize on the sides of the hummocks (*Sphagnum megellanicum*, M) will have a scale that is half of the scale of species found exclusively on the tops of the hummocks (*S. fuscum*, F) and those found only in the hollows (*S. angustifolium*, A).

then the groups of species that are most strongly controlled by each factor will have patterns of different scales. The vegetation, as a whole, has several scales of pattern, at least one for each set of species. Any method designed to investigate multispecies pattern must therefore be able to detect several different scales. The clarity of the detection of several scales of pattern is an important criterion to be used in evaluating the methods available.

Multiscale ordination

The first method proposed for multispecies pattern analysis is multiscale ordination (MSO) developed by Noy-Meir and Anderson (1971) and its concept remains one of the most sophisticated. The basis of the procedure is to calculate a variance-covariance matrix of the species for each of a range of block sizes and then the matrices are summed. The summed matrix is then subjected to principal components analysis (PCA) which gives several independent linear combinations of the species that explain as much of the variance as possible. For each new combination, its associated variance is partitioned by block size to produce a graph analogous to those in single-species methods (Chapter 3).

In the original version of the method (Noy-Meir & Anderson 1971), the variances and covariances were calculated using blocked quadrat variance (BQV); the modern version by Ver Hoef and Glenn-Lewin (1989) used TTLQV (Hill 1973) and its related covariance TTLQC (Greig-Smith 1983). We used three-term local quadrat variance (3TLQV) and its related covariance because it should be less affected by trends in the data (Dale & Zbigniewicz 1995, and see Chapter 3 for discussion of the advantages and disadvantages of those two methods).

Let the number of species be k; a $k \times k$ variance-covariance matrix, $\mathbf{C}(b)$, is calculated for each block size, b, from 1 to some maximum, M, using 3TLQV and 3TLQC, as described in Chapters 3 and 4. For clarity,

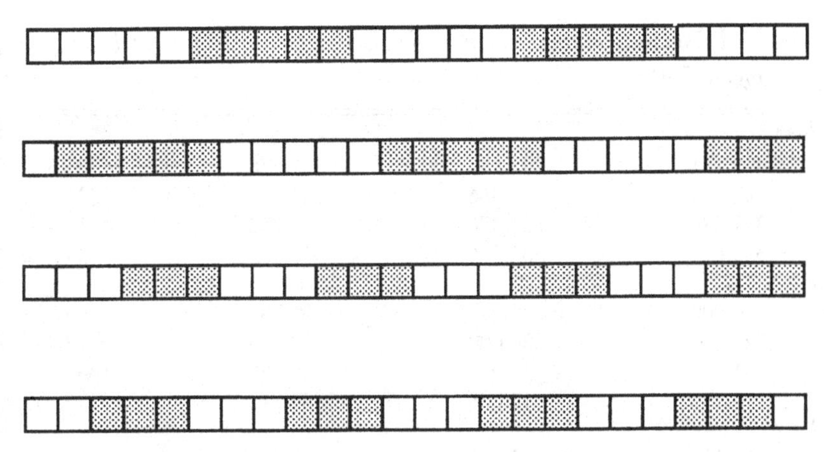

Figure 5.4 Artificial data for multiscale ordination (MSO) analysis. The shaded squares represent quadrats with high density. The first and second species have a scale of 5, but they are negatively associated. The third and fourth species have scale of three and are offset by one quadrat.

the procedure will be illustrated with artificial data for four species, as shown in Figure 5.4. Two of the species have pattern at scale five, but they are offset by four quadrats; they are negatively associated. The other two have pattern at scale three and are positively associated. Table 5.1 gives the variance-covariance matrices for block sizes 1 to 10. The M matrices for the range of block sizes are added together, as illustrated at the bottom of Table 5.1:

$$\mathbf{S} = \sum_{b=1}^{M} \mathbf{C}(b). \tag{5.1}$$

The sum is eigenanalyzed as in the familiar procedure of PCA. That procedure can be explained in several ways; for instance, we can say that it finds the linear combination of the k species that explains more of the variance than any other such combination. It then finds a second linear combination that maximizes the amount of the remaining variance explained and that is orthogonal to the first. It proceeds to find a third axis with the same properties, and so on. The variance associated with the first combination is called the first eigenvalue λ_1, the variance of the second is the second eigenvalue λ_2, and so on. Because the linear combinations are forced to be orthogonal, their covariances are all zero. Therefore, another way of explaining the process is to say that the original variance-covariance matrix $\mathbf{S}$ is transformed into a diagonal matrix of the eigenvalues, Λ:

Table 5.1. *Variance-covariance matrices for block sizes 1 to 10 (artificial data from Figure 5.4)*

$b=1$			
499.161	247.483	331.376	−335.570
247.483	494.966	331.376	−331.376
331.376	331.376	834.732	−419.463
−335.570	−331.376	−419.463	834.732
$b=2$			
1250.000	−625.000	−31.780	−6.356
−625.000	1250.000	19.068	−19.068
−31.780	19.068	3756.356	1887.712
−6.356	−19.068	1887.712	3756.356
$b=3$			
4162.386	−3232.021	37.101	−128.425
−3232.021	4140.982	142.694	−142.694
37.101	142.694	6095.891	2762.557
−128.425	−142.694	2762.557	6095.891
$b=4$			
8135.813	−6585.208	−45.415	−9.732
−6585.208	8153.114	22.708	−22.707
−45.415	22.708	1878.244	943.988
−9.732	−22.707	943.988	1878.244
$b=5$			
9000.000	−7027.972	61.189	−61.189
−7027.972	9055.944	75.175	−75.175
61.189	75.175	166.958	−83.916
−61.189	−75.175	−83.916	166.958
$b=6$			
5407.097	−4349.234	0.000	0.000
−4349.234	5382.803	0.000	0.000
0.000	0.000	0.000	0.000
0.000	0.000	0.000	0.000
$b=7$			
1785.714	−1392.857	44.005	−49.107
−1392.857	1785.714	49.107	−49.107
44.005	49.107	119.260	−59.949
−49.107	−49.107	−59.949	119.260
$b=8$			
313.064	−156.814	−6.769	0.000
−156.814	311.372	5.077	−5.077
−6.769	5.077	939.192	472.134
0.000	−5.077	472.134	939.192
$b=9$			
55.758	27.879	34.469	−38.524
27.879	55.758	38.524	−38.524
34.469	38.524	2031.630	920.519
−38.524	−38.524	920.519	2031.630

Table 5.1. (*cont.*)

b = 10			
0.000	0.000	0.000	0.000
0.000	0.000	0.000	0.000
0.000	0.000	751.384	377.767
0.000	0.000	377.767	751.384
sum over all block sizes			
30608.994	−23093.744	424.175	−628.903
−23093.744	30630.654	683.728	−683.728
424.175	683.728	16573.646	6801.350
−628.903	−683.728	6801.350	16573.646

Notes:
The first two species have scale of five and are negatively associated; see $b=5$ and $b=10$. Species three and four are positively associated with scale three; see $b=3$ and $b=6$.

$$\begin{bmatrix} \lambda_1 & & & & 0 \\ & \lambda_2 & & & \\ & & \cdot & & \\ & & & \cdot & \\ 0 & & & & \cdot \\ & & & & \lambda_k \end{bmatrix} \tag{5.2}$$

Table 5.2 shows the diagonal matrix of eigenvalues derived from the sum matrix for the artificial example.

There are several ways of explaining how this transformation is accomplished, but one way is to say that for each eigenvalue, λ_i, we are looking for an eigenvector, $\mathbf{u}_i$, such that $\mathbf{u}_i\mathbf{S}=\lambda_i\mathbf{u}_i$. We can, in fact, standardize the vector such that $\mathbf{u}_i\mathbf{u}_i^T=1$; $\mathbf{u}_i^T$ is the transpose of $\mathbf{u}_i$, so their product is a scalar. (The squares of the elements of $\mathbf{u}_i$ add to 1.) Most computer algorithms for performing eigenanalysis, such as Hotelling's procedure, will provide these vectors directly. Knowing vector $\mathbf{u}_i$, the eigenvalue can be partitioned by block size:

$$\begin{aligned}\lambda_i &= \mathbf{u}_i\mathbf{S}\mathbf{u}_i^T = \mathbf{u}_i\mathbf{C}(1)\mathbf{u}_i^T + \mathbf{u}_i\mathbf{C}(2)\mathbf{u}_i^T + \ldots + \mathbf{u}_i\mathbf{C}(M)\mathbf{u}_i^T \\ &= \lambda_i(1) + \lambda_i(2) + \ldots + \lambda_i(M).\end{aligned} \tag{5.3}$$

For an analysis of k species, k eigenvalues are produced, but the first few are the largest and may account for a large proportion of total variance and thus only they need to be considered further. In other applications of the procedure, such as PCA, the technique is used to

Table 5.2. *The diagonal matrix of eigenvalues derived from summed matrices of Table 5.1*

53714.48	0	0	0
0	23375.32	0	0
0	0	10299.65	0
0	0	0	6997.50

Notes:
The proportion of the total variance accounted for by the values are 57%, 25%, 11% and 7%.

reduce the number of dimensions that need to be considered. The largest eigenvalues are then each partitioned into the contributions of each block size and peaks or plateaux in the plot of variance as a function of block size are interpreted as corresponding to scales of pattern (*cf.* Ver Hoef & Glenn-Lewin 1989). In our example, the first eigenvalue accounts for 57% of the variance and its associated eigenvector is (0.71, −0.71,0,0). When the eigenvalue is partitioned, there is a variance peak at block size five (Table 5.3), reflecting the scale of pattern of the first two species. (Note that 0.71 is $1/\sqrt{2}$ and so the sum of the squares of the vector elements is 1.0.) The values for species 1 and 2 have opposite signs since they are negatively associated. The second eigenvalue accouts for 25% of the total variance. Its eigenvector is (0,0,0.71,0.71), picking up the third and fourth species and producing a variance peak at block size 3, as we would expect. The values have the same sign since the species are positively associated. The third and fourth eigenvectors are (0.29,0.28,0.65, −0.65) and (0.65,0.65, −0.28,0.29) with variance peaks at three and five. In a real analysis, we probably would not examine the third and fourth axes because they represent only a small proportion of the total variance, 11% and 7%.

As in single-species 3TLQV analysis, small-scale patterns may produce 'shoulders' in the variance plot, due to the large variance associated with pattern at large block sizes. Another feature of methods like 3TLQV is that pattern of scale B produces resonance peaks at block sizes approximately $3B$, $5B$, $7B$ . . ., with peak variances diminishing to 1/3, 1/5, 1/7. This phenomenon appears in the multispecies pattern analysis.

In evaluating the multiscale ordination technique, Ver Hoef and Glenn-Lewin (1989) suggest that, because larger block sizes tend to produce larger variances, the covariance matrices should be weighted prior to summing. Such a weighting procedure would be similar to sub-

Table 5.3. *Eigenvalues partitioned by block size and their associated intensities; maxima are underlined*

b	$\lambda_1(b)$	$J_1(b)$	$\lambda_2(b)$	$J_2(b)$	$\lambda_3(b)$	$J_3(b)$	$\lambda_4(b)$	$J_4(b)$
1	249.59	0.11	415.30	0.14	1660.01	0.22	338.68	0.10
2	1875.40	0.31	5643.93	0.53	1674.28	0.22	819.10	0.15
3	7384.40	0.61	<u>8858.38</u>	<u>0.66</u>	<u>3112.37</u>	<u>0.30</u>	1140.00	0.18
4	14729.84	0.86	2822.49	0.37	1037.57	0.17	1455.52	0.20
5	<u>16055.68</u>	<u>0.89</u>	83.31	0.06	630.46	0.13	<u>1620.42</u>	<u>0.22</u>
6	9743.94	0.70	0.15	0.00	167.32	0.07	878.49	0.16
7	3178.53	0.40	59.38	0.05	283.49	0.09	288.55	0.09
8	469.13	0.15	1411.28	0.26	418.43	0.11	203.98	0.08
9	27.95	0.04	2952.01	0.38	1001.83	0.17	192.98	0.07
10	0.02	0.00	1129.09	0.24	313.89	0.10	59.77	0.04

Notes:
λ_i is the eigenvalue (Equation 5.3) and $J_i(b)$ is the associated intensity (Equation 5.5).

stituting the intensity for the variance at each block size. As in single-species pattern, the intensity of the pattern at block size b, $I(b)$, is a function of the block size and the variance, $V(b)$:

$$I(b) = \sqrt{6b\ V(b)/(b^2+2)}. \qquad (5.4)$$

We therefore suggested weighting the variance-covariance matrices by the factor $6b/(b^2+2)$ before summation and eigenanalysis. The square-root is omitted, so that the matrices remain truly variance-covariance matrices and their sum can be partitioned (Dale & Zbigniewicz 1995). After eigenanalysis and partitioning, the values are converted back to the equivalent of the original variances by multiplying by $(b^2+2)/6b$. In our artifical example, the process of conversion and back conversion has little effect on the analysis because the scales are both small and close together.

For each eigenvalue, λ_i, from the analysis, there is a variance at each block size, $\lambda_i(b)$. Where there is a variance peak, it is useful in interpreting its importance to look at the associated intensity, $J_i(b)$:

$$J_i(b) = \sqrt{6b\ \lambda_i(b)/(b^2+2)}. \qquad (5.5)$$

In the example based in Figure 5.4, the first two axes have variance peaks of high intensity, 0.89 and 0.66, and the last two have low intensities, 0.30 and 0.22 (Table 5.3).

There is a second property of multispecies pattern that is important in evaluating the pattern: how much the various species contribute to the pattern. If one species dominates one of the eigenvectors, any pattern detected based on that eigenvector is not truly multispecies. A measure of species' contributions can be derived from the eigenvectors produced by the analysis. Recall that one explanation of the procedure is that from the k original variables, the x_j's, for each eigenvalue λ_i there is a new variable, y_i, that is a linear combination of the species densities with weights u_{ij}:

$$y_i = \sum_{j=1}^{k} u_{ij} x_j. \qquad (5.6)$$

We know that $\sum_{j=1}^{k} u_{ij}^2 = 1$, because $\mathbf{u}_i \mathbf{u}_i^T$ was set at 1. The variance of the squares of the weights, u_{ij}^2, can be used to evaluate the evenness of the species' contributions. The mean of the u_{ij}^2 is $1/k$ by definition. Let C_i be their coefficient of variation in eigenvector $\mathbf{u}_i$, variance over mean, expressed as a proportion not in percent. If evenness is at a maximum with all species having equal weights, C_i is 0, and if evenness is at a minimum with one weight of 1.0 and the other $k-1$ being 0, then C_i is $\sqrt{k-1}$. A measure of evenness is therefore:

$$E_i = 1 - C_i/\sqrt{k-1}. \qquad (5.7)$$

In the small example we have been using as an illustration, the first two axes have low evenness, $E = 0.43$, but the last two are more even with $E = 0.77$.

The weights, u_{ij}, can also be used to evaluate species association at the scale of pattern indicated by peaks in an eigenvalue's partitioning. Pairs of species which have large weights of the same sign are strongly positively associated and species which have large values of opposite sign are negatively associated at that scale (*cf.* Ver Hoef & Glenn-Lewin 1989). We have already commented on this feature of the artificial example; in the first eigenvector species one and two are of opposite signs whereas in the second eigenvector species three and four have the same sign. It is important to note, however, that the meaning of the signs does not carry through to the last two eigenvectors, which must be orthogonal to the first two and to each other.

We have applied this multiscale ordination to a range of data sets (*cf.* Dale & Zbigniewicz 1995), including the successional communities on the Robson moraines. On moraine three, the first eigenvalue represents 83% of the total variance, but it does not represent true multispecies pattern because it is very strongly dominated by a single species, *Picea engelmannii*, giving $E_i = 0.093$. If you glance at Figure 3.30, you will see why. The next two axes were also dominated by single species. The first three eigenvalues accounted for 95% of the total variance. On the younger moraines, the axes are also dominated by single species but the first eigenvalues were not as high.

In their study of a basalt glade prairie, Ver Hoef and Glenn-Lewin (1989) found much more even eigenvector loadings. An interesting hypothesis for future testing is whether the evenness of contributions to multispecies pattern is low in seral communities but is greater in climax vegetation. A last optional step in the analysis procedure is to plot the score of each quadrat for the first few axes along the length of the transect to portray the multispecies response. Noy-Meir and Anderson (1971) refer to such a diagram as a component profile. Ver Hoef and Glenn-Lewin (1989) recommend using a running average of the quadrat scores using a moving window of size equal to the major scale of pattern detected for that axis. For example, with their field data, the first eigenvalue had a peak at block size 50 (5 m) and so they plot moving averages of 50 quadrats. The 5 m scale of pattern was attributed to patches of crustose lichens on basalt outcrops, somewhat less than 5 m in size, separated by larger patches of other growth forms. The second eigenvalue had a similar peak at about 5m but it had a second peak at 9 m. This second scale is due to the segregation of regions between the lichen patches into two types, one dominated by cryptogams and the other dominated by graminoids. This situation is similar to that in Figure 5.3, with the crustose lichens (like '*M*' in the figure) having a scale of pattern that is half that of the other cryptogams and of the graminoids (like '*F*' and '*A*' in the figure).

Our evaluation of the MSO technique using artificial data is that it recovers most of the major features of the data reasonably well, even with presence/absence data which may violate the assumption of multivariate normality which underlies the eigenanalysis. One somewhat puzzling feature of the technique is that the results may depend on the maximum block size used (Dale & Zbigniewicz 1995). Ver Hoef and Glenn-Lewin (1989) also found much to recommend about this method: not only did it recover the known structure of fabricated data, but also, in field data, it revealed details of species associations over a range of scales.

Castro *et al.* (1986) used PCA for multispecies pattern analysis, but followed a different approach. They calculated covariance matrices for block sizes 1, 2, 4, 8, 16, and 32, and applied PCA to each matrix separately. The results were used to plot the first axis score as a function of position on the transect and to plot the first axis score of individual species as a function of block size. Because the matrices were analyzed separately, a high first axis score at one block size may have a different interpretation from a high first axis score at the next block size. Another problem of this method, compared to MSO, is that only the first axis is considered and, therefore, scales of pattern that may be associated with the second and third axes may be missed.

Semivariogram and fractal dimension

The analysis of fractal dimension as a means of detecting multispecies pattern is a method introduced into ecology from geostatistics. It is based on the empirical semivariogram which estimates the semivariance. For distance b, where $x_j(i)$ is the density of the jth species in the ith quadrat, the calculation is (Palmer 1988):

$$\gamma(b) = \sum_{j=1}^{k} \sum_{i=1}^{n-b} \left[x_j(i) - x_j(i+b)\right]^2 / 2(n-b). \tag{5.8}$$

Because the terms that contribute to γ are measures of dissimilarity at distance b, this technique is actually a multispecies version of spaced quadrat variance, PQV, described for the single-species case in Chapter 3 (*cf.* Ludwig & Goodall 1978). Ver Hoef *et al.* (1993) discuss the relationship between several spatial analysis methods and they show that PQV is a variogram estimator that can be used approximate TTLQV; therefore this multispecies spaced quadrat method is actually closely related to the MSO blocked quadrat method.

Having calculated $\gamma(b)$, the slope of $\log(\gamma)$ as a function of $\log(b)$ is $m(b)$. There is a choice of methods for calculating the slope, including linear regression over a small range of points or, more simply, following Palmer (1988):

$$m(b) = \left[\log \gamma(2b) - \log \gamma(b)\right] / \log(2). \tag{5.9}$$

The slope is then used to calculate the fractal dimension at scale b (Phillips 1985):

$$\mathfrak{D}(b) = \left[4 - m(b)\right] / 2. \tag{5.10}$$

When quadrats at distance b are very dissimilar and quadrats at distance $2b$ are similar, $m(b)$ will be large and negative, giving large positive values of $\mathfrak{D}$. When quadrats at distances b and $2b$ have the same similarity, $m(b)$ will be zero and $\mathfrak{D}$ will be 2.0. Large values of $\mathfrak{D}$ greater than 2.0 therefore indicate scales of pattern in the data. Because $m(b)$ can be negative, it is possible for $\mathfrak{D}$ to take values greater than two; for example, suppose the density data are in percent, and there are two species, the first has densities 99, 99, 99, 1, 1, 1, 99, 99, 99, ... and second has complementary densities 1, 1, 1, 99, 99, 99, 1, 1, 1, ... Under these conditions, $\gamma(3) = 9604$ and $\gamma(6) = 0$, giving $m(3) = -13.23$ [replacing log(0) with 0 by convention], and thus $\mathfrak{D} = 8.6$. The ability of $\mathfrak{D}$ to take such large values makes its interpretation as a fractional dimension difficult. One reason for this difficulty may be that the method was not originally designed for situations in which repeating spatial pattern would give a series of peaks and valleys in the variogram, but rather for variograms of phenomena for which increasing distance gives a continuing increasing difference (see Bell *et al.* 1993). Alternate methods for calculating the slope of the variogram may also improve the application of this method.

The strength of multispecies pattern as detected by the fractal method can be measured by the maximum value of $\mathfrak{D}$, since it will reflect the difference between the low similarity of quadrats at distance b and the high similarity at distance $2b$. If similarity is independent of distance, $m(b)$ will be zero, giving $\mathfrak{D}(b) = 2.0$; therefore, peak values of $\mathfrak{D}$ that are close to 2.0 indicate multispecies pattern that is very weak. For example, the Ellesmere Island transect data when treated with this kind of analysis produced peaks in $\mathfrak{D}$ that were close to 2.0 indicating that any multispecies pattern that was present was very indistinct. When we analyze the artificial data used to illustrate MSO with species patterns at scales three and five the plot of $\mathfrak{D}$ has peaks at 3 and 9 with values of 2.5 and 2.8, but the strongest peaks are at 15, 45 and 75 with values of 13.0!

Palmer (1988) applied the same method of fractal analysis to data sets from a range of plant communities including a suburban lawn and the trees and the understorey of a hardwood forest. As in our analysis of the Ellesmere data, he found fractal dimensions close to 2.0, suggesting weak spatial pattern in the communities. He also concluded that because fractal dimension is not a constant function of scale, patterns of spatial variation cannot be extrapolated from one scale to another.

Methods based on correspondence analysis

There are several methods available for multispecies pattern analysis based on correspondence analysis (CA). Galiano (1983) carried out that ordination procedure on data from transects consisting of 400 2 cm × 2 cm quadrats in a grassland in central Spain. The data used a three-category cover scale and so it was argued that CA was a more appropriate ordination procedure than PCA. Having performed the ordination, the quadrats' scores on the first ordination axis were analyzed by two-term new local variance (NLV). Before that analysis, the plot sequence was divided into two sequences: A, in which all the negative scores were replaced with zeros, and B, in which all the positive scores were replaced with zeros and the negative scores made positive. For example, the sequence 0.3, 0.9, 0.2, −0.4, −0.7, −0.5, 0.1, 0.6,... gives A = 0.3, 0.9, 0.2, 0, 0, 0, 0.1, 0.6, ... and B = 0, 0, 0, 0.4, 0.7, 0.5, 0, 0,.... Each of A and B are analyzed separately. The argument for this conversion is that the first CA axis will represent a major environmental gradient from moist to dry and NLV will then detect the sizes of the patches of plants associated with the two extremes of that gradient. This is more or less what the analysis found.

In general applications, in which CA would be followed by analysis using 3TLQV in order to detect the scale of pattern, the division of the transect into phases which are then analyzed separately would not be recommended (Gibson & Greig-Smith 1986). There are other two criticisms of the approach used by Galiano. The first is that NLV should be complemented by 3TLQV or PQV to examine the overall scale of pattern, not just the average size of the smaller phase. The second is that more than one axis should be examined since there may be sources of spatial pattern in the vegetation other than the moist to dry gradient. The interpretation of this procedure is made more complicated by the 'arch' or 'horsehoe' effect in CA: samples that are in fact arranged linearly along a gradient often appear in a horseshoe shape in the ordination diagram that shows the positions of the samples relative to the first two axes. This effect can be removed by a procedure known as detrending. If detrending is not used, the possible effects of this phenomenon on the use of the ordination procedure for pattern analysis must be investigated.

Gibson and Greig-Smith (1986) used a somewhat different method based on ordination by CA. They sampled dune grassland vegetation using three parallel transects of 64 5 cm × 5 cm quadrats. All 192 quadrats were subjected to detrended correspondence analysis (DCA). For each of the first two axes, the quadrats were assigned their scores for that axis and the scores were then analyzed with TTLQV and 3TLQV. TTLQV and 3TLQV were used to examine the scale of pattern of the microtopography, quantified by the relative height of the center of each quadrat.

Gibson and Greig-Smith extended their analysis by examining the correlation values derived from TTLQC between the microtopography and the DCA axis score. The analysis showed that microtopography had a scale of pattern of about 30 cm. The first DCA axis had a similar scale of pattern (40 cm) but was negatively correlated with microtopography. This correlation was interpreted as being related to the fact that the vegetation was comprised of hummocks that included the dominant grass *Arrhenatherum elatius*, alternating with hollows that contained the dominant herb *Hydrocotyle vulgaris*. The second DCA axis had a scale of about 15 cm, with a weak positive correlation with microtopography. The authors interpreted these results with reference to small tussocks including *Carex nigra* out of phase with other species group including *Equisetum variegatum*. Remember that our examination of artificial patterns showed that peaks in correlation were not good indications of pattern scale, only peaks in covariance. Using correlation to evaluate the relationship of scale already identified is a different procedure.

While Gibson and Greig-Smith used DCA and correlation analysis to examine the scale of the joint pattern of vegetation and an underlying environmental factor, we expect that in the future such joint pattern analysis will be based on the ordination canonical correspondence analysis (CCA) (ter Braak 1987). It includes both species data and environmental data in simultaneous ordination and therefore seems to be an obvious basis for a joint species-environment pattern study. Like CA, it also available in a detrended form.

Euclidean distance

A method suggested by Lepš (1990b) is based on a measure of dissimilarity, the average Euclidean distance, calculated from the species abundances, between adjacent blocks of *b* quadrats. When this distance measure is calculated for a range of block sizes, peaks in the plot of distance as a function of block size will correspond to scales of pattern in the whole community. The calculation of the average Euclidean distance at a range of block sizes can be achieved by adding together all the TTLQV curves of the individual species (Lepš 1990b). The 3TLQV equivalent is to be preferred over the TTLQV based method described by Lepš (1990b), for the reasons mentioned in previous discussions (see Chapter 3, 'Local Quadrat Variances'), including the fact that it is less susceptible to trends in the data. The calculation is then:

$$V_E(b) = \sum_{h=1}^{k} \sum_{i=1}^{n+1-3b} \left(\sum_{j=1}^{i+b-1} \left[x_h(j) - 2x_h(j+b) + x_h(j+2b) \right] \right)^2 / 8b(n+1-3b). \tag{5.11}$$

The consistency of this overall pattern can be measured by comparing the intensity of the pattern with the average density of species in quadrats where they are present. If D_j is the average of the nonzero densities of species j, and D_{avg} is $\Sigma D_j/k$, then a measure of total pattern consistency is:

$$T(b) = \sqrt{6b\ V_E(b)/k(b^2+2)}/D_{avg}. \tag{5.12}$$

For presence/absence data, D_{avg} is omitted, of course.

To illustrate the application of this method, we will use data from the successional communities on the proglacial deposits adjacent to the SE Lyell Glacier (52 N, 119 W) in the Canadian Rockies (Dale & MacIsaac 1989). The surfaces are at about 1600 m altitude and were uncovered by the retreat of the SE Lyell and Mons glaciers. The development of the vegetation begins with *Dryas drummondii* which forms an almost continuous carpet within 20 years. Shrubs, such as *Shepherdia canadensis* and *Salix* spp. appear at age 40 years and the tree canopy of *Picea engelmannii* starts to close around 90 years. At that point, the *Dryas* cover starts to decline. The terminal moraine (130 years) is vegetated by a *Picea* forest with a dense shrub stratum and almost no *Dryas*, but a ground cover of mosses.

The chronosequence was sampled at 11 locations, ranging in age from 6 years to 143 on the terminal moraine, and at a twelfth location in the adjacent old-growth forest, more than 400 years old. Each location was sampled with two transects of 300 contiguous 30 cm × 30 cm quadrats parallel to the moraine crests. Percentage cover was estimated for all species, with the cover of *Dryas* being divided into live and dead.

In applying the Euclidean distance method to data from these successional communities at SE Lyell, we found that the peak in V_E (variance based on Euclidean distance) often matched peaks in the variance related to the first eigenvalues derived from MSO, but the values of T (total consistency) associated with the peaks were quite low, averaging around 0.4. One of the strongest examples of multispecies pattern was at subsite 11 at Lyell, where the Euclidean distance method had a variance peak at block size 27 ($T = 0.62$) and the first axis of the MSO analysis, which accounted for 40% of the total variance, had a peak also at 27 with $E = 0.4$ and an intensity of 0.68. In other cases, the peaks did not match so well, indicating that the two methods are sensitive to different features of the data. Table 5.4 compares

Table 5.4. *A comparison of multiscale ordination (MSO) and Euclidean distance multispecies pattern analysis; transects in older vegetation at SE Lyell site*

Multiscale ordination				Euclidean distance Peak position (T)	
$\lambda_i/\Sigma\lambda$	E_i	Peak position	(intensity)		
Transect 8.1					
0.71	0.39	38 (0.42)	61 (0.36)	36 (0.28)	62 (0.24)
0.15	0.46	28 (0.20)	65 (0.14)		
Transect 8.2					
0.84	0.29	53 (0.47)	–	53 (0.25)	–
Transect 9.1					
0.76	0.31	26 (0.63)	–	26 (0.40)	–
0.15	0.33	16 (0.37)	–		
Transect 9.2					
0.48	0.47	11 (0.66)	33 (0.43)	33 (0.42)	57 (0.35)
0.39	0.49	21 (0.49)	61 (0.35)		
Transect 10.1					
0.48	0.54	33 (0.46)	–	40 (0.50)	–
0.26	0.41	45 (0.39)	–		
0.18	0.44	14 (0.48)	43 (0.22)	–	–
Transect 10.2					
0.47	0.56	15 (0.57)	55 (0.34)	23 (0.54)	53 (0.39)
0.22	0.56	18 (0.35)	26 (0.30)	42 (0.24)	
Transect 11.1					
0.48	0.40	27 (0.68)		27 (0.62)	
0.25		9 (0.48)	29 (0.31)		
Transect 11.2					
0.39	0.57	7 (0.52)	27 (0.38)	s59 (0.48)	
0.27	0.53	35 (0.43)			
0.18	0.62	11 (0.33)			
Transect 12.1					
0.44	0.63	12 (0.52)	31 (0.33)	13 (0.60)	31 (0.41)
0.23	0.66	14 (0.33)	27 (0.27)		
Transect 12.2					
0.50	0.53	53 (0.69)		60 (0.64)	
0.26	0.49	15 (0.51)			
0.15	0.60	14 (0.42)	43 (0.23)		

Notes:
Only those eigenvalues that account for 15% or more of the total variance are included. 's' indicates a shoulder rather than a peak in the variance plot.

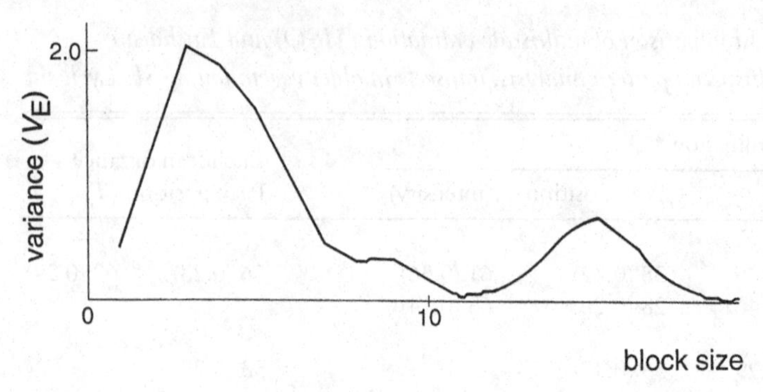

Figure 5.5 The variance based on Euclidean distance, V_E, for the artificial data in Figure 5.4. The scale of 5 is not very evident in the variance plot.

the results of MSO and Euclidean distance for transects on the older surfaces at the SE Lyell site.

The Euclidean distance method has the advantage that it summarizes the spatial pattern of many species in a single variance curve. The disadvantage is that any technique that produces only one curve may not be able to show that there is actually more than one scale of pattern. When the method is applied to our usual four-species artificial data set with scales of three and five, a curve with a strong peak at 3 is produced in which there is little evidence of the scale of 5 (Figure 5.5). The variance curve would have to be partitioned as described by Dale and Blundon (1990), removing the variance due to the pattern at scale three and the residual variance would then have a peak at scale five.

The satisfactory performance of the method reported by Lepš (1990b) was due, in part at least, to the fact that three of the five species used showed evidence of pattern at the same scale, about 2.4m, so that the analysis of the overall spatial pattern correctly reflected this characteristic.

Comments

In evaluating three of the methods so far described (MSO, Fractal dimension and Euclidean distance), we found that if the vegetation consists of several groups of species with different scales, neither the fractal dimension nor the Euclidean distance method can reliably pick out all scales of pattern (Dale & Zbigniewicz 1995). Their output is too much a summary in which important detail can become lost. We have to use MSO or DCA score pattern analysis because only those methods give several axes each of which can display more than one scale of pattern. The choice between the two will depend on the suitability of the data for the application PCA or DCA, and there is a wealth of literature on that and related topics (*cf.* Ludwig & Reynolds 1988; ter Braak & Prentice 1988; James & McCulloch 1990). Note, however, that in studies of artificial pattern, we found that MSO recovered the characteristics well, even with presence/absence data. We have also explored the possibility of using nonmetric multidimensional scaling (NMDS, Anderson 1971) as the ordination procedure from which to derive quadrat scores for several axes which are subsequently subjected to 3TLQV (unpublished). It also seems to perform well with artificial data.

Spectral analysis

In Chapter 3, we describe the use of spectral analysis to examine the spatial pattern of a single species. Like the other basic methods, it can be applied to quadrat ordination scores as well as to species density. Kenkel (1988b) applied spectral analysis to data from a hummock-hollow complex in a mire in Northern Ontario, by first applying CA. The first CA axis was highly correlated with microelevation and spectral analysis found similar scales of pattern in both variables. He also looked at the relationship between the two spectral series using cross-spectral analysis, which evaluates the relationship between two series. The similarity of the two series was evaluated by two spectra, the coherence spectrum which measures the similarity of amplitudes at particular frequencies and the phase spectrum which evaluates 'lead-lag' relationships. Cross-spectral analysis is described in greater detail in specialized texts on the subject of spectral analysis such as Jenkins and Watts (1969) or Koopmans (1974). In the mire study, the coherence values asssociated with the identified peaks were over 0.8 and their phase spectra were close to 0. These results show that the spatial patterns of the vegetation and of the microtopography are very similar and closely in phase. The hummock-hollow pattern is therefore thought to be related to the growth of *Sphagnum* on the branches of the low-growing *Chamaedaphne calyculata* shrubs.

Other field results

An early example of the use of MSO is by Williams *et al.* (1978) who used a BQV-based version of MSO to look at the influence of sheep on the spatial pattern of *Atriplex vesicaria*. Instead of looking at several species they looked at several ages of plants of the same species and concluded

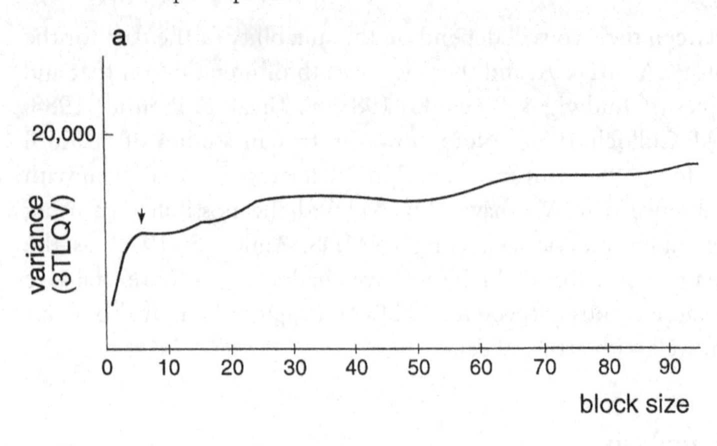

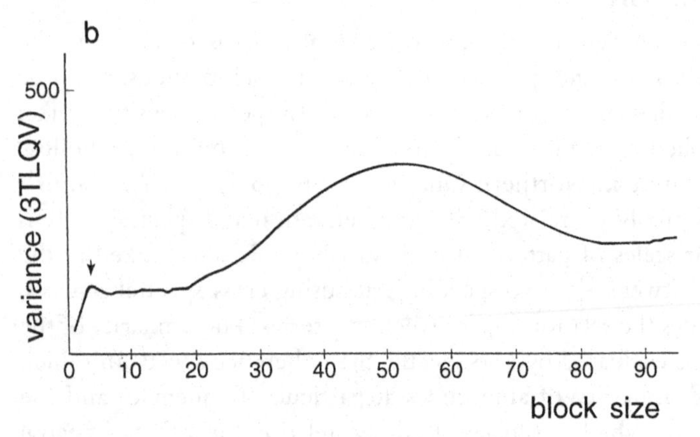

Figure 5.6 3TLQV performed on the first CA axis scores for the Ellesmere Island transects: *a* OWT, *b* CRT, *c* BMS and *d* BRS. All show evidence of small scale pattern (arrows).

that multiple pattern analysis provided an economical description of the total pattern of the age-structured population. They conclude that the plants' interactions with sheep are influenced by plant size, sex and location relative to other plants of the same species.

The Ellesmere Island sedge meadow data (Young 1994) were analyzed by performing 3TLQV on the individual species and most of them showed pattern at a scale below block size 20. That small-scale pattern does not show up clearly in the multispecies analysis. For instance, at site BMS, five of the eight most important species had pattern at block size six or seven, but there is no sign of matching scale in any of the three analyses using MSO, fractal dimension and Euclidean distance. This may be the result of

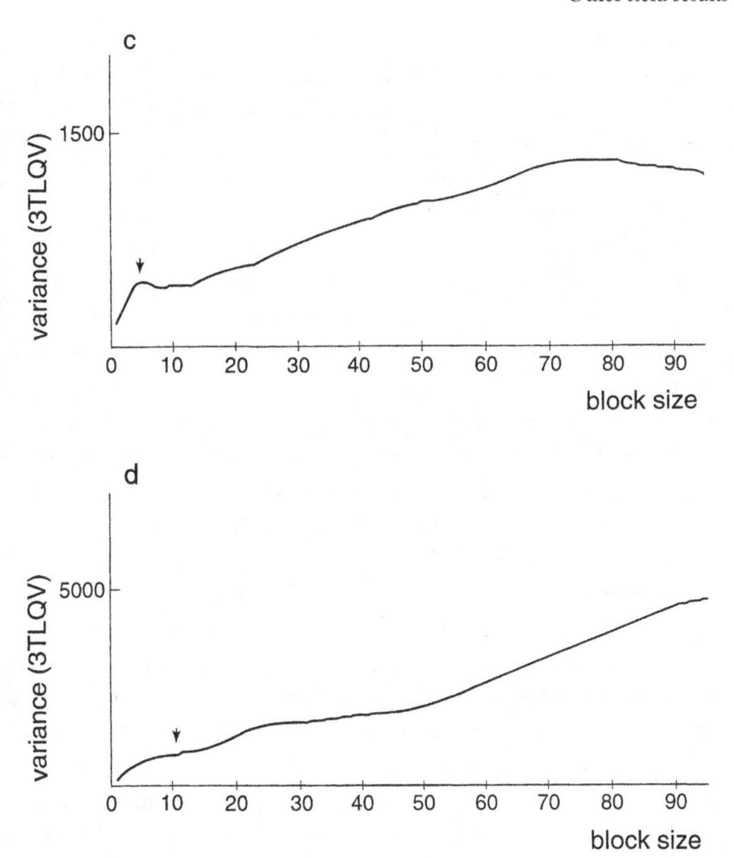

the individual patterns not being consistent in their phase relationships: neither constantly in phase nor constantly out of phase. It may be related, also, to nonstationarity in the patterns. At all four Ellesmere sites, the MSO produced three axes with the eigenvalue, λ, >0.10, usually with moderate values for E_i, the measure of evenness (average 0.59). In many instances, there was no evidence of pattern in the range of block sizes studied (up to 75 quadrats $=7.5$m), especially using the Euclidean distance method. The values of $\mathfrak{D}$ for peaks in the fractal dimension analysis were very close to 2.00, indicating little true pattern. What is especially interesting about this particular example is that when the data were subjected to CA and 3TLQV was carried out on the first CA axis scores, the pattern at small scales reappears, as is shown in Figure 5.6a–d. There was no clear agreement among the methods, not even among MSO, fractal dimension, and Euclidean distance which are closely related (Dale & Zbigniewicz 1995).

We applied these methods also to data from shrub communities in the Kluane Lake–Kloo Lake valley in the Yukon (61°N 138°W). This valley is the site of a collaborative project studying the effects of experimental treatments on the vertebrate and plant communities of the boreal forest (Krebs *et al.* 1995). The vegetation consists primarily of white spruce (*Picea glauca*) forest, varying from closed to very open, with many meadows and shrublands. In the shrublands sampled for this data set, the most prominent species were *Salix glauca, Arctostaphylos uvi-ursi, Festuca altraica, Betula glandulosa*, and *Picea glauca*. We sampled the vegetation at five sites using visual estimates of cover in 100m transects of 1001 10cm × 10cm quadrats.

While there was no agreement among the methods when applied to the Ellesmere Island data, the data from the Yukon shrub communities often gave good agreement between the peak positions in the Euclidean distance method, the fractal approach and one of the first two axes in MSO (Dale & Zbigniewicz 1995). In MSO, the evenness was often very low because of dominance of the axis by one or two species.

Maslov (1990) studied the multispecies pattern in Russian forest communities near Moscow, using a method like MSO, combining PCA and nested blocks. The results showed that the communities had several scales of pattern in the 40m to 70m transects studied. The axes were also interpreted using the ecological indicator values of the species. Each quadrat was giving an indicator value for each of several environmental variables such as light regime, moisture, and soil nitrogen (*cf.* Persson 1981). By calculating rank correlation coefficients between the derived variables and the axis scores, he showed that in the *Oxalis*-type mixed forest, most of the scales of pattern found on the first two axes were significantly related to the amount of nitrogen in the soil. He used a related approach (for details see Maslov 1990) to look at the influence of canopy species, and concluded that while most axes are interpretable with respect to a gradient of the influence of tree species, their influence works through the modification of the environmental factors.

Ver Hoef *et al.* (1989) examined the relationship between horizontal pattern in a grassland community and its vertical structure. The horizontal pattern was measured using the first four axes from MSO analysis and the vertical structure was characterized by vertical cover, vegetation height and center of vertical biomass in a 10cm-wide strip transect. The relationship between the two was evaluated using standardized cross-variograms (Chapter 4). Although both the first eigenvalue and the measures of vertical structure had scales of pattern at 3m, the relationship between the two characteristics was weak because they moved in and out of phase with each other along the length of the transect. This clearly is an important area of research for our understanding of the spatial structure of vegetation and it should receive more attention.

Species associations

In Chapter 4, we alluded to the spatial relationship of the plants of two species as being a kind of spatial pattern which is usually investigated as the association of the two species at a single spatial scale. We will concentrate in this chapter on multispecies aspects of species association, using two sampling methods, quadrats and point-contact samples.

The standard pairwise approach to the analysis of species association in a community is based upon the presence and absence of the two species in sampling units such as quadrats. The approach has been to test all pairs of species, using a 2×2 contingency table for each pair to determine whether the plants of the two species have a significant tendency to be found together or a significant tendency to be found apart. The cells in such 2×2 contingency tables are usually referred to as '*a*' (the number of quadrats with both present), '*b*' (A absent, B present), '*c*' (A present, B absent), and '*d*' (both absent).

		A: present	absent
B:	present	*a*	*b*
	absent	*c*	*d*

The significance of a 2×2 table's departure from the null hypothesis of equal proportions is tested by comparing a test statistic like X^2 or G with the χ^2_1 distribution and, if significant, the association is positive if $ad-bc$ is positive and negative when $ad-bc$ is negative.

This method of analysis can be used to investigate the dependence of species associations on scale when small quadrats are combined into blocks or when a range of quadrat sizes is used. For example, Økland (1994) used quadrats of five sizes, from $1/256\,\mathrm{m}^2$ to $1\,\mathrm{m}^2$, to investigate bryophyte associations in a Norwegian boreal forest. Very few of the associations detected were negative, probably resulting from a strong trend in diversity from under tree canopies to more open microsites between trees (Økland 1994). As quadrat size decreased from $1\,\mathrm{m}^2$ to

$1/64\text{m}^2$, there was no trend in the ratio of observed positive associations to the potential number, but it dropped markedly when the size decreased to $1/256\text{m}^2$. This result indicates that the strength of associations does not change with scale but that the smallest quadrats are too small to be representative of the community, with fewer than 2.5 species per sample (Økland 1994).

It has long been recognized that all these pairwise tests cannot be independent of each other because knowing that species pair A and B and pair B and C are both positively associated makes the positive association of A and C less surprising. There is, however, another problem in that, for biological reasons, the association between a particular pair of species may depend on the combination of other species in their neighborhood or on the environmental factors that determine that combination.

A simplistic version of this problem is illustrated by the following tables which show the joint occurrences of species A and B when species C is present, when it is absent, and when the presence or absence of species C is ignored.

	C present		C absent		C ignored	
A:	present	absent	present	absent	present	absent
B: present	100	50	50	100	150	150
B: present	50	100	100	50	150	150
test result:	significant		significant		not significant	
ad − *bc*:	positive		negative		zero	

This example shows that while A and B are positively associated when C is present, they are negatively associated in its absence. Ignoring species C leads to the erroneous conclusion that the two species occur independently of each other. We can imagine more complex examples in which the association of species A and B depends on the combinations of species C and D, or of C, D, and E, and so on. Therefore, in considering species association, we need to consider not only spatial scale, as discussed in Chapter 4, but also the effects of combinations of other species.

To test whether pairwise associations are independent of other species combinations, we must look at multispecies association and therefore the data cannot be reduced to a number of 2×2 contingency tables, one for each pair of species of which there are $\binom{k}{2}$. We have to use as a single 2^k table.

As in Dale *et al.* (1991), we will describe the method for six species, assuming that extensions to higher values of k are clear. Let the observed count in cell (i,j,k,l,m,n) be o_{ijklmn}, and its expected value, based on the hypothesis of complete independence, be e_{ijklmn}. The variables i to n are 0 when the corresponding species is absent and 1 when it is present. Based on the complete independence model, the expected value is calculated as:

$$e_{ijklmn} = \frac{o_{i+++++}o_{+j++++}o_{++k+++}o_{+++l++}o_{++++m+}o_{+++++n}}{o^5_{++++++}}, \qquad (5.13)$$

with the plus signs indicating summation over the variable in that position. Thus, o_{++++++} is the number of quadrats sampled. The test statistic for the whole table is:

$$G = 2\sum_i \sum_j \sum_k \sum_l \sum_m \sum_n o_{ijklmn} \log\left(o_{ijklmn}/e_{ijklmn}\right). \qquad (5.14)$$

This G statistic is compared with the χ^2 distribution on $2^k - k - 1$ degrees of freedom, ν, which is 57 for $k=6$. If the observed table differs significantly from the expected, the next step is to determine the extent to which each cell deviates from its expected value. This deviation can be measured by the Freeman–Tukey standardized residual:

$$d_{ijklmn} = \sqrt{o_{ijklmn}} + \sqrt{1 + o_{ijklmn}} - \sqrt{1 + 4e_{ijklmn}}. \qquad (5.15)$$

These residuals are used to determine which cells are the most aberrant and contribute most to the overall significance. Bishop *et al.* (1975) caution that there is no critical value that assures that a cell is different from its expected value with a known probability, and thus can be considered statistically significant, but Sokal and Rohlf (1981) suggest a value of $\sqrt{\nu\chi^2_1/2^k}$ as a guideline. For example, for $k=6$ and $\alpha=0.05$, the guideline value is 1.85.

The presentation of the results of this analysis will be difficult for large numbers of species, but, if there are relatively few, our 1991 method (Dale *et al.* 1991) seems to work well. A line of k boxes is drawn for the cell of each significant deviate and these are arranged in two columns, one for those in which $o>e$ and one for those with $o<e$. An empty box corresponds to a species being absent and the box is filled with a square if the species is present. Figure 5.7 gives an example.

One disadvantage of the method is that with many species the number of cells in the table increases quickly and even with moderately large data sets, the expected value for each cell falls rapidly, and it is well known that

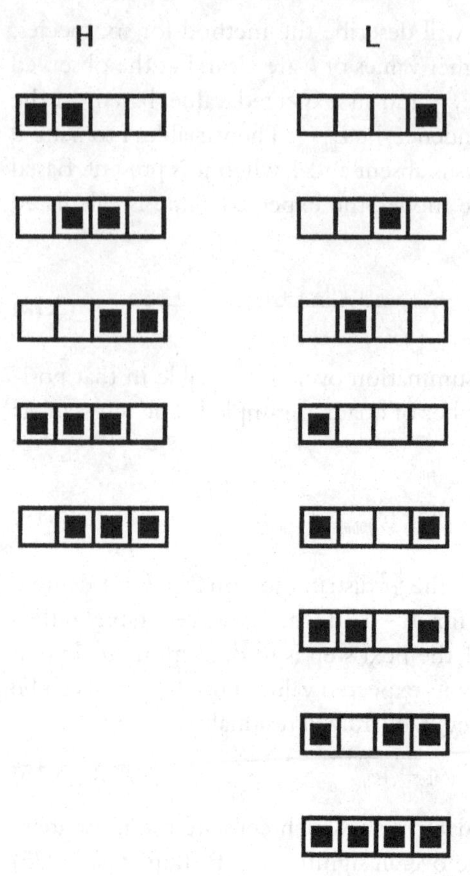

Figure 5.7 An example of how the box diagrams can be used to portray the results of 2^k testing. There are four species and the column labelled H shows combinations of presence (filled) and absence (empty) that were much more common than expected, the pairs 1 and 2, 2 and 3, and 3 and 4; the triplets 1, 2, and 3 and 2, 3, and 4. The column labelled L shows those combinations that had much lower frequency than expected, combinations with only one species present and combinations that have both species 1 and 4 present.

the test statistic's approach to the χ^2 distribution becomes poor when the expected values are small.

Another area of concern is the relationship between the number of species, the physical size of the plants, and the size of the sample unit. For instance, if the quadrat is small relative to the plants, there may be a practical upper limit to the number of different species that can be found in the quadrat because of the limit to the number of plants. (Think of sampling a forest canopy with 1 m × 1 m quadrats.) In those circumstances, the combinations that have many species included will have 0 in the observed counts, and the results of the test will be affected.

A third factor is the fact that if the quadrat data are collected in transects, the detection of association will be affected by the spatial autocorrelation in the data. Spatial autocorrelation decreases the effective sample size because ten adjacent quadrats may provide information equivalent to that in fewer than ten independent quadrats. For this reason, spatial autocorrelation leads to more apparently significant results and to avoid this effect, a deflation factor, Φ, is divided into the test statistic to decrease its value. Dale *et al.* (1991) discuss the problem and describe one way to calculate deflation factors. That method was based on the underlying model of a Markov chain. For multispecies testing, it may be that the amount of deflation necessary is small, again based on underlying Markov processes. Dale *et al.* (1991) describe a Monte Carlo procedure for deflating multispecies association test statistics. Willingness to apply the technique will depend on a researcher's assessment of the assumption of Markov processes. [Remember that Pielou (1977a) suggested that Markov models of two-species sequences were probably seldom tenable hypotheses.] It is our impression that the more species that are considered, provided at least some of them do not occur in long runs of quadrats, deflation values are small for multispecies association analysis. For example, in examining the Ellesmere Island data set, we found that as more species were considered, the degree of the Markov model needed to describe the spatial process decreased; in other words, as more species are considered, the combinations of species in adjacent quadrats become more independent. For this reason, the effect of spatial autocorrelation must be taken more seriously in pairwise tests than in multispecies testing.

For an illustration of the technique, we will use data sets from two sites, one from moraines one and three (dated 1801 and 1890) at Mt. Robson, and the other from proglacial deposits at the SE Lyell glacier, Alberta. The SE Lyell data were subdivided into two sets of 1200 quadrats (4 transects) each: those labelled 8 and 9 (dated at 1894 and 1885) and those labelled 10 and 11 (dated 1855 and 1840). We analyzed the eight most frequent species in each data set. In these seral communties, the deflation required for the multispecies test, Φ_a, averaged 1.35. In the data there are several examples of the species combination of only A and B being rare but the combination of A, B and C (or A, B, C, and D) being common, and there are examples where the reverse is true (either can be referred to as a reversal). There are several instances of species A and B being positively associated in the pairwise test, with the combination of species A and B only being significantly low in frequency. For instance in the first part of Figure 5.8, high-frequency combination one, which contains three species, reverses low-frequency combination seven, which contains two of the three. Low-frequency combination 26, with three species, reverses the high-frequency combination 15. An intriguing comparison is 'high 1' and 'high 2' with 'low 4' and 'low 6'; all four of the bryoids are a common combination whereas some combinations of three of the four are very rare and one is very common.

The major division between the species is that, in all cases, the first four are vascular plants and the last four are bryophytes or a lichen ('bryoids'). The morphology of the last four combines with the choice of quadrat size to make combinations of one, two, or even three of them in a quadrat, with no vascular plants, rare (Mt. Robson L1 to L7; Lyell 8 and 9 L1 to L5; Lyell 10 and 11 L1 to L3). Not surprisingly, only 'dominant' vascular plants like *Picea* or *Dryas* are commonly found alone in a quadrat, whereas the bryoids are very rarely found alone (Figure 5.8). There are many instances of high frequencies for combinations of a single vascular species with or without bryoids; this explains the predominance of negative associations in the tests of pairs of vascular plants (Tables 5.5 and 5.6). The large numbers of positive pairwise associations between bryoids must be caused by the high frequencies of combinations of one or two vascular plants with several bryoids. For instance, at Mt. Robson, *Hedysarum* (Table 5.6) was frequently found in association with three or all four (H5–H8) and rarely with only one or two (L8–L11), and most reversals contained *Tortella* and *Ditrichum*.

At the SE Lyell subsite 8 and 9, there is a clear division between quadrats in the more closed phase with *Picea* and/or *Shepherdia* and those in the *Dryas*-dominated open areas. Of the low-frequency combinations, nine have *Dryas* with *Picea* or *Shepherdia*. In spite of the overall negative associations between *Dryas* and *Shepherdia* and between *Dryas* and *Picea*, there are some high-frequency combinations containing *Dryas* and at least one of the other two: H20–H23. In fact, the combination of only *Dryas* and *Picea* has a high frequency (Figure 5.8b, H21).

The association between *Picea* and *Shepherdia* changes from positive at subsite 8 and 9 to negative at subsite 10 and 11 (Table 5.6), probably as the result of increased shading by the conifer. This is reflected by a change from ten high-frequency and two low-frequency combinations with both species at the first subsite to two high-frequency and eight

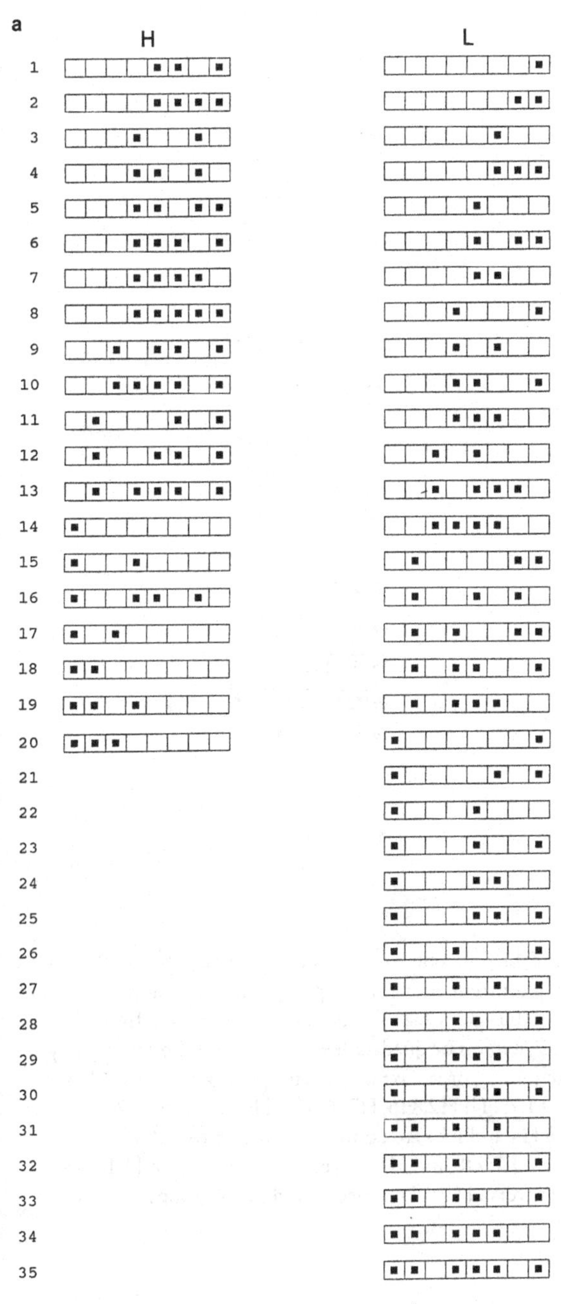

Figure 5.8 Box diagrams portraying the 2^k association analysis for the eight most important species in seral communities at *a* Mt. Robson, *b* SE Lyell subsites 8 and 9,

b

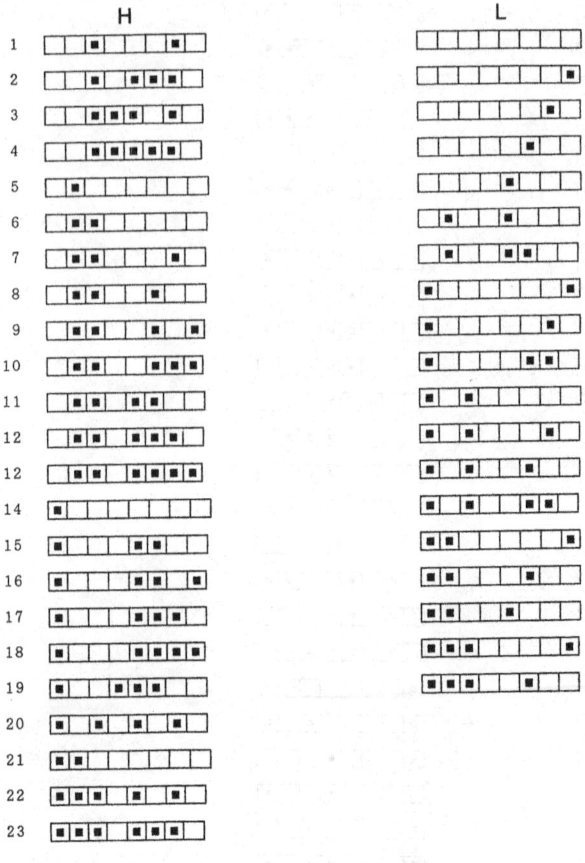

H

L

Figure 5.8 (cont.)
and *c* SE Lyell subsites 10 and 11. The column labelled H contains the combinations that occurred with much higher frequency than expected and the column labelled L shows those that occurred with much lower frequency than expected. In many instances, there are 'reversals', triplets or quadruplets of species of high frequency that contain pairs of low frequency, or *vice versa.* We will point out some of these: *a* H2 and H5 are reversals of L2; H1, H2, H6, H7, H9 and H12 reverse L7; L26, L27, L28 and L31 reverse H15. *b* H4 and L15 are complements; L15 to L19 are reversals of H21. *c* H8 and H10 reverse L7; H8 and H14 reverse L9; L5, L15, and L16 are reversals of H3; L17 to L19 reverse H12. The species and plant groups are given in Table 5.5, p. 156.

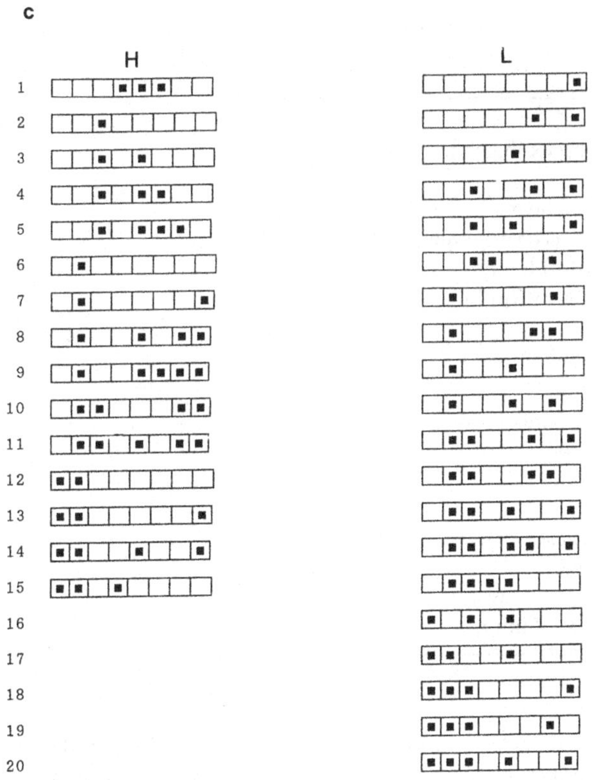

c

H L

1
2
3
4
5
6
7
8
9
10
11
12
13
14
15
16
17
18
19
20

low-frequency combinations at the older site. Between the younger and the older site, there also seems to be a decrease in the number of species present in the high-frequency combinations and an increase in the numbers in the low ones. This trend may be attributable to an increase in the sizes of the patches and in the scales of pattern, making it less likely that a single quadrat can contain many species.

The preceding discussion shows that analysis using the 2^k approach obviously can reveal details about the spatial relationships of species not available from the traditional analysis. We are not suggesting that the pair-wise analysis of species associations should be abandoned. In many instances, the average association of two species, which is what the results of 2×2 table testing represent, will be of interest and the 2^k method may not be tractable for large numbers of species. However, in other cases, the greater detail revealed by the multispecies method will prove informative, even if applied only to the subset of species that are most of interest.

Table 5.5. *Species and plant groups used in the analysis of field data*

Mt. Robson moraines
1. *Picea engelmannii* Parry ex Engelm.
2. *Dryas octopetala* L.
3. *Arctostaphylos rubra* (Rehder and Wils.) Fern.
4. *Hedysarum boreale* var *mackenzii* Nutt. (Rich.)
5. *Tortella inclinata* (Hedw.) Limpr.
6. *Ditrichum flexicaule* (Schwaegr.) Hampe
7. *Bryum* sp. Hedw.
8. *Cladonia* sp. (Hill) Hill
SE Lyell moraines
1. *Dryas drummondii* Richards
2. *Picea engelmannii* Parry *ex* Engelm.
3. *Shepherdia canadensis* (L.) Nutt.
4. *Salix glauca* L.
5. *Brachythecium groenlandicum* (C. Jens.) Schljak.
6. *Bryum caespiticium* Hedw.
7. *Drepanocladus uncinatus* (Hedw.) Warnst.
8. *Tortella tortuosa* (Hedw.) Limpr.

Before we leave the topic of species association as detected using data from strings of contiguous quadrats, we should emphasize a comparison with quadrat covariance analysis. Covariance analysis allowed us to look at the scales of association between two species, which could be positive at some scales and negative at others. Similarly, multispecies pattern analysis can be interpreted in terms of scales of association. In fact, Ver Hoef *et al.* (1989) refer to groupings of species detected by MSO as microassociations. These groupings are created, relative to each axis of the ordination, by the eigenvectors which give positive and negative weights or loadings to each species. For example, in their grassland study, the first eigenvector had a large positive weights for *Leontodon hispidus*, *Carex flacca*, and *Rhinanthus serotinus* and large negative weights for *Hypericum perforatum* and *Knautia arvensis*. Because the eigenvalue had a variance peak at 30 cm, we can interpret the positive and negative associations within that group of species as having a scale of 30 cm. On the other hand, the third eigenvector, which peaked at 3 m, had positive loadings for *Festuca pratensis* and *Linum catharcticum* and negative loadings for *Centaurea jacea* and *Carex flacca*; we would therefore interpret the positive and negative associations within that grouping of species as occurring at that larger scale.

Table 5.6. *Pairwise associations in seral communities based on quadrat data*

	Species						
Species	2	3	4	5	6	7	8
a) Mt. Robson moraines							
1	.	.	–	–	–	–	–
2	:	–	–	.	.	–	.
3		:	–	.	.	–	.
4			:	+	+	+	.
5				:	+	+	+
6					:	+	+
7						:	+
b) SE Lyell 8 and 9							
1	–	–	.	.	–	–	.
2	:	+	.	-	–	+	.
3		:	.	+	.	+	.
4			:	.	.	.	.
5				:	+	+	+
6					:	+	+
7						:	+
c) SE Lyell 10 and 11							
1	+	–	.	.	.	.	+
2	:	–	.	–	–	.	+
3		:	–	.	.	.	–
4			:	.	.	.	.
5				:	+	+	.
6					:	.	.
7						:	+

Notes:
'+' Indicates positive association and '–' indicates negative, '.' indicates that the test result was not significant. A blank indicates that no significant association was detected. See Table 5.5 for species names.

The second method for detecting species associations is contact sampling which was introduced by Yarranton (1966) for the study of a bryophyte-lichen community. At each of many regularly or randomly placed points, the species present at the point, the initial sample, is recorded together with the species that touches the initial one closest to the point, the contact neighbor. The method is said to examine the community from 'a plant's eye view' in that it investigates the plant's neighborhood as defined by physical contact (Turkington & Harper 1979a,b). When the distinction between the initial sample and its contact neighbor is preserved, the method can be used to examine asymmetric associations. That is, we can distinguish among three possibilities: (1) A is associated with B but not B with A; (2) B is associated with A but not A with B, and (3) A and B are associated with each other.

The original 'Chi-square' approach to the analysis of this kind of data has been shown to be wrong because the expected values based on the hypothesis of random pairing were calculated incorrectly (de Jong *et al.* 1980, 1983). Their 1983 paper gives the correct method of calculating expected values. They suggest that it is misleading to treat cases where the initial sample has no contact neighbor as if the plant were in contact with the pseudospecies 'no contact', because of the asymmetry it imposes on the count matrix. Their method uses an iterative technique to derive maximum likelihood expected values for contact frequencies based on the null hypothesis of random pairing. The de Jong method assumes that within-species contacts are disallowed, so that the main diagonal of the data matrix consists of zeros. The expected values can be calculated in two ways, with and without the 'proportionality hypothesis'. That hypothesis is the assumption that the probability of a species being a contact neighbor is proportional to the probability of its being an initial sample.

The proportionality hypothesis is unlikely to hold if the morphology of some plants makes it more probable that they will be initial samples rather than contact neighbors, such as a savanna tree, or more probably contact neighbors, such as climbing vines. Even in vegetation that is strictly two-dimensional, different probabilities might result from the shapes and sizes of the species patches: species that grow as large round patches will have a higher initial sample to contact neighbor ratio than those that form very irregular patches with a high perimeter to area ratio.

The usual method of analyzing this kind of data for k species is to make a $k \times k$ frequency table with the entry o_{ij} being the observed number of times that species i was the initial sample and species j was its contact neighbor. The expected value for that cell of the table, e_{ij}, is calculated based on the hypothesis of independence, and the deviation of the observed from the expected is measured by a test statistic for the whole table (Sokal & Rohlf 1981):

$$G = 2\sum\sum o_{ij}\ln(o_{ij}/e_{ij}). \tag{5.16}$$

The test statistic is then compared to the χ^2 distribution with an appropriate number of degrees of freedom. The test statistic, G, is calculated using expected values based on the assumption of randomness, first with the hypothesis of proportionality, G_p (k^2-1 df), and then without the hypothesis, G_n (k^2-k df). The proportionality hypothesis can be tested by comparing $G_h=G_p-G_n$ to the χ^2 distribution with $k-1$ degrees of freedom (de Jong *et al.* 1983).

Having tested the whole $k \times k$ sample-contact frequency table, with and without the proportionality hypothesis, and found it significant, we will want to know which pairwise associations are significant. de Jong *et al.* (1983) discuss the possibility of collapsing the table in order to test specific pairwise associations. If, however, the proportionality hypothesis has been found to be false, it is impossible to test a pairwise association without assuming that relationships of the two species of interest with others in the table are random (de Jong *et al.* 1983, Appendix 2). If the table as a whole is significantly nonrandom, no subsection of the table can be assumed to be random. It is certainly impossible to make that assumption while testing all possible pairs of species. The alternative is to calculate the Freeman–Tukey standardized residual for each cell:

$$z_{ij}=\sqrt{o_{ij}}+\sqrt{o_{ij}+1}-\sqrt{4e_{ij}+1}. \tag{5.17}$$

We studied a lichen community at Jonas Rockslide, a large rockslide in Jasper National park, 75 km south of the town of Jasper on Highway 93 (52°26′N, 117°24′W). It is 3.5 km long and 1 km wide, ranging in elevation from 1500 m to 2200 m. The rock is honey-colored quartzite sandstone and is in blocks of a variety of sizes, commonly up to 3 m × 3 m × 1 m, with flat surfaces. The slide faces southwest and, based on lichenometric evidence, it is at least 500 years old (John 1989).

Saxicolous lichens cover 87% of the exposed rock surface, and more than 100 species were identified, with growth forms including crustose (e.g., *Rhizocarpon*), foliose (e.g. *Umbilicaria, Melanelia*), and fruticose (e.g., *Pseudephebe*). Rock faces were selected for sampling with the criteria of low microtopographic relief, size larger than 1 m × 1 m, between 0° and 90° in slope angle and not closely sheltered by trees (see John 1989). Twenty-five rock faces were sampled using a rectangular grid of sample points with vertical spacing of 10 cm along the rock surface and horizontal spacing of 20 cm. The total sample was 2200 points. The species at each grid point was identified as well as the species touching the initial lichen closest to the grid point. Uncolonized rock was treated as if it were a species.

In this analysis of the rockslide community, only the 17 most common lichen species were used (those that occurred as an initial sample or a contact neighbor at least 70 times), with all the rest being lumped into an 'other species' category. Those 17 included bare rock treated as a species

Table 5.7. *Frequency of lichen species as initial samples or contact neighbors and ratios,* r, *of occurrence as initial sample and as contact neighbour*

Number	Species	*r*	$f(i)+s(i)$
1.	bare rock	0.66	822
2.	*Rhizocarpon bolanderi*	1.11	393
3.	*Spilonema revertens*	0.61	421
4.	*Umbilicaria hyperborea*	2.04	76
5.	*Rhizocarpon* sp. (grey)	1.02	261
6.	*Aspicilia cinerea*	1.62	160
7.	*Pseudephebe pubescens*	0.80	227
8.	*Melanelia sorediata*	0.94	91
9.	*M. stygia*	1.14	137
10.	*R. disporum*	1.33	100
11.	*R. grande*	1.49	117
12.	*R. geographicum*	1.05	123
13.	*M. granulosa*	0.92	71
14.	*Schaereria tenebrosa*	1.65	82
15.	*U. torrefacta*	1.51	188
16.	*Lecidea paupercula*	1.17	89
17.	*Lepraria neglecta*	0.51	80
18.	'other'	1.26	1084

Notes:
$f(i)+s(i)$ is the total number of times the species occurred in the frequency table.

both as an initial sample and as a contact neighbor (no contact). The proportionality hypothesis was tested as outlined above, but the result may be affected by small expected values.

In the lichen community, the table of sample-contact frequencies was very significantly nonrandom, both with and without the proportionality hypothesis: $G_p = 658.3$ on 323 df. and $G_n = 524.9$ on 306 df. The proportionality hypothesis is strongly rejected: $G_h = 133.4$ on 17 df. The pairwise species associations shown to be significant using the Freeman–Tukey residuals are shown in Table 5.8.

Because the lichen data were collected on a grid of sample points, we were concerned about possible effects of spatial autocorrelation. The effective sample size may be reduced because the probability of finding a particular species at a sample point is dependent on the species found at nearby points, due to their microhabitat ecology (John 1989). Therefore, more tests may give apparently significant results than the data justify. To evaluate spatial autocorrelation, we used join-count statistics.

Table 5.8. *Pairwise asymmetric species associations in the Jonas rockslide lichen community, based on point-contact sampling*

	Contact neighbor																	
Sample	1	2	3	4	5	6	7	8	9	10	11	12	13	14	15	16	17	18
1.	:	+	·	·	+	·	−	·	·	·	·	·	·	·	·	+	−	·
2.	+	:	+	·	·	−	−	+	·	·	·	·	·	−	·	·	·	−
3.	−	+	:	·	+	+	−	·	·	·	·	·	−	·	·	−	·	·
4.	−	·	·	:	·	·	+	+	·	·	·	·	·	·	·	·	·	·
5.	+	·	·	·	:	·	·	·	·	−	−	·	·	·	·	·	−	−
6.	·	·	+	·	−	:	·	·	·	·	·	·	·	·	·	·	·	·
7.	·	−	·	·	·	·	:	·	·	·	·	·	+	+	·	·	·	·
8.	−	·	+	·	·	+	·	:	·	·	·	·	·	·	·	·	·	·
9.	−	·	·	·	·	·	·	·	:	·	·	·	·	·	+	·	+	·
10.	·	·	·	·	·	·	·	·	·	:	·	·	·	·	·	·	·	·
11.	·	·	·	·	·	·	+	·	·	·	:	·	·	·	·	·	·	·
12.	·	·	·	·	·	·	·	·	·	·	·	:	·	·	·	·	·	−
13.	·	·	·	·	·	·	+	·	·	·	·	·	:	·	·	·	·	·
14.	·	·	·	·	·	·	·	·	·	·	·	·	·	:	·	·	·	−
15.	·	·	·	·	·	·	·	·	·	·	+	·	·	·	:	·	·	·
16.	+	·	−	·	−	·	·	·	·	·	·	·	·	·	·	:	·	·
17.	−	−	·	·	·	·	·	·	+	·	·	·	·	·	·	·	:	·
18.	·	−	−	−	−	−	·	−	·	·	·	·	−	·	·	·	+	:

Notes:
Species numbers as in Table 5.7. See legend to Table 5.6 for explanation of symbols.

Join-count statistics are used to determine whether the classes of points in a regular grid or other spatial structure are random or patchy in their distribution, by comparing the observed number of times that members of the same class are found at adjacent grid points with the number expected if the classes are randomly arranged (Upton & Fingleton 1985, Chapter 3). For rectangular grids, it is possible to compute the mean and variance based on the hypothesis of randomness which can then be used with a normal approximation to test whether observed numbers of adjacencies are significantly high or low (see Pielou 1977a, p. 146). The grids of sample points in the lichen study, however, were irregular because of the shapes of the rock faces, and so these calculations could not be used. We therefore used a Monte Carlo approach to the analysis.

On each rock face we counted the number of times that at any two adjacent sampling points were initial sample points of the same species, contact neighbors of the same species, and ordered pairs of initial sample and contact neighbor of the same two species. Adjacency was defined in the 'queen's move' sense, including horizontal, vertical and diagonal neighbors. The species were then reassigned at random to different points on the sampling grid, preserving the sample-contact pairs, and the three types of joins were counted again. Observed join counts that were equalled or exceeded in fewer than 50 of 1000 trials were considered to be evidence of significant autocorrelation. Since it is the sample-contact pairs that are of interest in a study of association, it is the join counts of the third kind that are the most important here. If these were consistently significant, the association findings would have to be re-evaluated.

In examining the lichen data for spatial autocorrelation, it was found that 11 of the 25 rocks had significant spatial autocorrelation in the initial sample species and 11 in the contact neighbor species. However, only 1 of the 25 showed significant autocorrelation in the sample-contact pairs (1 in 20 is expected based on randomness). Therefore, while there is spatial autocorrelation in the data, it does not greatly decrease the effective sample size for species association analysis and the procedure and interpretation do not have to be modified.

The second vegetation type that was studied was the successional communities on moraines at the Robson Glacier in British Columbia, already described. In total 200 point samples were used on each of moraines 8, 7, 5, 4, 3, and 1. The points were located using random numbers and a rectangular reference grid along the top of the moraine. For each sampling point the species of the first plant hit by the sampling point and the next different species in contact with it and nearest to the sample point were recorded. In cases where a species hit by the sample point did not touch another plant, or where the only individuals contacted were members of the same species, a 'no contact' was recorded. 'No contact' samples were recorded to provide information on the tendency of a particular species to occur as isolated clumps or individuals.

In analyzing the successional communities, we used the same methods, with one modification. We wanted to lump together some taxonomic groups (*Salix* species) and some life-form groups such as lichens and mosses (bryoids). When this is done, the contact neighbor can be a member of the same class as the sample, so that the elements on the main diagonal are not all zero. The diagonal entries for single species remain zero because self-contact was not allowed. Therefore, we used Deming–Stephan iterative proportional fitting to derive the expected

Table 5.9. *Results of goodness-of-fit tests on the successional communities at Mt. Robson*

Moraine	G_p	G_n	G_h
8	73.6	42.7	30.9
7	130.5	98.9	31.6
5	180.2	105.3	75.0
4	167.25	121.4	45.9
3	219.4	177.5	41.9
1	178.0	161.0	17.0

Notes:
G_p is the test statistic calculated with the proportionality hypothesis, G_n is the test statistic calculated without the hypothesis and G_h is the statistic testing the proportionality hypothesis itself.

values (Bishop *et al.* 1975). This was done both with and without the proportionality hypothesis, and the difference between the two test statistics, $G_h = G_p - G_n$, was used to test that hypothesis.

For the successional community data, 'bare rock, no contact' was not used as a species in the analysis. Instead, we examined the frequency of 'no contact' for each species or class, using the standard goodness-of-fit test, to determine whether species tend to occur as isolated patches. Because of small numbers, we combined pairs of transects based on surface age: young, middle, and old.

The frequency tables of contact sampling of the successional vegetation were significantly nonrandom on all six moraines, both with and without the proportionality hypothesis (Table 5.9). The proportionality hypothesis was rejected for all six moraines.

The results of pairwise tests of association are given in Table 5.10. In general, there is positive association between *Hedysarum* and *Dryas*, with bryoids associated with *Hedysarum*, and there are positive associations among *Picea*, *Salix* and 'other', with negative associations between these two groups.

Species with significantly many or significantly few cases of no contact are given in Table 5.11. Since the tests for the species are more or less independent, the number of tests that would result in apparently significant results from randomness alone is one. Since eight were found, we interpret them as reflecting real biological effects. On the youngest surfaces, there are three categories with more no contacts and none with

Table 5.10. *Species associations based on initial sample:contact neighbor data in successional communities*

Moraine	Positive associations			Negative associations		
8	*Dryas*	→	*Hedysarum*			
	other	→	other			
7	*Picea*	→	*Salix*	*Picea*	→	bryoid
	Salix	→	*Picea*	*Hedysarum*	→	*Salix*
	Salix	→	*Hedysarum*	*Salix*	→	bryoid
	Hedysarum	→	bryoid	bryoid	→	*Salix*
	Dryas	→	bryoid	bryoid	→	other
	bryoid	→	bryoid	other	→	other
	other	→	*Salix*			
5	*Picea*	→	*Salix*	*Picea*	→	bryoid
	Salix	→	*Picea*	*Hedysarum*	→	*Salix*
	Hedysarum	→	bryoid	*Hedysarum*	→	*Picea*
	Dryas	→	bryoid	*Salix*	→	bryoid
	other	→	*Salix*	*Dryas*	→	other
4	*Salix*	→	*Picea*	*Picea*	→	bryoid
	Picea	→	*Salix*	*Hedysarum*	→	*Salix*
	Picea	→	other	*Salix*	→	bryoid
	Hedysarum	→	bryoid	bryoid	→	*Salix*
	bryoid	→	bryoid	bryoid	→	other
3	*Picea*	→	*Salix*	*Picea*	→	bryoid
	Salix	→	*Picea*	*Salix*	→	bryoid
	Salix	→	other	bryoid	→	*Salix*
	Hedysarum	→	bryoid	bryoid	→	*Picea*
	bryoid	→	bryoid	*Hedysarum*	→	*Picea*
				Hedysarum	→	*Salix*
				Hedysarum	→	other
				other	→	bryoid
1	*Picea*	→	*Salix*	*Picea*	→	bryoid
	Salix	→	*Picea*	*Salix*	→	bryoid
	Hedysarum	→	bryoid	*Hedysarum*	→	other
	Picea	→	other	bryoid	→	*Salix*
	bryoid	→	bryoid	bryoid	→	other
	Salix	→	other	other	→	bryoid
	other	→	other			

Notes:
Dryas → *Hedysarum* in the positive column means that *Hedysarum* was found as a contact neighbour of *Dryas* significantly more often than expected.

Table 5.11. *Tests of clump isolation: plant categories that have significantly more or fewer no contacts than expected based on G test*

Moraines	More 'no contacts'	Fewer 'no contacts'
7 and 8	*Hedysarum*, *Dryas*, other	
4 and 5		*Picea*
1 and 3		*Hedysarum*, *Salix*, *Picea*, other

fewer, whereas on the oldest surfaces, there are no cases of too many no contacts and three or four with too few. The difference can be related to the sparseness and density of vegetation in the developing community.

Concluding remarks

In looking at species association, whether the data are from point-contact sampling or from quadrats, it is clear that the plants of different species do not occur independently of each other. Quadrats represent small subsections of the area occupied by a community and the combinations of species found in those small areas are not random; some combinations are very common and some are very rare.

These nonrandom combinations are an important feature of the structure of a plant community, but this spatial heterogeneity has implications for its dynamics as well. One version is the 'spatial segregation hypothesis'; it states that 'finite dispersal and spatially local interactions lead to spatial structure that enhances ecological stability (resilience) and biodiversity' (Pacala 1997). This hypothesis is especially important because it explains the coexistence of similar species which lack other methods of reducing the intensity of competition such as resource partitioning. Pacala (1997) suggests that an informal appreciation of the hypothesis can be gained by the inspection of a species-rich lawn. Estimate the abundance of one species, A, first by counting the number of individuals of A in small circles of a constant diameter centered on plants of species B and second in similar circles randomly placed. If the species are spatially segregated, the first estimate will be considerably less than the second. In the terminology of this chapter, the nonrandom arrangement of the plants of the two species shows that they are negatively associated.

Nonrandom combinations of species are an important precondition for the existence of multispecies pattern. In many instances, however, the high-frequency combinations are not particularly distinct but grade into each other by the loss or gain of a single species. For example, in the high (H) combinations of Figure 5.8a, numbers 5, 6, 7, and 8 are just sets of four or five of the last five species. Therefore, if we are picturing the high-frequency combinations as each representing a color of tile in a mosaic, many of the colors may be very similar so that the phases of the mosaic are not very distinct. On the other hand, the presence or absence of a particular species can have a big effect; in the same figure, high-frequency combinations in which spruce is present (14–20) seem to have fewer species present than high-frequency combinations in which it is absent (1–13). This phenomenon is related to the fact that, in MSO, the first axis was dominated by this single species, and thus was not truly multispecies at all.

Even given distinct high-frequency combinations of species in quadrats, the existence of spatial pattern and our ability to detect it depend on how the combinations are arranged in space. For instance, given combinations of species that we can label A, B, C, D and E, with closeness in the alphabet reflecting compositional similarity, if a string of quadrats looks like 'ABECADBDEACEADABCEAECE . . .', there is little spatial pattern to detect. Only if the phases of the mosaic show more order such as 'AABCCCDDEEDCBBAABCDDDCCBBCCDEEE . . .' will pattern be detectable.

As in other parts of this book, the message is that a single method of analysis cannot tell you all you need or want to know about the data. The 2^k association analysis will tell you whether the combinations of species are nonrandom and what the common combinations are. If several of them are single species, of course, the chances of finding meaningful multispecies pattern are greatly reduced. MSO does its best to find multispecies pattern in the data, and the association analysis can help you interpret what pattern is found. The use of a complementary approach such as examining the CA scores of the quadrats (using the first two or three CA axes) and applying 3TLQV to them may also provide useful insights into the spatial structure of the vegetation. Further developments in the area of combined analysis for the evaluation of multispecies pattern await us.

Recommendations

1. For the analysis of multispecies pattern, we recommend the use of a multiscale ordination procedure, MSO, based on 3TLQV and 3TLQC, using the squared intensity weighting. It is a good idea, also,

to evaluate the multispecies nature of the pattern by examining the evenness of the species' contributions.

2. In the detection of species association, using either quadrats or point sampling, any interpretation of the results should be based on the fact that all pairwise tests are not independent. The 2^k method can be used to study the complex structure of multispecies association for particular small groups of species.

6 · *Two-dimensional analysis of spatial pattern*

Introduction

The preceding three chapters have examined various aspects of the study of spatial pattern in one dimension using data acquired by several sampling methods including strings of contiguous quadrats. In fact, we know that in real vegetation, spatial pattern exists in at least two, if not three, dimensions. We know, also, that spatial pattern may be anisotropic, exhibiting different characteristics in different directions. Depending on the application, we may want to retain that anisotropy in the analysis, or we may want to average over all possible directions to look at the overall spatial pattern.

There are several approaches to two-dimensional analysis available, most of which are adaptations of methods initially developed for one-dimensional analysis. It is interesting that one of the earliest methods, blocked quadrat variance (BQV), was originally proposed in a two-dimensional version (Greig-Smith 1952), for the analysis of spatial pattern in grids of quadrats.

The practical problem associated with two-dimensional analysis is the collection of data. Studies of pattern in one dimension have usually sampled the vegetation with strings of small contiguous square quadrats from as few as 36 (Usher 1983) to 1001 (Dale and Zbigniewicz 1995). The amount of work necessary for one-dimensional studies is often great, and to extend them into two dimensions would, in many instances, be impractical. Most of the two-dimensional analysis methods described here may be most useful when applied to data collected by means such as the digitizing of images or the direct acquisition of digital images. For instance, one of the methods described below (Dale 1990) was developed to compare the scale of pattern of different vegetation types at Kluane based on LANDSAT imagery.

Having just said how impractical traditional quadrats may be for two-dimensional studies, we will now describe some data that we will use for illustration in this chapter that were collected in just that way. Stadt (1993) studied four forest sites dominated by *Pinus contorta* (lodgepole pine) in the Rockies using 20×20 grids of 5 m×5 m quadrats. All living trees were classified by species, height and diameter class and ocular estimates were made of the cover of understorey species. What is especially interesting about this data set from an ecological point of view is that the quadrats which Stadt sampled in 1989 were the same ones that had been sampled in the same way by Hnatiuk in 1967 (Hnatiuk 1969). Three of the sites are in Jasper National Park (Athabasca, Whirlpool and Sunwapta) and one (Hector) is in the adjacent Banff National Park.

Blocked quadrat variance

The original pattern analysis method is blocked quadrat variance and, when applied in two dimensions, it requires that the dimensions of the grid in two quadrats (or grid units) be powers of two in length, such as 64×64. The quadrats are then combined into exclusive blocks of two, four, eight, and so on and the variance is calculated in the usual way for each block size. Peaks in the variance are interpreted as being related to the scale of pattern, but the scale here is actual area rather than linear dimension. We listed some of the drawbacks of the one-dimensional version of this technique in Chapter 3. The two-dimensional version shares those problems with an additional consideration: the blocks with areas that are even powers of two are square, but those with areas that are odd powers of two are rectangular (e.g. $4\times8=32=2^5$) and the orientation of those rectangles may affect the outcome of the analysis (see Figure 6.1 cf. Pielou 1977a, p.141). For these reasons, as in the one-dimensional case, BQV is not recommended.

Spatial autocorrelation and paired quadrat variance

In examining quadrats in a grid, their spatial relationships lead us to expect that adjacent quadrats will be more similar in species composition than ones that are not adjacent. If there is a repeating spatial pattern in the data, similarity should first decrease with increasing distance and then increase again. The distance at which similarity reaches a minimum (or a maximum) can be converted into information about the scale of pattern because the scale of pattern is approximately the average distance

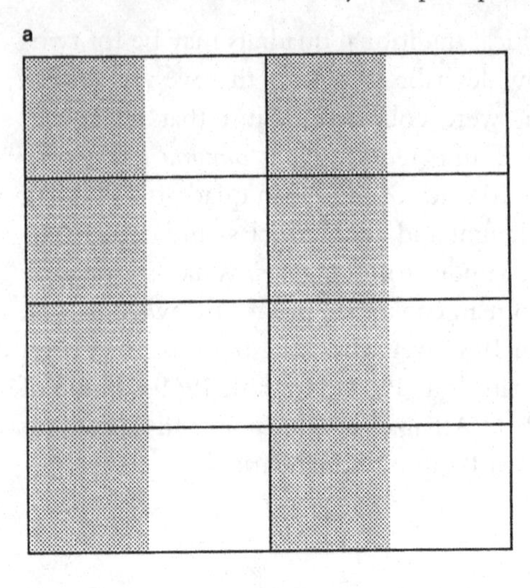

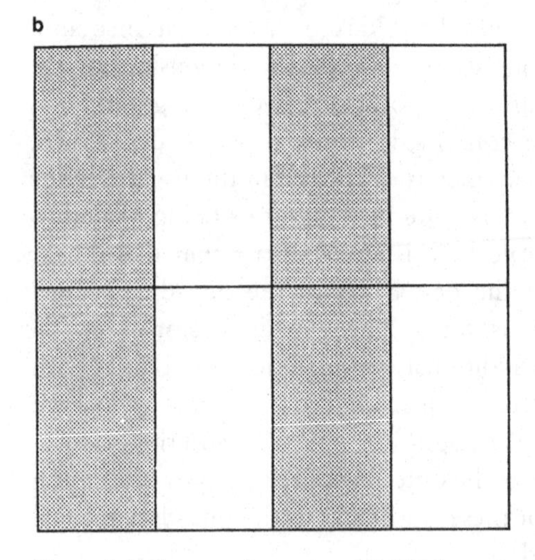

Figure 6.1 In two-dimensional BQV, the orientation of the rectangular blocks affects the outcome. The dark areas are patches and the light areas are gaps. In part *a*, all eight blocks have the same density and the variance is low; in part *b*, some blocks are dense and some empty so that the variance is high.

between the most unlike quadrats or half the distance between those that are most similar. This approach to the study of pattern is the analysis of spatial autocorrelation because it examines correlation within the data set as a function of distance. As we saw in Chapter 3, there is a close relationship between the analysis of spatial autocorrelation and the paired quadrat variance (PQV) approach, with one being almost the additive inverse of

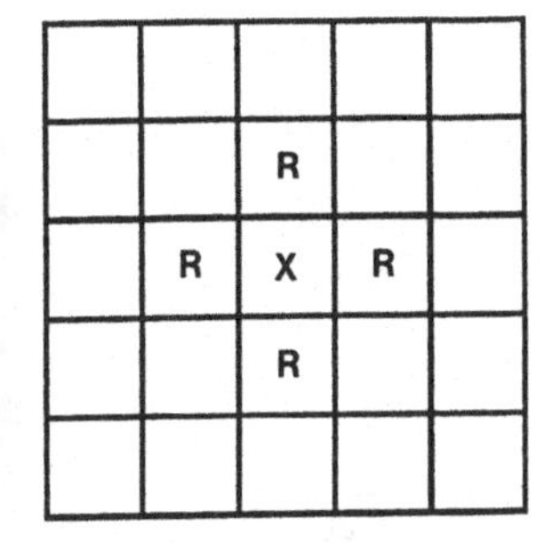

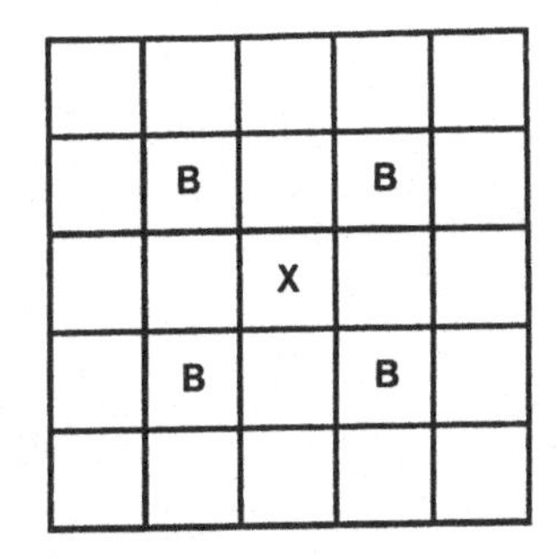

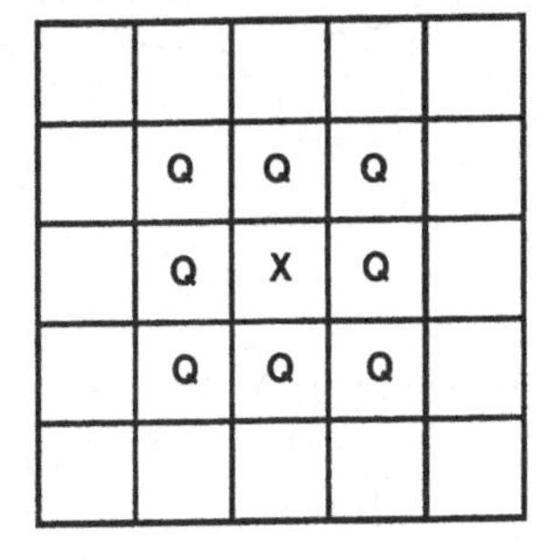

Figure 6.2 Three kinds of neighbors of the central quadrat 'X'. 'R' quadrats are rook's move neighbors. 'B' quadrats are bishop's move neighbors. 'Q' quadrats are queen's move neighbors.

the other. That relationship holds equally well in two-dimensional studies.

Sokal and Oden (1978a,b) discuss a variety of methods and provide examples of the analysis of spatial autocorrelation in biological studies. They do not explicitly apply their techniques to the detection of scale of spatial pattern, but it is easy to do so. Just as peaks in a variance plot are related to pattern scale, minima in the autocorrelation plot indicate the scale of pattern.

In evaluating spatial autocorrelation, there are three different ways of considering which quadats of a grid are adjacent. These are usually described with reference to the moves of chess pieces (*cf.* Sokal & Oden 1978a). If only horizontal and vertical neighbors are considered to be adjacent, this is referred to as the 'rook's move' definition of adjacency (see Figure 6.2). If only diagonal neighbors are considered as adjacent, this is the 'bishop's move' definition of adjacency (Figure 6.2). If all eight surrounding neighbors are considered to be adjacent, this is 'queen's move' adjacency (Figure 6.2). These adjacencies can then be used as a

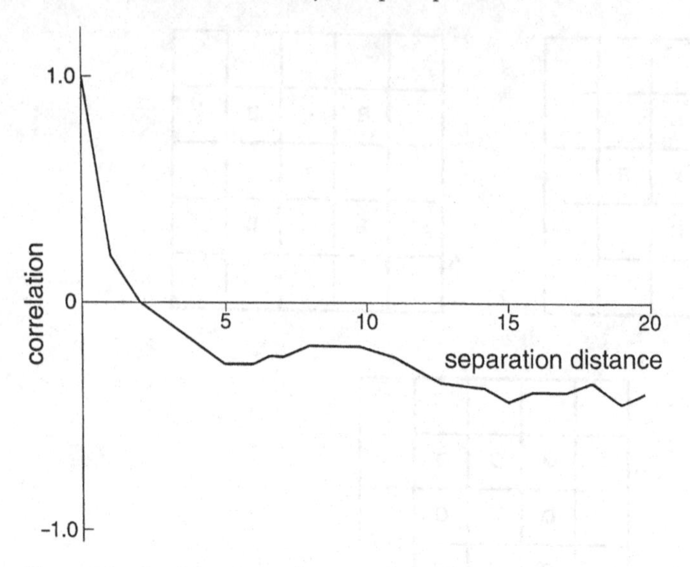

Figure 6.3 Spatial autocorrelation analysis of the Athabasca site smallest diameter class. There is evidence of pattern at a scale of about 5 quadrats or 25 m. The correlation of -0.24 is highly significant when tested without correcting for spatial autocorrelation.

framework for the analysis of spatial structure: first-order neighbors are reachable by one move, second-order neighbors by two, and so on.

In queen's move adjacency, there is a certain ambiguity in the calculation of the separation between quadrats: a quadrat that is four to the right and four up is separated by that path by eight rook's moves but only four diagonally (bishop's moves). To avoid this ambiguity, the rook's move definition may be preferable and also because quadrats that are adjacent by that definition are fully contiguous.

None of the three definitions solve the '$\sqrt{2}$ problem' which is that quadrats that are M moves away may actually be M units away or $M\sqrt{2}$ units or $M/\sqrt{2}$ in actual distance, depending on the definition. By the queen's move, for example, each quadrat has 32 neighbors with a separation of 4 moves, which range in actual distance from 4 to $4\sqrt{2}$; the rook definition gives 16 such neighbors at distances from $2\sqrt{2}=4/\sqrt{2}$ to 4. We will look at one solution to the problem later in this section.

The autocorrelation analysis proceeds by calculating the correlation coefficient between all quadrats separated by r moves for $r=1,2,3\ldots$ and then plotting that correlation as a function of separation. Negative peaks in that plot indicate the scale of pattern and positive peaks show the distance between areas of high similarity. Figure 6.3 shows the autocorrelation plot for the smallest class of trees at the Athabasca site in 1989; there is evidence of pattern at a scale of five quadrats or 25 m.

An alternative calculation that is similar to the autocorrelation approach is to calculate the variance associated with quadrats separated by r moves for $r = 1, 2, 3 \ldots$ and then to plot variance as a function of r. This is a two-dimensional version of PQV and, as usual, peaks in that plot indicate the scale of pattern. Because it uses pairs of individual quadrats, rather than blocks, one drawback in its application is nondiminishing resonance peaks.

Because of the difficulties of interpreting distance as measured by moves, instead of basing the analysis on the autocorrelation framework of adjacent quadrats, PQV analysis can be based on the actual measured distance between quadrats. Based on moves, PQV is calculated as:

$$V_{\mathrm{M}}(r) = \sum (x_{ij} - x_{pq})^2 / n_r, \tag{6.1}$$

where x_{ij} is the density of the species of interest at grid position (i,j), $r = |i-p| + |j-q|$, so that x_{ij} and x_{pq} are separated by exactly r moves, and n_r is the number of such pairs. Based on actual distance, the calculation is:

$$V_{\mathrm{D}}(d) = \sum (x_{ij} - x_{pq})^2 / n_d, \tag{6.2}$$

with $d = \sqrt{(i-p)^2 + (j-q)^2}$, so that x_{ij} and x_{pq} are separated by distance of approximately d (usually an integer), and n_d is the number of such pairs. It is important in implementing this analysis that the pairs of quadrats are chosen in an unbiassed manner; for example from an initial quadrat, quadrats south and east should be examined as well as quadrats south and west.

As an illustration of the methods, Figure 6.4 shows that analysis of all trees at Stadt's Athabasca site. V_{M} and V_{D} give similar results, of course, with the curves diverging due to the different measures of distance. On the whole, for most purposes, the distance-based method is to be preferred because the interpretation as physical distance is more straightforward. The move-based approach to autocorrelation has advantages in situations where the samples are not in a regular grid and not contiguous. Under those circumstances, it may be important to identify which samples are immediate neighbors, which are neighbors of immediate neighbors, and so on. Related material will be discussed in Chapter 7. Interpretation of either move- or distance-based PQV should be based on the fact that the methods average over all directions, so that a peak

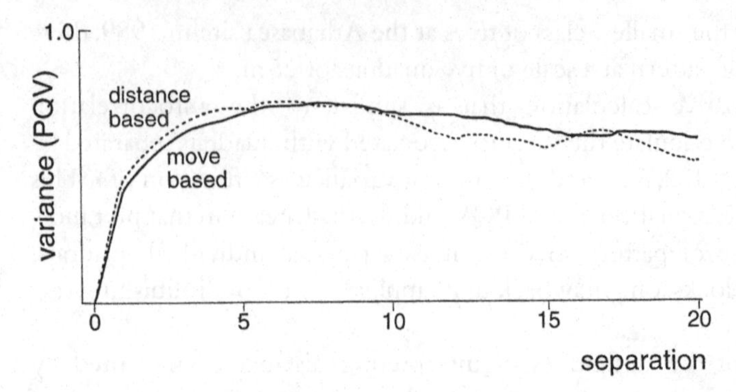

Figure 6.4 Two-dimensional PQV analysis of all trees at the Athabasca site. Both the move-based (solid line) and distance-based (broken line) methods are shown. There is evidence of pattern at a scale of 6 quadrats or 30 m.

variance at scale 10 could result from anisotropic pattern that had a scale of 7 in the North-to-South direction and 13 in the East-to-West direction. The following methods make an evaluation of anisotropy possible.

Two-dimensional spectral analysis

A two-dimensional version of spectral analysis was described by Bartlett (1964) and by Renshaw and Ford (1984); what follows is based on their work. They put forward the technique as an improvement over BQV, particularly because the new method made possible an evaluation of anisotropy. The basic calculations are very similar to those for the one-dimensional case (Chapter 3).

Consider a grid that is $m \times n$ units in size, with species density x_{ij} at grid position (i,j). The periodogram for frequencies p, in the direction of the first axis of the grid, and q, in the direction of the second axis, is I_{pq}:

$$I_{pq} = mn(c_{pq}^2 + s_{pq}^2), \tag{6.3}$$

where:

$$c_{pq} = \sum_{i=1}^{m} \sum_{j=1}^{n} x_{ij}[\cos 2\pi(ip/m + jq/n)] \tag{6.4}$$

and

$$s_{pq} = \sum_{i=1}^{m} \sum_{j=1}^{n} x_{ij}[\sin 2\pi(ip/m + jq/n)]. \tag{6.5}$$

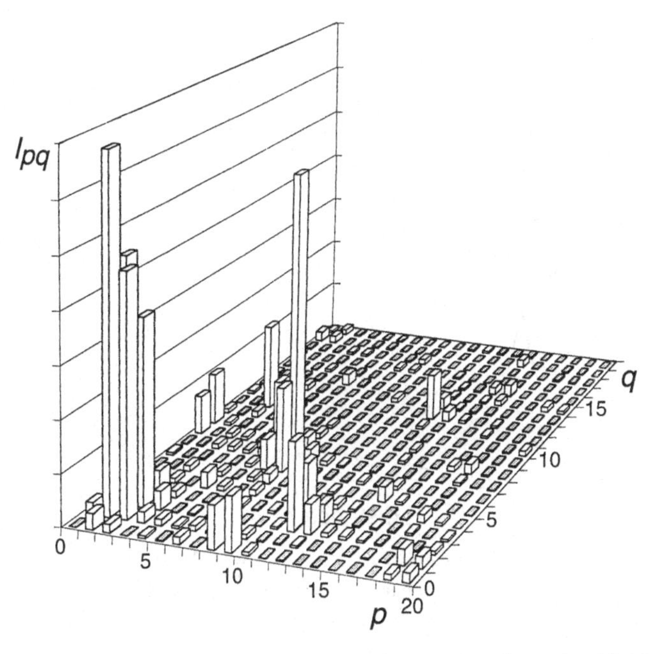

Figure 6.5 Results of two-dimensional spectral analysis of artificial data. The height of the column represents the magnitude of I_{pq} for combination of p and q. There are obvious peaks at (2,2) and (8,8).

The results of this analysis can be presented as a three-dimensional plot with two axes representing the values of p and q, and the third representing the magnitude of I_{pq} (see Figure 6.5). The significance of individual values of I_{pq} can be tested by comparing the proportion of the total variance that I_{pq} represents with the critical value of χ_2^2/mn (Renshaw & Ford 1984).

The three-dimensional representation can be collapsed into summary figures into two ways, referred to as the R-spectrum and the Θ-spectrum. The R-spectrum combines all I_{pq} for which the values of $r=\sqrt{p^2+q^2}$ are the same and plots I as a function of r (Figure 6.6). The Θ-spectrum combines all I_{pq} for which the values of $\theta = \tan^{-1}(p/q)$ are the same and plots I as a function of θ (Figure 6.7). Individual components of either spectrum can also be tested for statistical significance by comparison with the χ^2 distribution. The critical value for a component which contains N periodogram elements is $(1/2N)\chi_{2N}^2$ (Renshaw & Ford 1984).

In a companion paper, Ford and Renshaw (1984) investigated the development of spatial pattern in populations of *Calluna vulgaris*

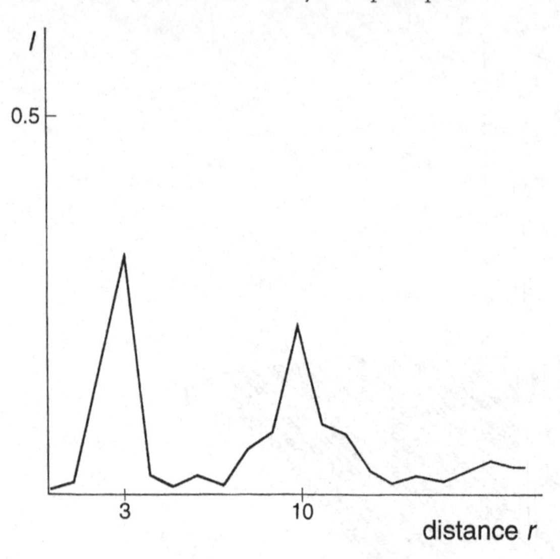

Figure 6.6 Results of spectral analysis summarized by distance classes: the R-spectrum. I is shown as a function of separation $\sqrt{r=p^2+q^2}$. The peaks at 3 and 10 correspond to the peaks in Figure 6.5 at (2,2) and (8,8).

(heather), shown in Figure 1.1, and of *Epilobium angustifolium* (fireweed). They found strong anisotropy in both populations, based on the Θ-spectrum, which they related to the process of colonization and the plants' morphology.

Following their approach, we split the *Calluna* data into the left and right halves and analyzed them separately using the distance-based version of PQV, Equation 6.2. Our analysis of both halves showed a scale of pattern around 0.8 m, which is somewhat smaller than the scale in the single one-dimensional transect illustrated in Figure 1.1 (1.2 m). Ford and Renshaw found that the *R*-spectrum of spectral analysis gave similar results, showing wavelengths of 2 m and 1.67 m which convert to scales of 1 m and 0.83 m. They interpreted this scale as representing the interaction between the size and spacing of the bushes.

Newbery *et al.* (1986) used two-dimensional spectral analysis in an interesting study of Kerangas forest, examining the spatial pattern of the 64 most common tree species. Some species exhibited pattern and some did not. The scale of pattern in the former group of species matched the size of gaps produced by wind throw, and the authors concluded that the most strongly patterned species were the shade-intolerant ones that require gaps for recruitment. On the other hand, they found little evi-

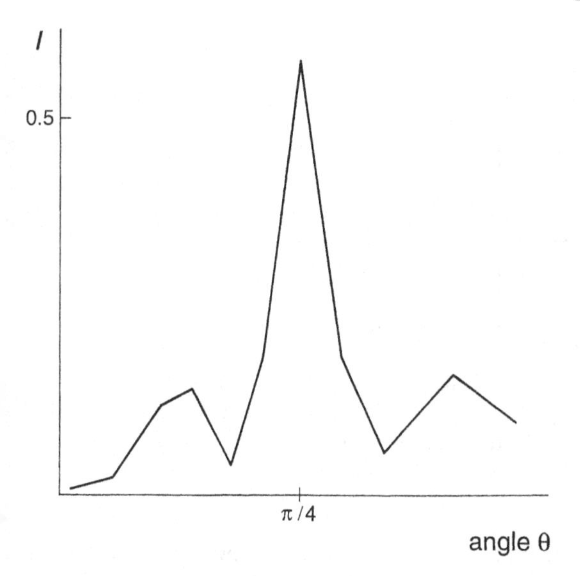

Figure 6.7 Results of spectral analysis summarized by directional classes: the Θ-spectrum. I is shown as a function of the angle, θ. There is a strong peak at $\theta = \pi/4$ or 45°, corresponding to the peaks at (2,2) and (8,8) in Figure 6.5.

dence that species distributions were influenced by the topography of slight ridges and shallow valleys.

The spectral analysis method is clearly one way in which anisotropy can be examined. One cautionary comment is that the analysis requires that the anisotropy is stationary, that is, the same in the entire region sampled. Curving patterns of high and low densities do not give clear results.

Two-dimensional local quadrat variances

For the analysis of two-dimensional pattern, we can develop methods based directly on the one-dimensional method TTLQV, which was described in Chapter 3 (cf. Hill 1973). That method calculates a variance for each block size:

$$V_2(b) = \sum_{j=1}^{n+1-2b} \left(\sum_{i=j}^{j+b-1} x_i - \sum_{i=j}^{j+b-1} x_{i+b} \right)^2 \Bigg/ 2b(n+1-2b), \tag{6.6}$$

with the positions of peaks in the variance plot being interpreted as reflecting pattern at that scale.

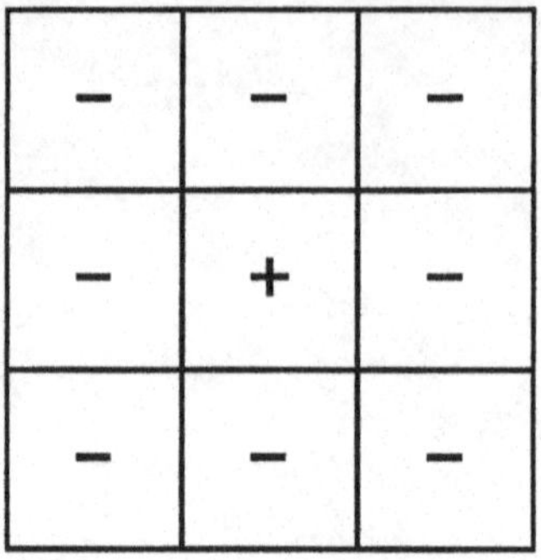

Figure 6.8 The template for the calculation of nine-term local quadrat variance (9TLQV): the data in the '+' block are summed and multiplied by 8; the data in the '−' blocks are subtracted and the result squared. 9TLQV averages the squared differences over all possible positions of the template.

In extending this method to two dimensions, one possibility would be to look at the average squared difference between the total density in a $b \times b$ square and the average total density in the eight $b \times b$ squares that surround it, its Queen's move neighboring blocks (Figure 6.8). This approach can be referred to as a nine-term local quadrat variance (9TLQV). Another possibility is to use a 2×2 arrangement of a $b \times b$ blocks (Figure 6.9), given a four-term variance (4TLQV). We will examine the four-term version of two-dimensional analysis first.

Four-term local quadrat variance

In the $m \times n$ grid of quadrats, let $s_b(i,j)$ be the total of the observations in the $b \times b$ block of quadrats starting at position (i,j):

$$s_b(i,j) = \sum_{p=1}^{i+b-1} \sum_{q=j}^{j+b-1} x_{pq}, \tag{6.7}$$

where x_{pq} is the density of the quadrat in the pth column and qth row of the data (see Figure 6.10). There are four different ways of calculating the variance in this square of four blocks, since the value of any of the four can be multiplied by three and the other totals subtracted. The four terms are, of course, not independent; they are:

$$v_1(b) = \sum_{i}^{n+1-2b} \sum_{j}^{m+1-2b} \frac{[3s_b(i,j) - s_b(i+b,j) - s_b(i,j+b) - s_b(i+b,j+b)]^2}{8(n+1-2b)\,(m+1-2b)b^3}, \tag{6.8}$$

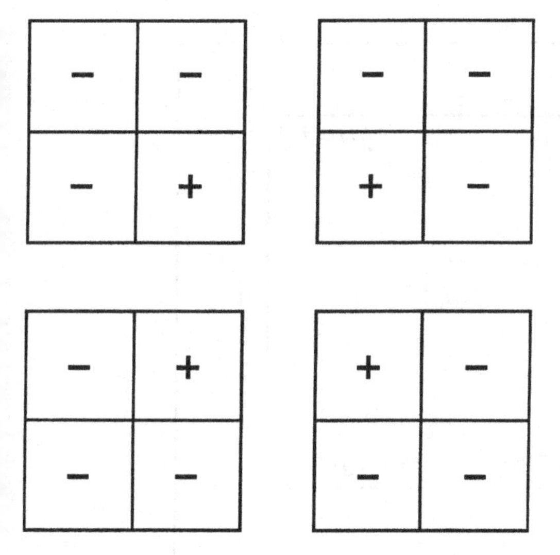

Figure 6.9 The template for the calculation of 4TLQV: the data in the '+' block are summed and multiplied by 3; the data in the '–' blocks are subtracted and the result squared. 4TLQV averages the squared differences over all possible positions of the template and over the four orientations of the template.

$$v_2(b) = \sum_{i}^{n+1-2b} \sum_{j}^{m+1-2b} \frac{[3s_b(i+b,j) - s_b(i,j) - s_b(i,j+b) - s_b(i+b,j+b)]^2}{8(n+1-2b)(m+1-2b)b^3}, \quad (6.9)$$

$$v_3(b) = \sum_{i}^{n+1-2b} \sum_{j}^{m+1-2b} \frac{[3s_b(i,j+b) - s_b(i+b,j) - s_b(i,j) - s_b(i+b,j+b)]^2}{8(n+1-2b)(m+1-2b)b^3}, \quad (6.10)$$

$$v_4(b) = \sum_{i}^{n+1-2b} \sum_{j}^{m+1-2b} \frac{[3s_b(i+b,j+b) - s_b(i+b,j) - s_b(i,j+b) - s_b(i,j)]^2}{8(n+1-2b)(m+1-2b)b^3}. \quad (6.11)$$

Then the overall two-dimensional variance, based on the four blocks, is $V_4(b)$:

$$V_4(b) = [v_1(b) + v_2(b) + v_3(b) + v_4(b)]/4 \quad (6.12)$$

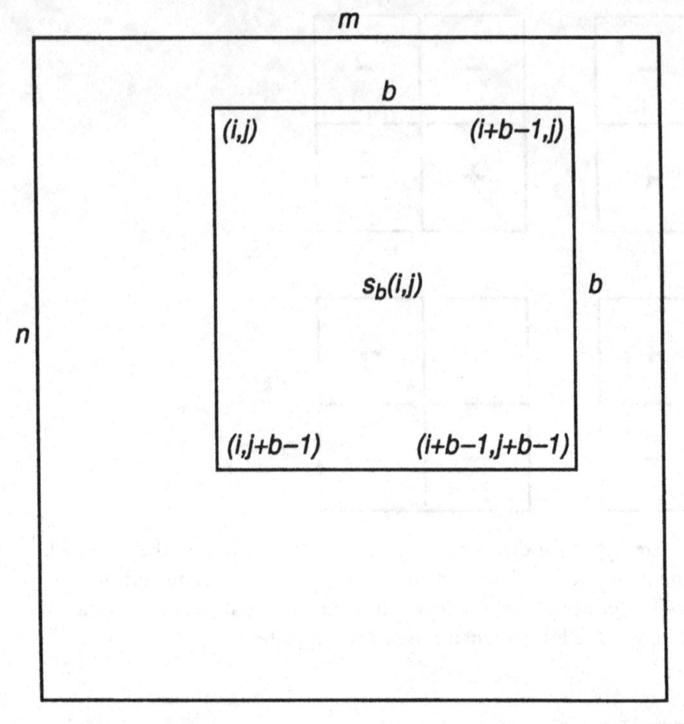

Figure 6.10 The definition of $s_b(i,j)$: it is the sum of quadrats in a $b \times b$ block starting at quadrat (i,j) and ending at quadrats $(i,j+b-1)$, $(i+b-1,j)$ and $(i+b-1,j+b-1)$.

(*cf.* Dale & Powell 1994). Peaks in the plot of V_4 as a function of b correspond to scales of pattern in the vegetation. Another version is to create terms such as:

$$D_{ij} = [3\,s_b(i,j) - s_b(i+b,j) - s_b(i,j+b) - s_b(i+b,j+b)]^2. \tag{6.13}$$

There are four such terms possible for each block size and each initial position (i,j). The other three are:

$$D_{i+b,j} = [3s_b(i+b,j) - s_b(i,j) - s_b(i,j+b) - s_b(i+b,j+b)]^2, \tag{6.14}$$

$$D_{i,j+b,} = [3s_b(i,j+b) - s_b(i+b,j) - s_b(i,j) - s_b(i+b,j+b)]^2, \tag{6.15}$$

$$D_{i+b,,j+b} = [3s_b(i+b,j+b) - s_b(i+b,j) - s_b(i,j+b) - s_b(i,j)]^2. \tag{6.16}$$

For example, if the four values for a particular block size are the following:

2 4
6 5

$D_{ij} = (3\times2-4-6-5)^2 = 81,$
$D_{i+b,j} = (3\times4-2-6-5)^2 = 1,$
$D_{i,j+b} = (3\times6-2-4-5)^2 = 49,$
$D_{i+b,j+b} = (3\times5-2-4-6)^2 = 9.$

The average of the four is 35.

Then:

$$V_4(b) = \sum_{i=1}^{n+1-2b} \sum_{j=1}^{n+1-2b} \frac{D_{ij} + D_{i+b,j} + D_{i,j+b} + D_{i+b,j+b}}{32b^3(n+1-2b)(m+1-2b)}. \tag{6.17}$$

Because V_4 is the two-dimensional equivalent of TTLQV, most of the comments made about that method in Chapter 3, concerning the effects of resonance peaks and the detection of shoulders in the plot of V_2, apply to V_4 as well. That maximum value of $V_4(b)$ is achieved when the data consist of solid stripes B units wide consisting alternately of dense quadrats, d's, and of empty quadrats, 0's, oriented parallel to either the x- or the y-axis. In that case, $V_4(b) = d^2(B^2+2)/6$ which for presence/absence data with $d=1$ simplifies to $(B^2+2)/6$. A measure of the consistency of the pattern is, as usual:

$$I_4(B) = \sqrt{6BV_4(B)/(B^2+2)}. \tag{6.18}$$

An earlier version of this method (Dale 1990) presented a slightly different formulation, with $18b^2$ in the denominator of the variance where we have $8b^3$ above (Equations 6.8–6.11). The reason for the modification is that the new version gives exactly the same values as the one-dimensional TTLQV when the pattern is parallel to one of the axes. Figures 6.11 and 6.12 illustrate the fact that each term added to the two-dimensional variance is $(2b)^2$ times the terms added to the one-dimensional version. This modification allows the one-dimensional findings to be applied directly in the two-dimensional case, including the rate at which the variance peak drifts away from the true scale of pattern with increasing block size.

As in TTLQV, pattern at scale B will produce resonance peaks at $3B$, $5B$, $7B$, and so on; with the earlier formulation, these variance peaks were at the same height as the peak at b, rather than diminishing to 1/3, 1/5, 1/7,... as they do with the new version. As we discussed in Chapter

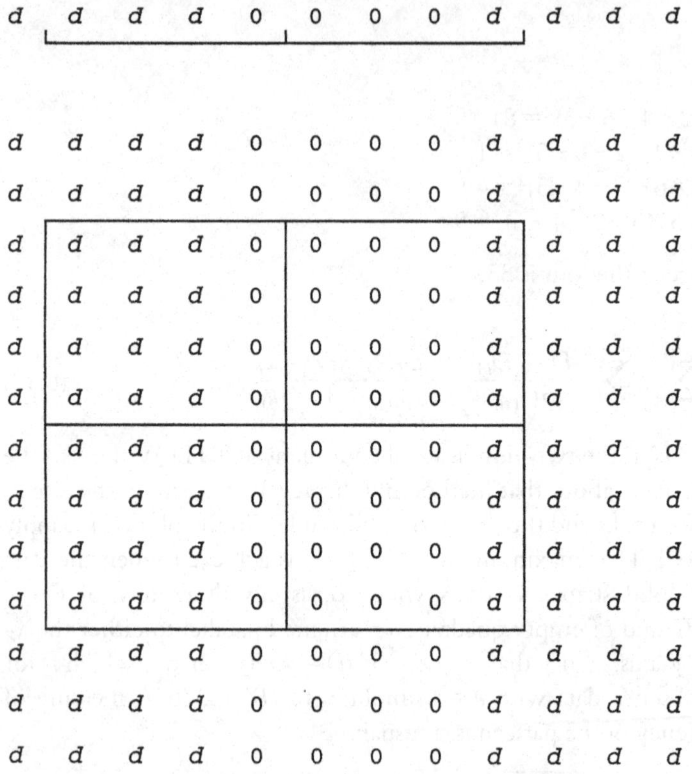

```
d d d d 0 0 0 0 d d d d

d d d d 0 0 0 0 d d d d
d d d d 0 0 0 0 d d d d
d d d d 0 0 0 0 d d d d
d d d d 0 0 0 0 d d d d
d d d d 0 0 0 0 d d d d
d d d d 0 0 0 0 d d d d
d d d d 0 0 0 0 d d d d
d d d d 0 0 0 0 d d d d
d d d d 0 0 0 0 d d d d
d d d d 0 0 0 0 d d d d
d d d d 0 0 0 0 d d d d
d d d d 0 0 0 0 d d d d
d d d d 0 0 0 0 d d d d
```

Figure 6.11 Comparison of TTLQV and 4TLQV computation. Scale and block size are the same: $B=4$ and $b=4$. TTLQV averages terms such as $(3d-1d)^2=(2d)^2$; 4TLQV averages terms such as $(3\times 12d-12d-4d-4d)^2=(16d)^2$ and $(3\times 4d-12d-12d-4d)^2=(16d)^2$. The 4TLQV term is $(2b)^2$ multiplied by the TTLQV term.

3, there is a tradeoff between the disadvantages of peak drift and non-diminishing resonance peaks determined by the factor of b in the divisor. In many applications, block size precision may be less important than the ease of interpretation, and that is the reason why the new version is recommended here.

Dale (1990) showed that the method detects the characteristics of most artificial patterns tested. The change in the divisor does not change that conclusion. The method is not totally independent of the orientation of the pattern with respect to the sampling grid (*cf.* Dale 1990). For instance, a checker-board pattern at 45° to the axes is interpreted by the method as stripes of varying width.

The usefulness of the V_4 analysis is enhanced when TTLQV is carried d d d d d 0 0 0 0 0 d d d d

d d d d d 0 0 0 0 0 d d d d
d d d d d 0 0 0 0 0 d d d d
d d d d d 0 0 0 0 0 d d d d
d d d d d 0 0 0 0 0 d d d d
d d d d d 0 0 0 0 0 d d d d
d d d d d 0 0 0 0 0 d d d d
d d d d d 0 0 0 0 0 d d d d
d d d d d 0 0 0 0 0 d d d d
d d d d d 0 0 0 0 0 d d d d
d d d d d 0 0 0 0 0 d d d d
d d d d d 0 0 0 0 0 d d d d
d d d d d 0 0 0 0 0 d d d d
d d d d d 0 0 0 0 0 d d d d
d d d d d 0 0 0 0 0 d d d d
d d d d d 0 0 0 0 0 d d d d

Figure 6.12 Comparison of TTLQV and 4TLQV computation. Scale and block size are the same: $B=5$ and $b=5$. TTLQV averages terms such as $(4d-1d)^2=(3d)^2$; 4TLQV averages terms such as $(3\times 20d-20d-5d-5d)^2=(30d)^2$ and $(3\times 5d-20d-20d-5d)^2=(30d)^2$. The 4TLQV term is $(2b)^2$ multiplied by the TTLQV term.

out separately on the rows (parallel to the x-axis) and columns (parallel to the y-axis) of the grid to produce variances V_x and V_y. The one-dimensional analyses help in the interpretation of the two-dimensional results.

The method was applied in a landscape-scale study of a valley approximately 30km long and 10km across between Kluane and Kloo Lakes in the Yukon. Knowing that the valley was heterogeneous in its vegetation, we wanted to test whether there were large differences in the scale of the vegetation's pattern. This concern was particularly important because we were applying experimental treatments to 1 km units of the valley (Krebs *et al.* 1995). The data were derived from a LANDSAT image in which each pixel represents a 30m × 30m square. The pixels were classified by their spectral properties related to the kind of vegetation in them. An analysis was carried out for each of the most common classes in 2km × 2km squares by converting the data into 66 × 66 matrix of 1's (pixels in the class) and 0's (pixels not in the class).

In the original analyses of the data, most of the 68 2km × 2km squares analyzed show a single scale of pattern in each vegetation type in each square. The scale of pattern was fairly consistently in the range of 11–23 pixels (330–690m) suggesting that there may be a single cause or set of causes for the observed patterns. This consistency gives some reassurance that different parts of the valley would not respond differently to experimental treatments because of large differences in the scale of their spatial pattern. The values of I_4 associated with the variance peaks were close to 0.30 for common vegetation classes and near to 0.20 for the less common ones, indicating that the patterns are somewhat diffuse.

A minor drawback of the method just described is that the horizontal and vertical adjacency of blocks (centers b units apart) is treated the same as diagonally adjacent blocks (centers $\sqrt{2}\,b$ units apart), creating a certain amount of uncertainty about the conversion of the block size that gives a variance peak to physical distance. The major drawback of the method is the fact that, used alone, it cannot detect or evaluate anisotropy.

Having shown that the four-block approach to spatial pattern analysis is equivalent to one-dimensional TTLQV, we will examine the nine-block method which is comparable to 3TLQV. Using $s_b(i,j)$ as defined above, define:

$$\begin{aligned} T_b(i,j) = {} & s_b(i-b,j-b) + s_b(i-b,j) + s_b(i-b,j+b) \\ & + s_b(i,j-b) - 8s_b(i,j) + s_b(i,j+b) \\ & + s_b(i+b,j-b) + s_b(i+b,jb) + s_b(i+b,j+b). \end{aligned} \tag{6.19}$$

Then:

$$V_9(b) = \sum_{i=b+1}^{n+1-2b} \sum_{j=b+1}^{m+1-2b} \frac{[T_b(i,j)]^2}{72b^3(n+1-3b)(m+1-3b)}. \tag{6.20}$$

The $72b^3$ in the divisor comes from the fact that the one-dimensional version, 3TLQV, has $8b$ in the divisor and, as Figure 6.13 shows, each term added to the two-dimensional version is $(3b)^2$ times the one-dimensional term and $8b \times 9b^2 = 72b^3$. The properties of V_9 will follow those of V_3. As usual, a simple measure of the pattern's intensity is:

$$I_9(b) = \sqrt{6bV_9(b)/(b^2+2)}. \tag{6.21}$$

d d d d 0 0 0 0 d d d d 0 0

d d d d 0 0 0 0 d d d d 0 0
d d d d 0 0 0 0 d d d d 0 0
d d d d 0 0 0 0 d d d d 0 0
d d d d 0 0 0 0 d d d d 0 0
d d d d 0 0 0 0 d d d d 0 0
d d d d 0 0 0 0 d d d d 0 0
d d d d 0 0 0 0 d d d d 0 0
d d d d 0 0 0 0 d d d d 0 0
d d d d 0 0 0 0 d d d d 0 0
d d d d 0 0 0 0 d d d d 0 0
d d d d 0 0 0 0 d d d d 0 0
d d d d 0 0 0 0 d d d d 0 0
d d d d 0 0 0 0 d d d d 0 0
d d d d 0 0 0 0 d d d d 0 0
d d d d 0 0 0 0 d d d d 0 0
d d d d 0 0 0 0 d d d d 0 0

Figure 6.13 Comparison of 3TLQV and 9TLQV computation. Scale and block size are the same: $B=4$ and $b=4$. 3TLQV averages terms such as $(3d-2\times 1d+3d)^2 = (4d)^2$; 9TLQV averages terms such as $(8\times 4d-6\times 12d-2\times 4d)^2=(48d)^2$ and $(8\times 12d-6\times 4d-2\times 12d)^2=(48d)^2$. The 9TLQV term is $(3b)^2$ multiplied by the 3TLQV term.

If large grids of quadrats are being analyzed, the nine-term method will have the same advantages over the four-block version as 3TLQV does over TTLQV. The 4TLQV method allows the examination of larger block sizes if we are willing to break Ludwig and Reynold's rule that the block size examined should not exceed 10% of n or m, the lengths of the sides of the grid (Ludwig & Reynolds 1988).

As an illustration, we analyzed two 100 pixel $\times$ 100 pixel grids from the Yukon LANDSAT data, from near the center of the valley. The first indicated a scale of pattern around seven pixel widths (200 m) for the most common vegetation types and the second at a somewhat larger scale averaging 11 (330 m). In these data, there was good agreement between the results of the 9TLQV analysis and the 4TLQV analysis. Given that the change in the divisor of the four-term method tends to shift the variance peak downward, the results also agree with the overall findings of Dale (1990) which showed a general feature of two-dimensional pattern at a scale of 330–690 m.

Random paired quadrat frequency

To examine anisotropy, we have developed a method based on the random paired quadrat frequency method (RPQF) method (Goodall 1978). It was specifically developed to examine spatial pattern in closed mosaics of saxicolous lichens, and it will be presented initially in those terms. The lichen mosaics can be treated as two-dimensional grids of numbers or symbols representing the different species.

This method is designed to deal with mosaic data which have been converted to a grid of quadrats, with each quadrat assigned to the dominant species in it. Let us consider first the analysis of spatial pattern of a single species. Pairs of quadrats are chosen randomly and we record their distance apart parallel to the x-axis, d_x, and their distance parallel to the y-axis, d_y. We then count the frequency with which quadrats with separation (d_x, d_y) both have the species of interest and the frequency with which they do not. Let $f(1,u,v)$ be the frequency with which pairs with $d_x = u$ and $d_y = v$ both have the species and let $f(0,u,v)$ be the frequency with which they do not. If q is the maximum distance of interest between quadrats parallel to either axis, the result is a $q \times q \times 2$ table. For each cell of the $q \times q$ target species half of the table, we can calculate an expected value, e_{uv}, as the product of the totals for row u, column v and the target species plane divided by the square of the overall total. We then compare the expected value with the observed frequencies, $o_{uv} = f(1,u,v)$, by calculating for each cell the Freeman–Tukey standardized residual:

$$z_{uv} = \sqrt{o_{uv}} + \sqrt{o_{uv} + 1} - \sqrt{4e_{uv} + 1}. \tag{6.22}$$

These can be printed out as a table or portrayed graphically by circles or squares with positions determined by the coordinates u and v, the size determined by the absolute value of z_{uv}, and solid or hollow according to whether they are positive or negative (see Figure 6.14). A threshhold

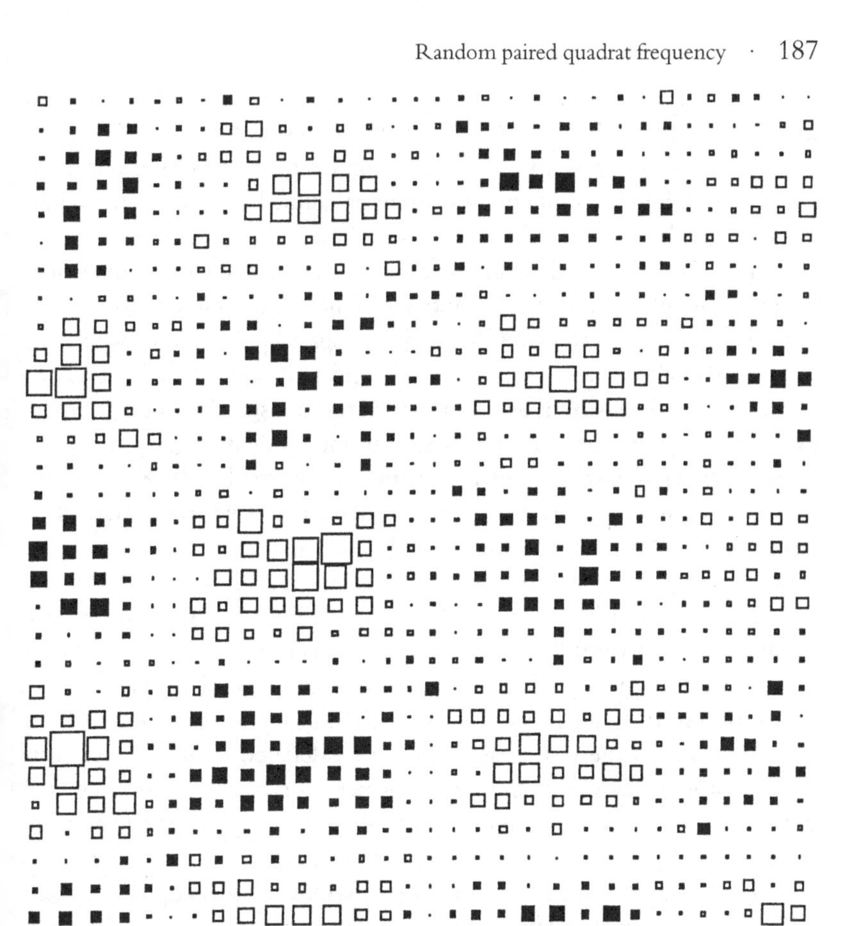

Figure 6.14 Application of the random paired quadrat frequency (RPQF) method to artificial data consisting of angled stripes of 1's and 0's each 10 quadrats wide. The resulting diagram shows correctly the angle and the spacing of the stripes and also the feature of resonance.

value of 2.0 can be used to determine which differ most from expectation although we cannot, strictly speaking, call them significant (*cf.* Bishop *et al.* 1975). We can also calculate a goodness-of-fit statistic to test for overall departure from randomness in the table.

If it is the intention to look at all species, rather than just one at a time, $f(1,u,v)$ becomes a count of the frequency with which quadrats of separation (u,v) are the same species and $f(0,u,v)$ becomes a count of the frequency that they are different. We analyze the 'same species' plane of the table as we analyzed the 'target species' plane in the first version.

A third version would be based not on the identities of the species in the mosaic, but on the shapes of the thalli or tiles of the mosaic. This can be accomplished by setting up a rectangular grid as before, but with entries of 1 when the unit contains an interthalline boundary and 0 when it does not. The analysis then proceeds as above.

This RPQF method avoids the $\sqrt{2}$ problem of two-dimensional TTLQV, because the actual x- and y-distances are retained, and it also both handles and portrays anisotropy in the spatial pattern very well.

As stated above, the observed and expected values can be used to perform a goodness-of-fit test using the G test statistic. Almost any deviation from expectation will give a significant result if large numbers of pairs of grid elements are used. For example, if the grid is 100×100, there are 10 grid units and 5×10^7 pairs of them. It is certainly practical to sample 100000 pairs. If we limit d_x and d_y to a maximum of 30, then the observed frequency table has 900 cells, giving an average expected value per cell of about 100. Such large numbers will make the method very sensitive to departures from expectation, but the significant results need to be treated with caution, because of the large amount of spatial autocorrelation in the data. It is not possible to evaluate, with current knowledge, how much the test statistic should be deflated to account for this spatial autocorrelation. The amount of deflation necessary will change with the relationship between the sizes of the tiles of the mosaic and the size of the grid units used (cf. Dale *et al.* 1991).

For small grids, it might be practical to sample all pairs up to a certain distance, but as grids become larger, an 'all pairs of quadrats' approach becomes impractical. Notice that for large grids, only a very small proportion of quadrat pairs may be sampled; in the numerical example above only 1 in 500.

Dale (1995) shows that the random pair method detects the essential features of artificial data sets, including anisotropy.

As a field example, let us look at lichen mosaics from three sites in the Subalpine and Alpine zones of the Canadian Rockies in Yoho National Park (N.P.) and Glacier/Mt. Revelstoke N.P. The sites had many species in common, including several species of *Rhizocarpon* (*R. geographicum*, *R. polycarpum*, *R. grande*, *R. disporum*), *Lecidea auriculata*, *Lecidea paupercula*, and *Aspicilia cinerea*. The data were converted from photographs into 50×70 grids. Figure 6.15 shows the analysis of a single species, *Rhizocarpon polycarpum*; it displays anisotropy, with greater similarity horizontally than vertically. Figure 6.16 shows the analysis of all species in the same mosaic from which Figure 6.15 was derived; it is much less anisotropic than the

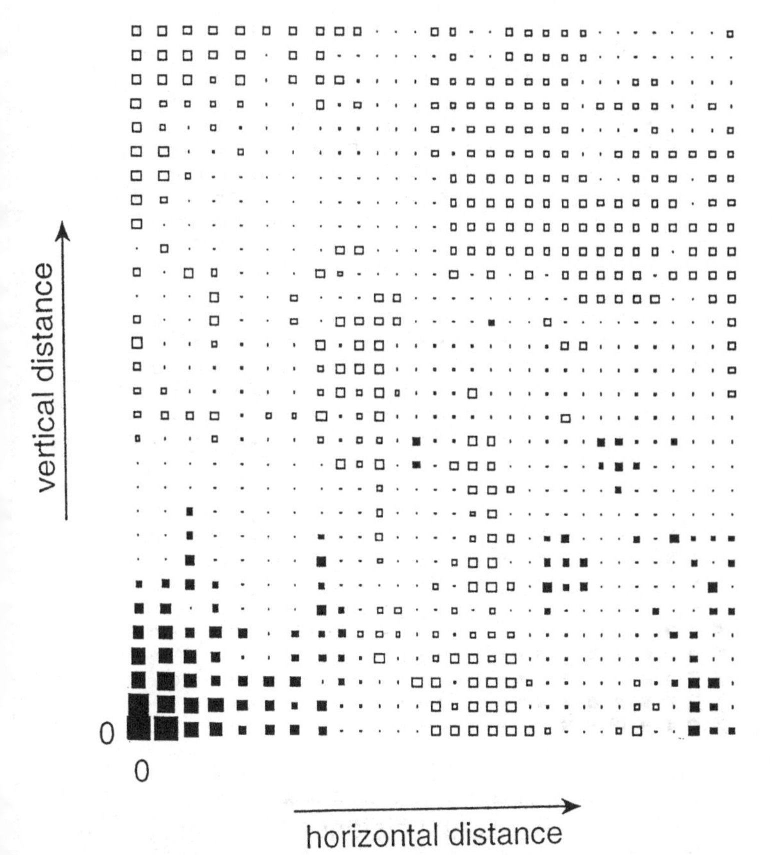

Figure 6.15 RPQF analysis of a single species, *Rhizocarpon polycarpum*, in a lichen mosaic showing anisotropy.

single-species pattern. One explanation is that the anisotropy of another single species, *R. geographicum*, resulted from strong vertical similarity so that when both species are included, the two directions of anisotropy cancel.

The RPQF approach has several advantages: it is simple to use, the output is easy to interpret, and the $\sqrt{2}$ problem is avoided. Like all paired quadrat techniques, however, pattern at scale B will give large positive residuals at distance $2B$, $4B$, $6B$, and so on as a result of resonance (*cf.* Figure 6.14). This resonance may cause confusion if there are several scales of pattern. The main advantage of the method is that it gives a clear portrayal of anisotropy.

As a second example of this method, we will use the brousse tigrée illustrated in Figure 1.2. We converted about a fifth of the figure into a

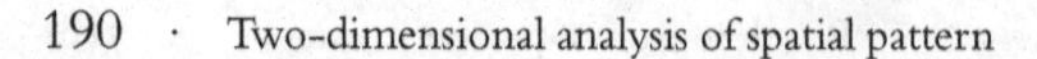

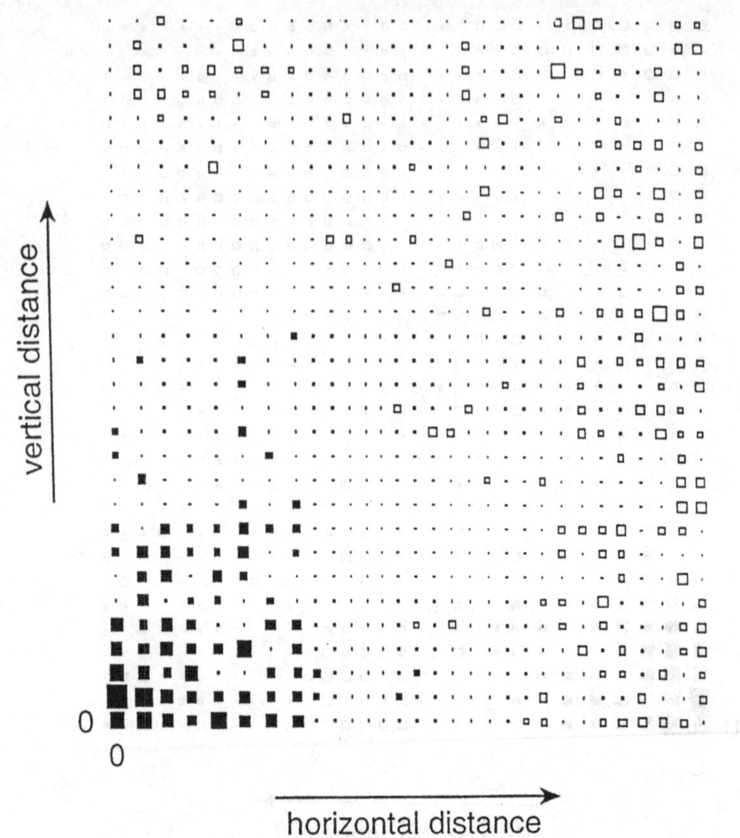

Figure 6.16 RPQF analysis of all species in the same lichen mosaic; the anisotropy is greatly reduced.

40 × 40 matrix of 1's and 0's (Figure 6.17). Figure 6.18 shows the results of the analysis, in which the effects of anisotropy are very obvious at a scale of three quadrats.

Variogram

In Chapter 3, we described the close relationship between the paired quadrat variance technique for studying spatial pattern in one dimension and the geostatistical approach of estimating the semivariogram or variogram. It is no surprise that the variogram approach has been used in two dimensions. The variogram can be calculated in different directions to determine whether the spatial structure is isotropic (Rossi *et al.* 1992).

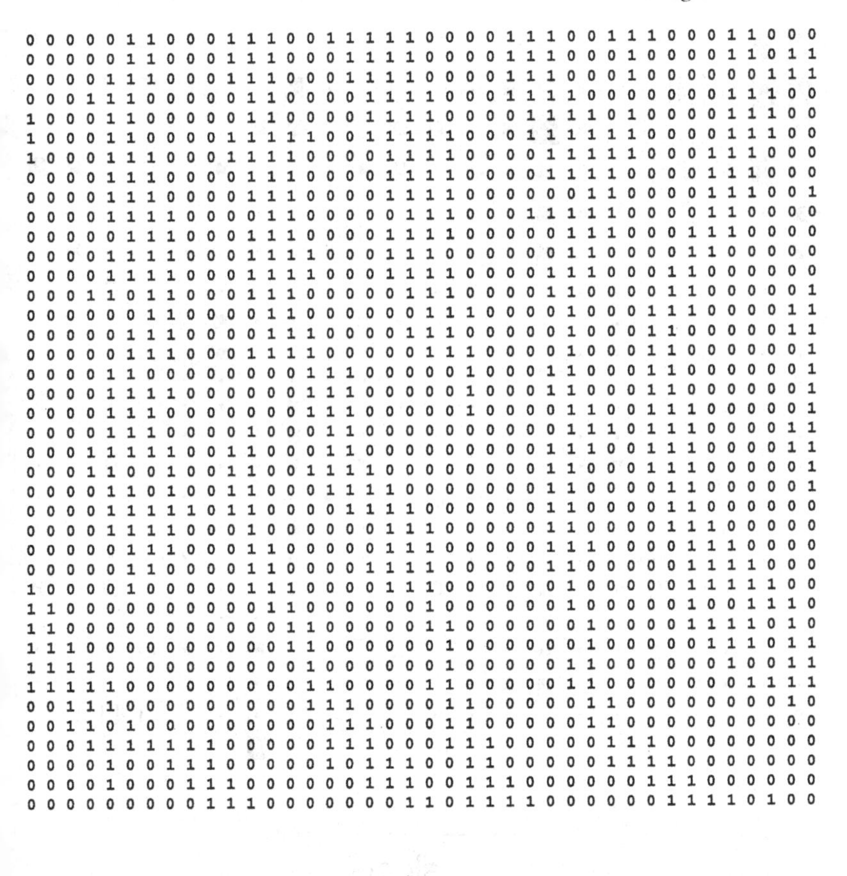

```
0 0 0 0 0 1 1 0 0 0 1 1 1 0 0 1 1 1 1 1 0 0 0 0 1 1 1 0 0 1 1 1 0 0 0 1 1 0 0 0
0 0 0 0 0 1 1 0 0 0 1 1 1 0 0 0 1 1 1 1 0 0 0 0 1 1 1 0 0 0 1 0 0 0 0 1 1 0 1 1
0 0 0 0 1 1 1 0 0 0 1 1 1 0 0 0 1 1 1 1 0 0 0 0 1 1 1 0 0 0 1 0 0 0 0 0 0 1 1 1
0 0 0 1 1 1 0 0 0 0 0 1 1 0 0 0 0 1 1 1 1 0 0 0 1 1 1 1 0 0 0 0 0 0 0 1 1 1 0 0
1 0 0 0 1 1 0 0 0 0 0 1 1 0 0 0 0 1 1 1 1 0 0 0 0 1 1 1 1 0 1 0 0 0 0 1 1 1 0 0
1 0 0 0 1 1 0 0 0 0 1 1 1 1 1 0 0 1 1 1 1 1 0 0 0 1 1 1 1 1 1 0 0 0 0 1 1 1 0 0
1 0 0 0 1 1 1 0 0 0 1 1 1 1 0 0 0 0 1 1 1 1 0 0 0 0 1 1 1 1 1 0 0 0 1 1 1 0 0 0
0 0 0 0 1 1 1 0 0 0 0 1 1 1 0 0 0 0 1 1 1 1 0 0 0 0 1 1 1 1 0 0 0 0 1 1 1 0 0 0
0 0 0 0 1 1 1 0 0 0 0 1 1 1 0 0 0 0 1 1 1 1 0 0 0 0 0 0 1 1 0 0 0 0 1 1 1 0 0 1
0 0 0 0 1 1 1 1 0 0 0 0 1 1 0 0 0 0 0 1 1 1 0 0 0 1 1 1 1 1 0 0 0 0 1 1 0 0 0 0
0 0 0 0 0 1 1 1 0 0 0 1 1 1 0 0 0 1 1 1 1 0 0 0 0 0 1 1 1 0 0 0 1 1 1 0 0 0 0 0
0 0 0 0 1 1 1 1 0 0 0 1 1 1 1 0 0 0 1 1 1 0 0 0 0 0 1 1 0 0 0 0 1 1 0 0 0 0 0 0
0 0 0 0 1 1 1 1 0 0 0 1 1 1 1 0 0 0 1 1 1 1 0 0 0 0 1 1 1 0 0 0 1 1 0 0 0 0 0 0
0 0 0 1 1 0 1 1 0 0 0 1 1 1 0 0 0 0 0 1 1 1 0 0 0 0 1 1 0 0 0 0 1 1 0 0 0 0 0 1
0 0 0 0 0 0 1 1 0 0 0 0 1 1 0 0 0 0 0 0 1 1 1 0 0 0 0 1 0 0 0 1 1 1 0 0 0 0 1 1
0 0 0 0 0 1 1 1 0 0 0 0 1 1 1 0 0 0 0 1 1 1 0 0 0 0 0 1 0 0 0 1 1 0 0 0 0 0 1 1
0 0 0 0 0 1 1 1 0 0 0 1 1 1 1 0 0 0 0 0 1 1 1 0 0 0 0 1 0 0 0 1 1 0 0 0 0 0 0 1
0 0 0 0 1 1 0 0 0 0 0 0 0 0 1 1 1 0 0 0 0 1 0 0 0 1 1 0 0 0 1 1 0 0 0 0 0 0 0 1
0 0 0 0 1 1 1 1 0 0 0 0 0 0 1 1 1 0 0 0 0 0 1 0 0 0 1 1 0 0 0 1 1 0 0 0 0 0 0 1
0 0 0 0 1 1 1 0 0 0 0 0 0 0 1 1 1 0 0 0 0 0 1 0 0 0 0 1 1 0 0 1 1 1 0 0 0 0 0 1
0 0 0 0 1 1 1 0 0 0 0 1 0 0 0 1 1 0 0 0 0 0 0 0 0 0 0 1 1 1 0 1 1 1 0 0 0 0 1 1
0 0 0 1 1 1 1 1 0 0 1 1 0 0 0 1 1 0 0 0 0 0 0 0 0 0 1 1 1 0 0 1 1 1 0 0 0 0 1 1
0 0 0 1 1 0 0 1 0 0 1 1 0 0 1 1 1 1 0 0 0 0 0 0 0 0 1 1 0 0 0 1 1 1 0 0 0 0 0 1
0 0 0 0 1 1 0 1 0 0 1 1 0 0 0 1 1 1 1 0 0 0 0 0 0 0 1 1 0 0 0 0 1 1 0 0 0 0 0 1
0 0 0 0 1 1 1 1 1 0 1 1 0 0 0 0 1 1 1 1 0 0 0 0 0 0 1 1 0 0 0 0 1 1 0 0 0 0 0 0
0 0 0 0 1 1 1 1 0 0 0 1 0 0 0 0 0 0 1 1 1 0 0 0 0 0 1 1 0 0 0 0 1 1 1 0 0 0 0 0
0 0 0 0 0 1 1 1 0 0 0 1 1 0 0 0 0 1 1 1 0 0 0 0 0 1 1 1 0 0 0 0 1 1 1 0 0 0 0 0
0 0 0 0 0 1 1 0 0 0 0 1 1 0 0 0 0 1 1 1 1 0 0 0 0 0 1 1 0 0 0 0 0 1 1 1 1 0 0 0
1 0 0 0 0 1 0 0 0 0 0 1 1 1 0 0 0 0 1 1 1 0 0 0 0 0 0 1 0 0 0 0 0 1 1 1 1 1 0 0
1 1 0 0 0 0 0 0 0 0 0 0 1 1 0 0 0 0 0 0 1 0 0 0 0 0 0 1 0 0 0 0 0 1 0 0 1 1 1 0
1 1 0 0 0 0 0 0 0 0 0 0 0 1 1 0 0 0 0 0 1 1 0 0 0 0 0 0 1 0 0 0 0 1 1 1 1 0 1 0
1 1 1 0 0 0 0 0 0 0 0 0 0 1 1 0 0 0 0 0 0 1 0 0 0 0 0 0 1 0 0 0 0 0 1 1 1 0 1 1
1 1 1 1 0 0 0 0 0 0 0 0 0 0 1 0 0 0 0 0 0 1 0 0 0 0 0 1 1 0 0 0 0 0 0 1 0 0 1 1
1 1 1 1 1 0 0 0 0 0 0 0 0 0 1 1 0 0 0 0 1 1 0 0 0 0 0 1 1 0 0 0 0 0 0 0 1 1 1 1
0 0 1 1 1 0 0 0 0 0 0 0 0 0 1 1 1 0 0 0 0 1 1 0 0 0 0 0 1 1 0 0 0 0 0 0 0 0 1 0
0 0 1 1 0 1 0 0 0 0 0 0 0 0 0 1 1 1 0 0 0 1 1 0 0 0 0 0 1 1 0 0 0 0 0 0 0 0 0 0
0 0 0 1 1 1 1 1 1 1 0 0 0 0 0 1 1 1 0 0 0 1 1 1 0 0 0 0 0 1 1 1 0 0 0 0 0 0 0 0
0 0 0 0 1 0 0 1 1 1 0 0 0 0 0 1 0 1 1 1 0 0 1 1 0 0 0 0 0 1 1 1 1 0 0 0 0 0 0 0
0 0 0 0 1 0 0 0 1 1 1 0 0 0 0 0 0 1 1 1 0 0 1 1 1 0 0 0 0 0 0 1 1 1 0 0 0 0 0 0
0 0 0 0 0 0 0 0 0 1 1 1 0 0 0 0 0 0 0 1 1 0 1 1 1 1 0 0 0 0 0 0 1 1 1 1 0 1 0 0
```


Figure 6.17 A 40 × 40 grid of brousse tigrée data used for analysis as 1's and 0's.

In their discussion of spatial pattern and its importance in ecological analysis, Legendre and Fortin (1989) use the example of a temperate forest in Québec, containing 28 tree species, which was sampled with 200 regularly spaced quadrats. To demonstrate the evaluation of anisotropy, they calculated directional variograms at 45° and at 90°, each with a window of 22°. The two angles gave different scales of pattern, 445 m and 685 m, indicating that the pattern was anisotropic. When the variogram was calculated for all directions, while there was an indication of a scale of pattern just less than 400 m, the pattern at 685 m produced a larger variance.

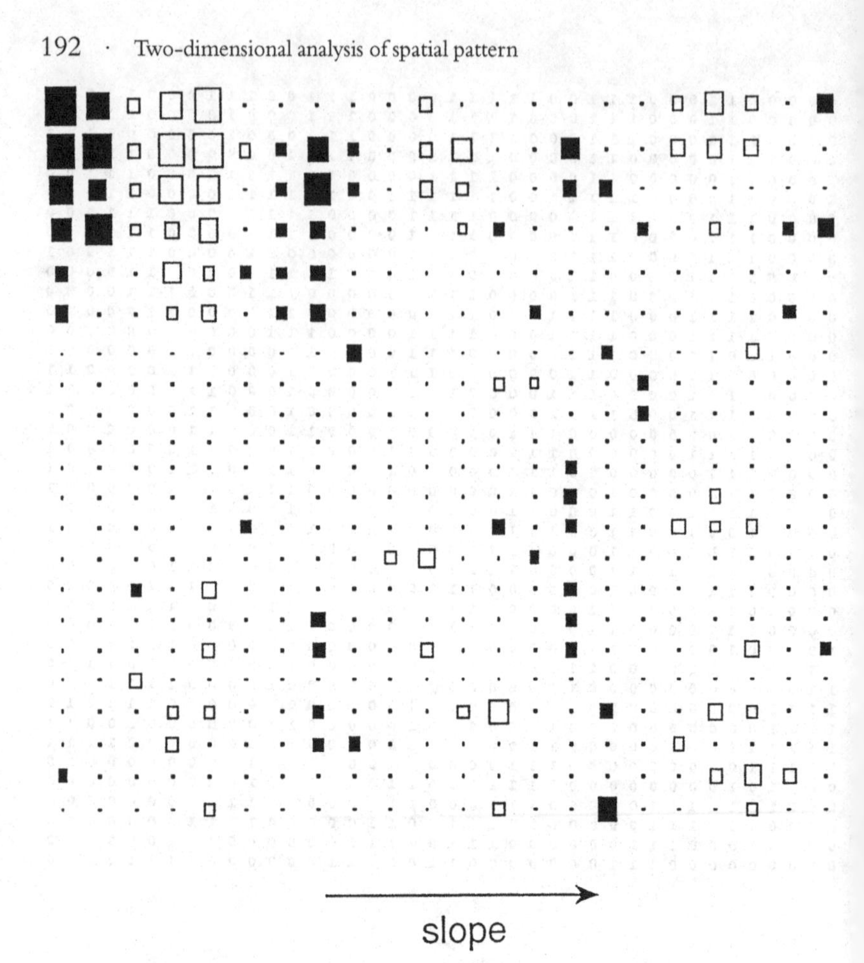

Figure 6.18 RPQF analysis of the brousse tigrée data showing both the scale of the pattern and its strong anisotropy.

For a visual evaluation of anisotropy, Rossi *et al.* (1992) suggest drawing a contour diagram of the variogram value as a function of displacement in the 0° and 180° direction and in the 90° and 270° direction.

Covariation

Many of the single-species methods described above, such as RPQF and 4TLQV, can be modified to deal with two different kinds of plants, just as the one-dimensional single species methods from Chapter 3 were modified for two species in Chapter 4. There are at least two approaches to the relationship between two kinds of plants in a plane. The first is to describe the spatial arrangement of each kind separately and then to describe how the two separate patterns are combined. The second is to describe the spatial arrangement of all the plants of either kind and then to label the plants' positions as belonging to the two different kinds. In either case, it is of interest to find out whether the plants of the different kinds are segregated or aggregated. When the data collected are the counts or densities of the two species in quadrats, we will examine whether those measures for the two species covary positively or negatively.

One feature of the concepts of segregation and aggregation is that the phenomena are scale dependent. Hurlbert (1990) points out that, '(The) degree of aggregation in nature is always strongly a function of spatial scale.' Therefore, we should not ask whether the two kinds of plants are segregated or aggregated; we should ask at what scale or scales they are segregated and aggregated. For quadrat data, the question will be translated into the question of what block sizes maximize or minimize the two species' covariance.

If the data are in the form of mapped plant locations, they can be converted into the equivalent of grids of contiguous quadrats by setting up a matrix in which the elements of the matrix are the numbers of plants of each kind in a corresponding square of the map (Figure 6.19). The use of smaller and thus more numerous quadrats will lose less information in this conversion. In most applications, the matrices will be sparse, consisting mainly of 0's with a sprinkling of 1's. With data on the densities of species in a grid of contiguous quadrats, as in the Stadt data set, the data are used directly.

Paired quadrat covariance (PQC)

The study area is represented by a square $n \times n$ or rectangular $n \times m$ grid of quadrats. Let x_{ij} be the number of individuals or the density of species A in the quadrat (i,j) of the grid and y_{ij} is the measure of the density of species B. Then we define the covariance as:

$$C_D(d) = \sum (x_{ij} - x_{pq})(y_{ij} - y_{pq})/n_d, \tag{6.23}$$

with $d = \sqrt{(i-p)^2 + (j-q)^2}$, so that quadrats (i,j) and (p,q) are separated by distance of approximately d and n_d is the number of such pairs. The term 'approximately d' can refer to the nearest integer less than the real

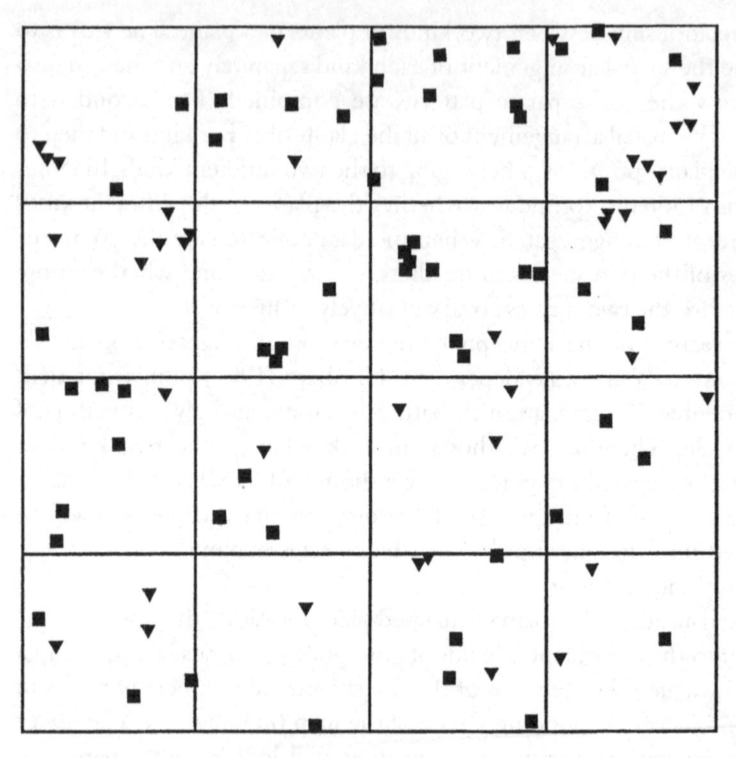

Figure 6.19 The conversion of mapped data to quadrats. The plants are *Salix arctica* on Ellesmere Island, the triangles are male plants and the squares are female plants (data from M. H. Jones, personal communication). The counts are as follows:

male

3	2	0	8
6	0	1	4
1	1	3	1
3	1	3	2

female

2	4	8	4
1	4	9	3
6	3	0	3
3	1	4	1

The sexes are negatively correlated in the quadrats.

distance. A move-based measure of distance can also be used in calculating covariance as we did above for variance, but we do not include it here. Figure 6.20 shows the covariance of all regenerating stems, including seedlings and all size classes of trees at Stadt's Athabasca site in 1989. There is evidence of negative covariance or segregation at a scale of five quadrats or 25 m, probably reflecting the size of open areas in which regeneration is more successful.

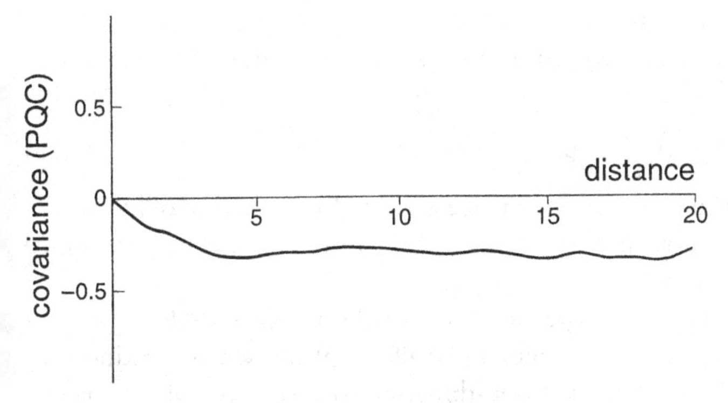

Figure 6.20 Paired quadrat covariance (PQC) analysis of regenerating stems and trees at Stadt's Athabasca site. The two classes are segregated at a scale of 5 quadrats.

Four-term local quadrat covariance

The next method we examine is a modification of 4TLQV described above, now in a covariance form to deal with two species. Define $s_b(i,j)$ as the sum for species A in the $b \times b$ square of quadrats starting at (i,j) (Figure 6.10):

$$s_b(i,j) = \sum_{g=1}^{i+b-1} \sum_{h=j}^{j+b-1} x_{gh}. \tag{6.24}$$

The equivalent for species B is:

$$t_b(i,j) = \sum_{g=1}^{i+b-1} \sum_{h=j}^{j+b-1} y_{gh}. \tag{6.25}$$

We then use terms such as:

$$D_{ij} = [3s_b(i,j) - s_b(i+b,j) - s_b(i,j+b) - s_b(i+b,j+b)]^2. \tag{6.26}$$

There are four such terms possible for each block size and each initial position (i,j), as described above.

Then:

$$V_A(b) = \sum_{i=1}^{n+1-2b} \sum_{j=1}^{n+1-2b} \frac{D_{ij} + D_{i+b,j} + D_{i,j+b} + D_{i+b,j+b}}{32b^3(n+1-2b)^2} \tag{6.27}$$

The divisor of this calculation is modified from that in Dale and Powell (1994) with $32b^3$ replacing $72b^2$, for reasons explained above to bring the calculation into line with TTLQV. The net effect of this change is not large, but it tends to shift the peak downward because of the division by an extra factor of b.

The variance of species B, V_B, and that of the combined data, V_{A+B}, are calculated in the same way for block size b. The covariance of A and B at scale b is:

$$\mathrm{Cov}_{AB}(b) = \left[V_{A+B}(b) - V_A(b) - V_B(b)\right]/2, \tag{6.28}$$

(*cf.* Kershaw 1961). When covariance is plotted as a function of block size, peaks, both positive and negative, can be interpreted as the approximate scales of covariance.

We suggest looking at the plot of the total variance $V_T = V_{A+B}$, since it will reflect the spatial arrangement of all the plants, whatever kind they are. We can also examine the one-dimensional variances and covariances parallel to the x-axis and parallel to the y-axis, to help interpret the pattern if it is anisotropic.

To evaluate the significance of the pattern detected, we can calculate the expected value of V_4 on the assumption that the plants are randomly arranged in space. On that assumption, for small values of b, we need to derive the expected value of terms such as $(3\xi - \eta - \zeta - \varphi)^2$ where the ξ, η, ζ, and φ are treated as independent random binomial variables. Where m is the number of plants, the size of the grid is $n \times n$, and the block size is b, then:

$$E[(3\xi - \eta - \zeta - \varphi)^2] = 12\ \mathrm{Var}(\xi) = mp(1-p) \tag{6.29}$$

where $p = b^2/n^2$.

For $b = 1$, $E[V(1)] = 2(1 - 1/n^2)m/3n^2$, which is approximately $2m/3n^2$. As b increases, ξ is increasingly negatively correlated with $(\eta + \zeta + \varphi)$. At the maximum value of b, which is $n/2$, $E[V(n/2)]$ can be derived by calculating $E[3\xi - (m - \xi)]^2 = E(16\xi^2 - 8\xi m + m^2) = 16mp(1-p) + m^2(4p-1)^2$ where $p = 1/4$. This gives $E[V(n/2)] = 2m/3n^2$. Computer trials confirm that the average value of V_4 remains near this value from $b = 1$ to $b = n/2$.

We know that the expected value of the covariance is zero if the plants of the two species are arranged independently of each other. There is, however, little point in pursuing an analytical approach because the distribution of the covariance will change with the arrangement of the plants and we do not really expect that to be random. If the plants are known to be patchily distributed, it is preferable to determine whether any apparent aggregation of the different kinds of plants is due solely to the overall patchiness, or whether there is aggregation in addition to the overall patchiness. This question can be addressed using a randomization approach. Keeping the positions of the plants constant, the 'labels' of the plants are randomly permuted a number of times with the variances and covariance calculated for each relabelling. Then, 100 or more trials can be used to judge whether the observed results are significantly high or low for the positions of the plants. Dale and Powell (1994) found that these methods were successful in recognizing the characteristics of artificially constructed data.

We can illustrate the method using data from a large hay field at the Wagner Natural Area near Edmonton, Alberta. The field was seeded with grasses and alfalfa more than 15 years before the study and it had been mowed for hay in each intervening year. Several weed species, including *Solidago canadensis*, were invading the field from the edges. The plant is attacked by several kinds of insect herbivores including gall-formers (*cf.* Felt 1940, Hartnett and Abrahamson 1979). Twelve 2m × 2m plots were placed at the edge of the field and all *Solidago* stems were mapped in each plot with each classified as having a gall or not.

The mapped data were converted to 50 × 50 grids of 4cm × 4cm units and analyzed using the method just described.

All 12 plots had evidence of spatial scale between block sizes 14 and 21 (0.56m to 0.84m) and ten were judged to be significant compared to random dispersion. Only three covariance plots showed segregation at small scales (block sizes 1–3) and nine showed aggregation at larger scales (13–22).

The variances for all plants and the covariances calculated parallel to the x-axis and to the y-axis provide little evidence that the spatial pattern was anisotropic. The variance profiles were usually similar in the two directions, as were the covariances which were often close to zero.

The randomization test showed that in most cases there was either no significant aggregation of galled and ungalled plants within the overall patchiness, or that the two kinds of plants were actually segregated when tested in that way. Therefore, much of the apparent aggregation of the galled and ungalled plants is attributable to the overall patchiness of the plants.

The second example of this kind of analysis uses data from the literature, specifically the Lansing Woods data given in Diggle (1983) and discussed by Upton and Fingleton (1985). These data are the mapped positions of trees of various species in a hardwood forest in Clinton County, Michigan, U.S.A. As with the *Solidago* data, the positions were converted to a 50 × 50 grid of counts. Upton and Fingleton (1985) remark that maples (*Acer* sp.) and hickories (*Carya* sp.) are complementary, with maple gaps matching hickory patches and *vice versa*. Figure

6.21a shows that the segregation of the two species has a maximum at a scale of 12 quadrats (68 m). In contrast, red oaks and hickories show aggregation at a scale of 8 quadrats (45 m) as illustrated in Figure 6.21b. Black oak (*Quercus nigra*) with hickory, on the other hand, has a very different pattern, with aggregation increasing with scale (Figure 6.21c). This example shows that, within a single community, and a small set of species, there is a wide range of covariance responses, indicating that interspecific association changes over distance in different ways depending on the pair of species.

Plant–environment correlation

Many of the methods we have been describing for examining the spatial covariation of two kinds of plants in two dimensions can obviously be used to investigate the relationship between a species and an environmental factor. Reed *et al.* (1993) used a somewhat different technique to examine the scale dependence of correlation between vegetation and environmental factors. In a woodland in North Carolina, they sampled the vegetation using a grid of 256 16 m × 16 m cells, each containing eight nested quadrats in one corner from 0.0156 m to 256 m^2 in size. They also collected 289 soil samples from the corners of the cells and analyzed them for a number of characteristics including pH, bulk density and the available amounts of a range of elements. In the absence of directly measured soil variables for the small nested quadrats, those values were interpolated using the geostatistical technique of kriging (David 1977; Bailey & Gatrell 1995). The vegetation data were subjected to ordination, detrended correspondence analysis (DCA), and the environmental variables' coefficients of multiple determination with the first DCA axes were calculated. These values tended to be larger for the larger quadrats, which the authors interpreted as indicating that the correlation of vegetation and environment increases with increasing grain size or scale.

Cross-variogram

In Chapter 4, we introduced the general concept of the cross-variogram (Equation 4.5) which is the geostatistical equivalent of a covariance that is calculated over a range of distances. In that chapter, the concept was applied in one dimension, but it can obviously be used equally well in two dimensions. As with the variogram, in two dimensions it can be used either with averaging over all directions or in a directional manner. Rossi *et al.* (1992) provide an interesting example of the application of the

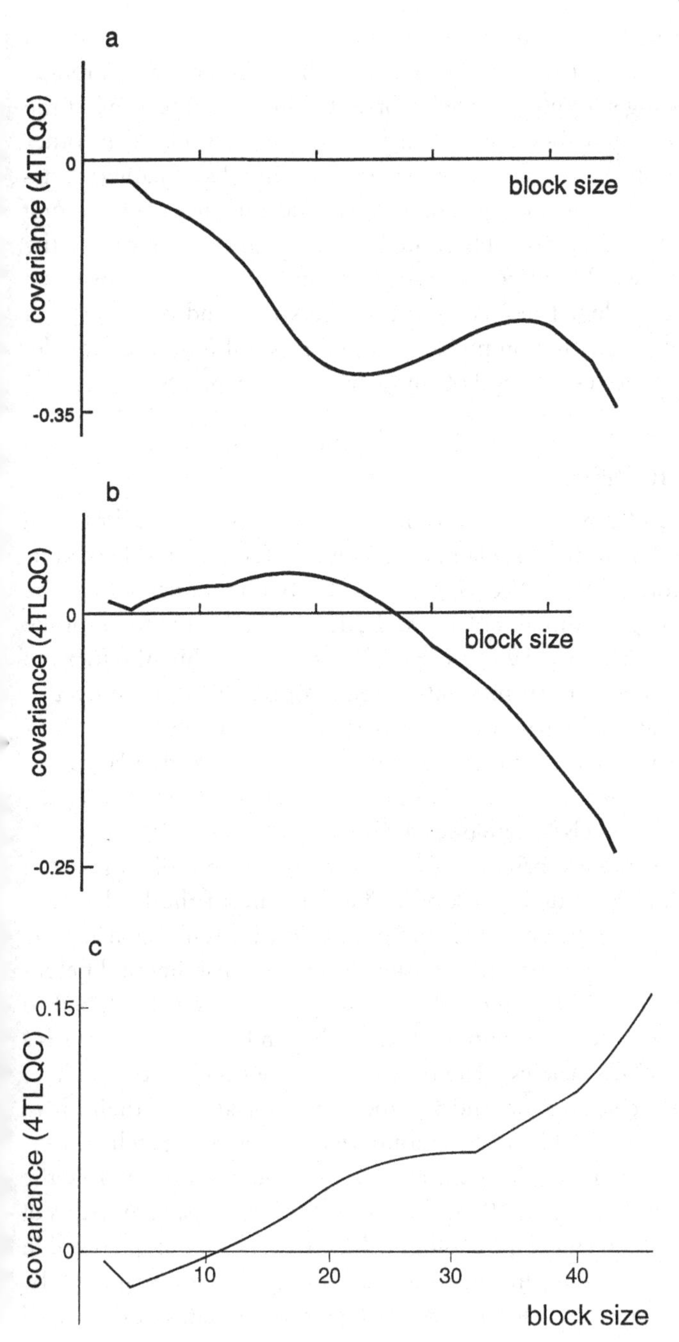

Figure 6.21 4TLQC analysis of Lansing Woods data. *a* Maples and hickories. *b* Red oak and hickory. *c* Black oak and hickory. There is a complete range of responses in the one data set.

directional cross-variogram in the study of the ecology of two carabid beetles. The cross-variogram showed that their pattern was strongly anisotropic, being strongly positively correlated in some directions at the same scales at which they were strongly negatively correlated in other directions. This pattern was interpreted as indicating that if habitat partitioning is the cause of the pattern, its spatial dependence had a strong directional component, probably related to a trend in soil moisture in the area sampled. It would not be surprising if a similar study of the plants in the same area produced results indicating the same kind of patterns. A spatial gradient in an environmental factor such as soil moisture provides an obvious and interesting kind of anisotropy (see Chapter 8).

Landscape metrics

The field of 'landscape ecology' was originally envisioned as dealing with a human-based definition of a landscape, that is systems on a scale of kilometers or more (Forman & Godron 1986), but its concepts can be applied to ecological mosaics of any scale, depending on the focus of the study (Wiens & Milne 1989; Kotliar & Wiens 1990). While the focus of this book is spatial pattern in plant communities, there is some overlap with quantification in landscape ecology, and our discussion of two-dimensional pattern should include some reference to related landscape metrics. We will not give the technical details of the landcape metrics, which can be found elsewhere (see McGarigal & Marks 1995).

Landscape ecology is based on the underlying concept of a mosaic of discernible patches of different kinds. The structure of the landscape is then determined by the characteristics of individual patches such as type, size, and shape, and by spatial relationships among the patches of different types (Turner *et al.* 1991). A number of metrics has been developed to quantify the characteristics of individual patches and of the landscape as a whole. Many of the metrics relate to basic quantities such as the number of patches of a certain kind and the mean and variance of their areas. Other measures are related to the amount of edge between patches in the landscape and the amount of contrast between patches separated by the edges (McGarigal & Marks 1995). One aspect of spatial pattern that is of great importance in landscape ecology, which we have not emphasized in our discussion of spatial pattern, is the shape of the patches. Shape can be measured, in part, by metrics such as perimeter to area ratio or the fractal dimension of the boundary.

Part of the reason that shape is emphasized in landscape ecology is that organisms may be sensitive to the effects of edges between different kinds of patches and may be restricted to what is called the 'core' area of their preferred patch type. How organisms are affected by the ratio of edge to core has important management implications in fragmented landscapes (Turner 1989). Similar concerns about the movement of organisms between patches of their preferred type has led to the introduction of a number of metrics related to the distances to the nearest patches of the same type. The ability of organisms to move between patches and thus recolonize after local extinction will depend, at least in part, on the spatial arrangement of the patches (Turner 1989).

In quantitative landscape ecology, as in spatial pattern analysis, there is a need to evaluate how heterogeneity changes with scale. A recent development in landscape ecology is the application of lacunarity analysis which is a scale-dependent measure of heterogeneity or the 'texture' of an object (Plotnick *et al.* 1993, 1996). The easiest way to explain this concept is to revert to one dimension. Consider a string of 1' and 0's of length n. For each block size b, count the number of 1's in all possible positions of a single block of size b and derive the frequency distribution. Where $\bar{x}(b)$ is the mean of this distribution and $s^2(b)$ is the sample variance, the lacunarity at scale b is (Plotnick *et al.* 1996):

$$\Lambda(b) = 1 + s^2(b)/\bar{x}^2(b). \tag{6.30}$$

This index is one more than the square of the coefficient of variation and higher values indicate clumping. It is therefore also closely related to the variance to mean ratio and to a variety of familiar measures such as Morisita's index (Chapter 3). As with quadrat variance methods, the characteristics of the spatial pattern are based on the shape of the plot of the index $\Lambda(b)$ as a function of b, usually in a log-log form. Fractal patterns produce straight lacunarity plots because the fractals are self-similar at all scales; patterned data produce lacunarity plots with distinct breaks in the slope corresponding to the scales of pattern (Plotnick *et al.* 1996).

The basic method we have outlined can be extended to quantitative data and into more spatial dimensions. Plotnick *et al.* (1996) suggest that one strength of this approach is that it can be applied to real patterns which may not be fractal, by determining scale-dependent changes in spatial structure. They believe the method will have wide applicability in fields concerned with spatial pattern description.

There is clearly overlap between the spatial characteristics studied in landscape ecology and in pattern analysis in plant ecology. Both are concerned with the relationship between pattern and process and with the quantification of that pattern. Both are concerned with the concept and importance of spatial scale (Turner *et al.* 1991). For example, Cullinan and Thomas (1993) evaluate some pattern analysis methods for determining the scale of pattern in a landscape, including TTLQV, spectral analysis, and the semivariogram. They concluded that more than one method should be used, because no single method consistently gave reliable estimates of scale.

On the other hand, spatial pattern analysis and landscape ecology have differences in emphasis, arising in part from the underlying motivation. If the focus is the conservation of birds of the interior of old-growth forests in an exploited landscape, the emphasis on edges, core area and fragment shape is crucial. The motivation to examine those characteristics is less strong in studies of processes in natural vegetation. In the future, however, the questions and methods of the two areas will undoubtably converge further.

Other methods

Other approaches to the analysis of spatial pattern in two dimensions are possible. For instance, we can speculate on how a two-dimensional version of multiscale ordination (MSO) might work but it has not yet been introduced to our knowledge. A two-dimensional equivalent of MSO would calculate a variance–covariance matrix for each block size, as in the one-dimensional case, but would use 4TLQV and 4TLQC or the nine-term equivalents. These matrices would be summed over block sizes, eigenanalyzed with the resulting variance repartitioned by block size. As with the methods that are its basis, this version of MSO does not deal with anisotropy.

Another approach to examining the scale of multispecies pattern in two dimensions is illustrated in Legendre and Fortin (1989) based on Oden and Sokal (1986) and Sokal (1986). A matrix of ecological multivariate dissimilarities among the samples is calculated, call it **X**. A series of matrices, **Y**, is calculated, one for each distance class, consisting of 1's for pairs of samples in that distance class and 0's elsewhere. A normalized Mantel (1967) statistic is calculated for **X** with each **Y** and the value of the statistic is tested and plotted against distance in the usual way. In the Québec forest example described above, Legendre and Fortin found that the statistic reached a maximum at distance class two and declined thereafter, indicating spatial pattern at a scale of 50 m (see our discussion of the concept of scale in multispecies pattern in the introduction to Chapter 5).

A two-dimensional version of new local variance (NLV) would average the absolute values of adjacent 4TLQV or 9TLQV terms. The purpose would be to detect the size of the smaller phase. The details of this approach would require careful consideration before the results could be reliably interpreted.

We can also imagine a revised version of the RPQF method described earlier: rather than focusing our attention on the pairs of cells that both were dominated by species A (same species), we could count and analyze the number of pairs of cells in which one belonged to species A and one belonged to species B. The rest of the analysis would proceed as described. If each quadrat could contain many species, the quadrat pairs could be classified as similar or dissimilar in composition and then counted and analyzed in the usual way. There are many variants of this technique possible.

An interesting and very promising approach is to include spatial information in the analysis of ecological data in order to partial out the variation that can be attributed to spatial structure (Borcard *et al.* 1992). The underlying concept is that the observed variation is caused by environmental factors and by spatial structure, but there is also variation that can be explained by neither. It is important to recognize that environmental variance and spatial variance overlap, and therefore the explained variation is divided into three parts: nonspatial environmental variation, variation due to spatial structure alone, and variation due to spatial structuring shared by the environmental data. The procedure is to use a canonical ordination technique like canonical correspondence analysis (CCA, ter Braak 1986, 1987), which ordinates species and environmental data together, including the spatial information as one kind of environmental data. The spatial data are included not just as simple (x,y) coordinates, but with sufficient polynomial terms for a cubic trend surface regression:

$$z = b_1x + b_2y + b_3x^2 + b_4xy + b_5y^2 + b_6x^3 + b_7x^2y + b_8xy^2 + b_9y^3. \qquad (6.30)$$

This ensures that more complex spatial structures such as patches and gaps can be accounted for (Borcard *et al.* 1992).

Among the several data sets used to illustrate the technique, the authors included the data previously analyzed by Legendre and Fortin (1989). The data are from 200 regularly spaced quadrats and include species abundance of 28 species of tree (12 retained in this example), 6 geomorphological variables and the spatial locations of the samples. They found that 18% of the total variation was accounted for by space, 11% by the environmental variables, and 8% by a combination of space and environment, leaving 63% unexplained (Borcard *et al.* 1992). The high proportion of unexplained variation may be related to the environmental factors chosen in the study. Almost 50% of the explained variation was due to spatial structure.

Økland and Eilertsen (1994) applied the same analysis to understorey data from coniferous forest patches in southeastern Norway. They subdivided the data in two ways: by forest type, pine or spruce dominated, and by plant type, vascular or cryptogam. In the four subgroups, the average amount of the total variation explained was about 10% space, 20% environment, 10% space and environment, and 60% unexplained. Although the proportion of the variation explained is similar in this study to that in Borcard *et al.* (1992), here about 50% of the explained variation is due to environmental factors rather than to space. Økland and Eilertsen (1994) interpret this result as supporting the view that topography is an important factor in differentiating between forest vegetation in geologically homogeneous areas.

The method developed by Borcard *et al.* (1992) is a modification of established ordination methods to include spatial information. In a similar way, classification methods can be modified to include spatial information by allowing the clustering of groups only if they are spatially contiguous (Legendre & Fortin 1989). The approach of including spatial information in ordination or classification analysis is an area of spatial pattern analysis where there are more interesting techniques to be explored and facts to be discovered.

Concluding remarks

Most of the vegetation that we discussed exists in at least three dimensions, but we often chose to analyze it in fewer dimensions, chiefly for practical reasons. Advances in technology, particularly in the area of computing, have made the task of analysis so much easier over the last few decades; it will be interesting to see whether technology can be equally helpful in the collection of data for the study of two-dimensional spatial pattern.

Recommendations

1. Fully mapped data of some kind probably provides the greatest flexibility for different kinds of analysis. Spaced samples can also

provide valuable insights, as can be seen from some of the examples discussed in this chapter.

2. Measures of spatial autocorrelation are complementary to the two-dimensional version of PQV analysis. As in one dimension, resonance peaks occur with PQV.
3. Two-dimensional spectral analysis permits the evaluation of anisotropy, but the anisotropy must be stationary.
4. The two-dimensional versions of TTLQV and 3TLQV seem to be reliable methods, but they do not detect anisotropy. In converting mapped data to grids of quadrats for this kind of analysis, more than one grid orientation should be used.
5. The frequency of like or unlike species in random pairs of quadrats gives a good visual portrayal of anisotropy in the two-dimensional pattern.
6. The two-dimensional quadrat variance methods can be converted into covariance methods to look at the scales of segregation and aggregation of two kinds of plants. These methods have the same strengths and weaknesses as those from which they were derived. The cross-variogram can also be used for two species.
7. In spite of differences in history and motivation, the methods of pattern analysis and quantitative landscape ecology have many similarities.
8. Many kinds of multivariate analysis of ecological data, such as ordination and classification, may benefit from including the spatial relationships of the samples in the analysis.

7 · *Point patterns*

Introduction

In Chapter 1, one of the first topics introduced was the distinction between treating the information on the spatial arrangement of plants as dimensionless points in a plane or as a mosaic of patches filling the plane. In this chapter, we will examine a number of methods that evaluate certain properties related to spatial pattern using the positions of individual plants in a plane. Several reviews of the analysis of spatial point patterns are available (Diggle 1983; Upton & Fingleton 1985; Cressie 1991), and it is not the intention to repeat a great deal of the material covered in those books. Instead, those methods that parallel the approaches described elsewhere in this book, but using points rather than density or presence, will be emphasized. In general, the kind of data that will be used here is mapped plant positions within a defined study area or plot. Considerations of the shape of the study plot to be used are discussed in Chapter 2.

There are several considerations to be included in our examination and evaluation of methods based on the positions of individual plants. The first is that for an investigation of spatial pattern, techniques that merely distinguish among the three possibilities of clumped, more-or-less random, and overdispersed are not really of interest for the purposes of this book. We want to get more out the analysis; for example, if the plants are overdispersed, what is their spacing, how uniform is the spacing, is the spacing the same between plants of different kinds? If the plants are clumped, what size are the clumps, what size are the gaps between them? If there are two or more kinds of plants, are the different kinds segregated or are they aggregated? How does the segregation or aggregation relate to the overall pattern of the plants?

The second consideration is that it is desirable for methods to deal equally well with clumps of points in an otherwise empty plane and with definite gaps or holes in a plane that is otherwise densely populated (that is a 'positive-negative' pair of spatial patterns as in Figure 7.1). Because of the way that the quadrat variance methods are formulated in terms of density differences (Chapter 3), spatial patterns in which patches and gaps are reversed give similar results (cf. Pielou 1977a). In analyzing point patterns, it will not always be true that positive-negative pairs of patterns like those in Figure 7.1 give the same result. It would be best to find a method that gives the same evaluation of the scale of spatial pattern in both the positive and negative versions, but that also provides a way of distinguishing between them.

Univariate point patterns

A large number of methods have been described that can be used to quantify the characteristics of spatial point patterns such as the mapped positions of plants (cf. Diggle 1983). Many of them concentrate on determining whether the plants are clumped or overdispersed as opposed to being randomly arranged in the plane. As we pointed out in Chapter 1, spatial dispersion is scale dependent: the same set of points can appear overdispersed at one scale and clumped at another (cf. Figure 1.10). In this section, one focus will be the investigation of the scale of spatial pattern: given that the plants are clumped, we want to determine the size of the clumps and their spacing.

The size of the clumps combines with the overall density of plants to determine the local density of plants. It is the local density that is of ecological importance, since, if all else is equal, plants with the highest local density of neighbors grow more slowly and experience the highest rates of mortality (Mithen *et al.* 1984; Silander & Pacala 1985, 1990).

Neighbor distance methods

The literature on the analysis of point patterns includes a large number of methods based on the distance of each plant to its nearest neighbor. One of the most famous is the Clark and Evans (1954) test that distinguishes random dispersion from clumped or overdispersed based on nearest-neighbor distances. From the point of view of analyzing spatial pattern, the draw-back of many of these methods is that while they can distinguish the kind of dispersion, having found clumping, for instance, they give no information on the size or spacing of the clumps. In other words,

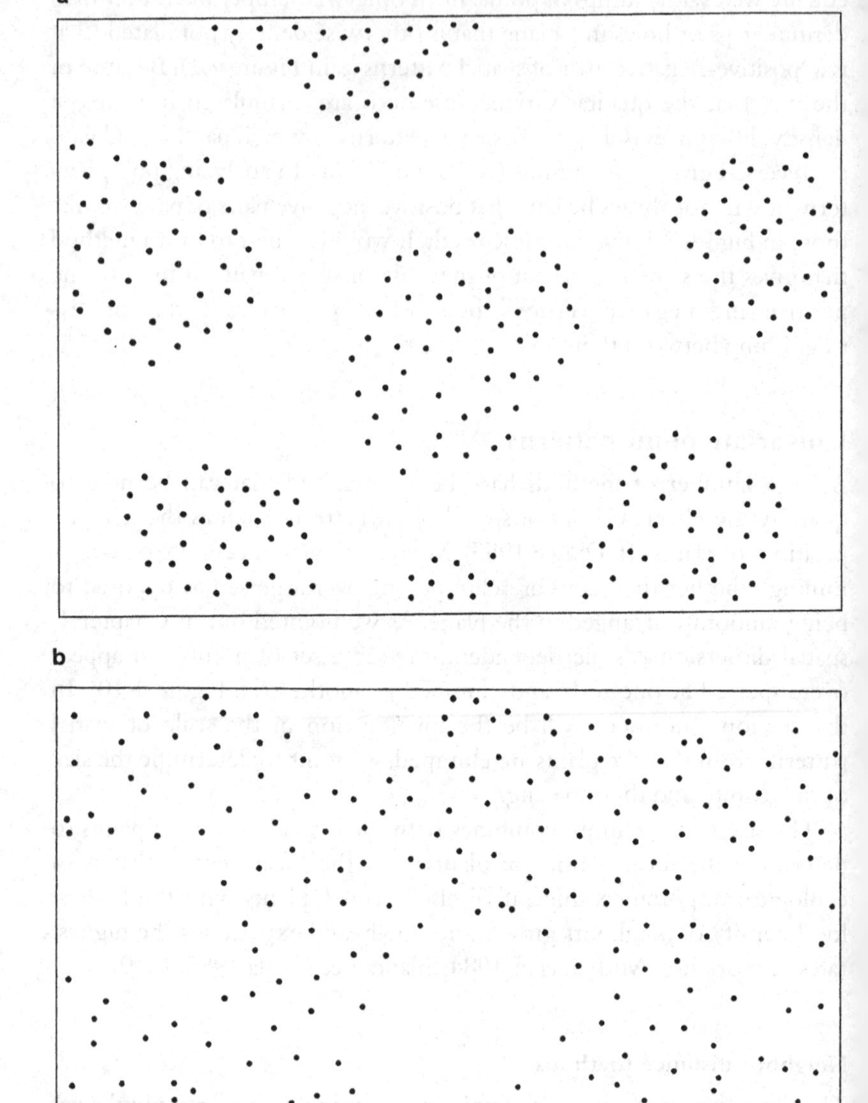

Figure 7.1 Spatial pattern that consists of: *a* clumps of points in an otherwise empty plane, or *b* a plane that is otherwise densely populated by plants except for definite gaps or holes.

they are related to the intensity of pattern rather than to the scale (Pielou 1977a). It is easy to see why this is so by considering that very different spatial patterns can have the same distribution of nearest neighbor distances, for example when all nearest neighbor distances are the same (Figure 7.2). For this reason, the usefulness of nearest neighbor distances for spatial pattern analysis is very limited.

If the distances from plants to their single nearest neighbor cannot be used to analyze or characterize spatial point pattern fully, an obvious extension is to look at the first and second nearest neighbors or first, second, and third nearest neighbors (cf. Thompson 1956). Clearly, the characteristics of the pattern will be captured more fully as more neighbors are used.

Plant-to-all-plants distances

The most extreme extension of nearest neighbor analysis uses the distances between all possible pairs of plants and the method is therefore called plant-to-all-plants distance analysis (Galiano 1982b). It looks at the frequency distribution of the distances between all pairs of plants in the plot. Where t is one of a given range of distances (e.g. 0–1 cm, 1–2 cm, 2–3 cm, . . .), let $f(t)$ be the number of pairs of plants, i and j, for which the distance between them, d_{ij}, is in the range of t. $f(t)$ is then plotted as a function of t and the plot is interpreted: large increases in $f(t)$ indicate overdispersion at scale t and decreases indicate clumping. For example, if the plants occur in clumps of diameter d separated by gaps also of size d, then there will be an excess of distances smaller than d, due to distances between plants within clumps. There will also be a deficit of distances of length d and greater because, for most plants, points at distance d will be outside their own clump. There will also be an excess of interplant distances just under $3d$ due to pairs of plants in adjacent clumps (see Figure 7.3; cf. Galiano 1982b).

Based on the null hypothesis of complete spatial randomness, the expected value of $f(t)$ increases linearly with t. Galiano (1982b) suggests converting the frequencies to 'conditioned probabilities' by calculating the number of distances within each distance class per unit area. Another modification suggested by Galiano (1982b) is that to avoid edge effects, the distances to all other plants should be calculated only for those plants in the center of the sample plot, specifically those further from the edge than the maximum distance examined. Unfortunately, if the plot is circular, and the maximum distance examined is 2/3 of the plot's radius, the result is that less than half the plants are fully used.

a

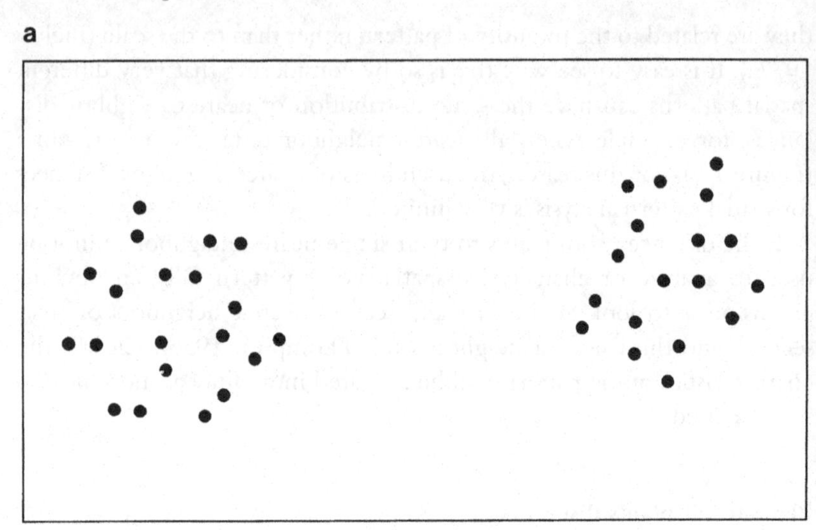

b

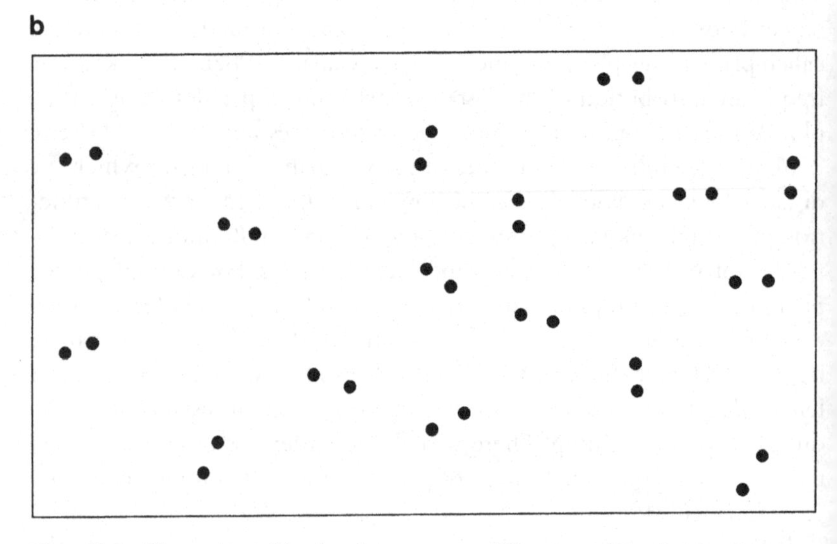

Figure 7.2 Illustration of the fact that two very different spatial point patterns can have the same distribution of nearest neighbor distances. Here all nearest neighbor distances are the same because the plants occur in pairs.

Figure 7.4 shows the arrangement of plants in clusters and Figure 7.5 shows the plant-to-all-plants analysis, $f(t)$ as a function of t when all plants are used. The method does not give as clear a result for the 'negative' of that pattern, where there are distinct gaps in a general backgound of points; as Figure 7.6 shows, the results are not readily distinguishable from those for points placed at random. Figures 7.5 and 7.6 show the raw fre-

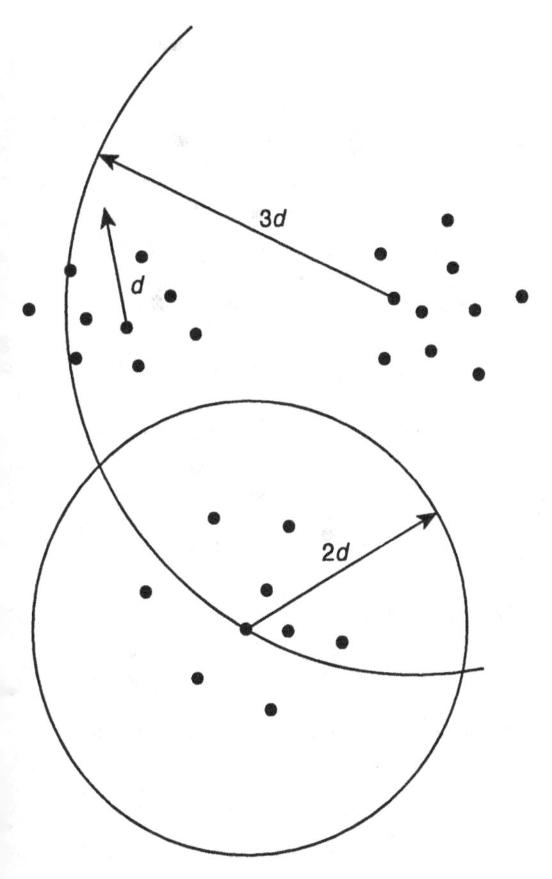

Figure 7.3 When the plants are in clumps of diameter *d* separated by gaps of size *d*, there are many interplant distances less than *d*. There are very few of about 2*d* because that distance goes beyond the plant's patch and more again at distance 3*d*, because circles of that size can reach adjacent patches.

quencies as a function of *t* with no edge correction, but the results are similar when only the plants in the center of the plot are used, following Galiano's suggestions. In either case, the results do not give a clear picture of the characteristics of the spatial pattern.

Second-order analysis

Although described as an extension of nearest neighbor methods, the plant-to-all-plants analysis is closely related to another approach referred to as a second-order statistic, introduced earlier by Ripley (1976, 1977). It is one of the more commonly used methods for studying the spatial

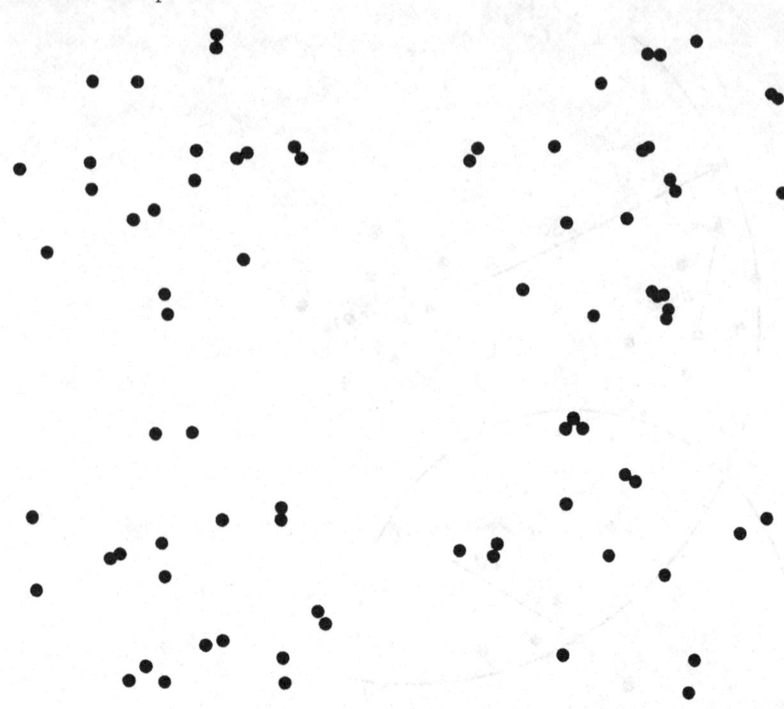

Figure 7.4 An example of plants in clusters for which the plant-to-all-plants analysis is illustrated in Figure 7.5 (artificial data).

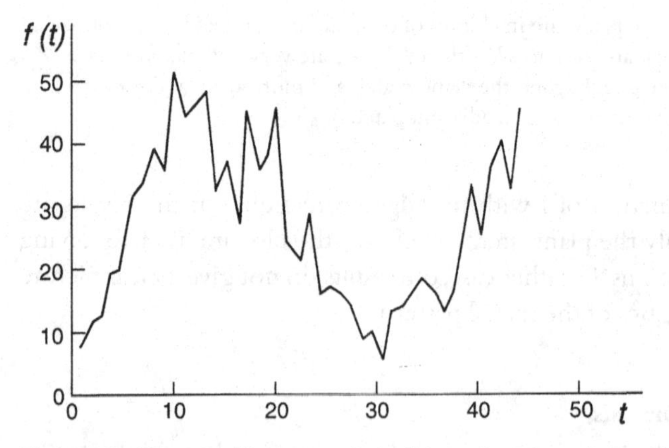

Figure 7.5 Plant-to-all-plant analysis: $f(t)$ as a function of t for all plants when the plants are in clusters (Figure 7.4).

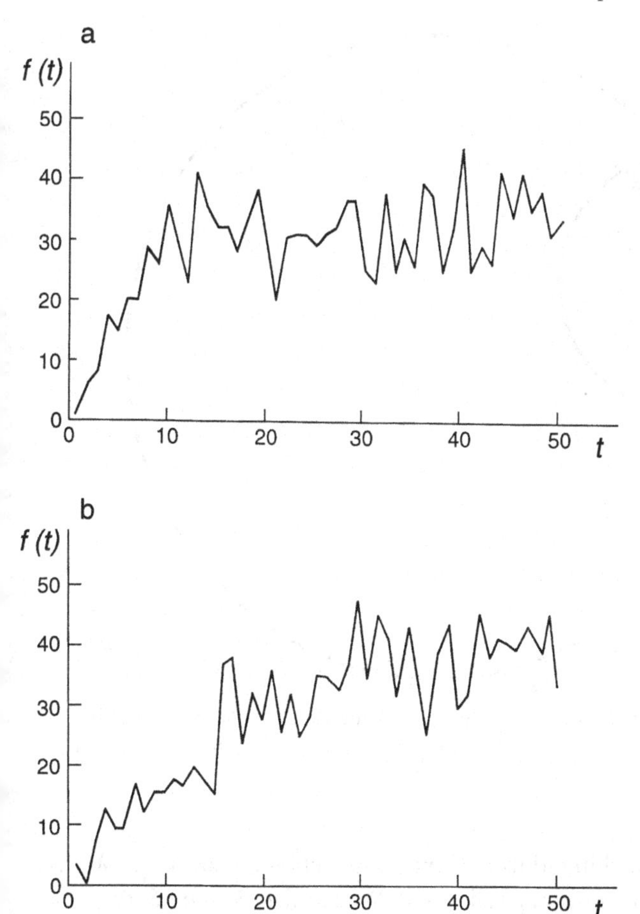

Figure 7.6 (*a*) $f(t)$ as a function of t when the points are placed randomly except for four gaps or holes. It is not readily distinguishable from the result when the plants are completely random, (*b*). All plants were used.

pattern of mapped points and one of the better ones available (Andersen 1992). The method is also based on the distances between pairs of points, because it counts the number of points within a certain distance, t, of each point, with t taking a range of values. The process is essentially the same as counting the number of points in circles of radius t centered on the n points. It can be considered to be an examination of the cumulative frequency distribution of the plant-to-all-plants technique.

Formally, let d_{ij} be the distance between points i and j and let I_t take the value 1 if the distance between i and j is less than t, and the value 0 otherwise. Where λ is the density of plants per unit area, the expected number

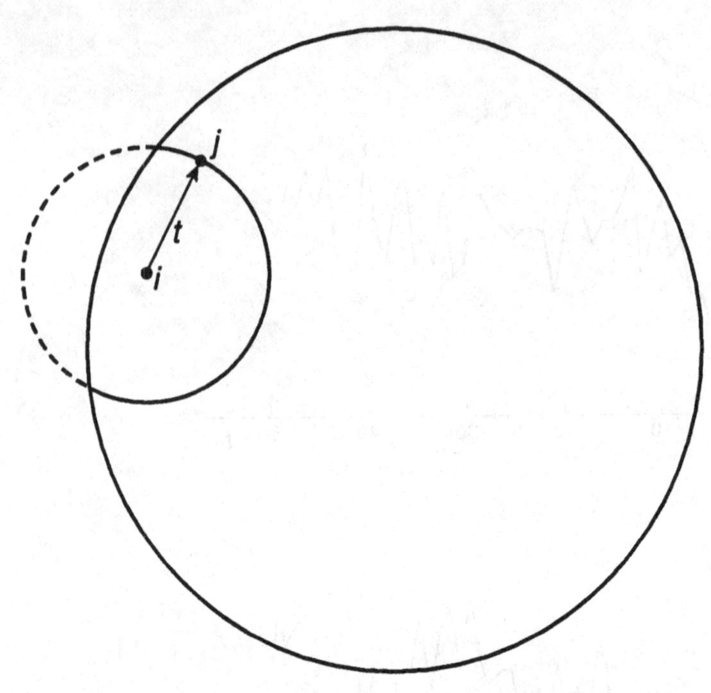

Figure 7.7 Edge correction technique: the reciprocal of the proportion of the circle centered on point i going through point j that is within the sample plot is added to the frequency count. Here $w_{ij} = 1.77$.

of other plants within radius t of a randomly chosen plant is just λ multiplied by some function of t, call it $K(t)$. We estimate $K(t)$ by $\hat{K}(t)$:

$$\hat{K}(t) = A \sum_{i}^{n} \sum_{\neq j}^{n} w_{ij} I_t(i,j)/n^2, \tag{7.1}$$

where A is the area of the plot, and w_{ij} is a weighting factor used to reduce the problem of edge effects. If the circle centered on i with radius t lies totally within the study plot then $w_{ij} = 1$, otherwise it is the reciprocal of the proportion of that circle's circumference that lies within the plot (Figure 7.7). The purpose of the weighting factor is to remove edge effects, where large circles centered on points near the edge contain fewer points than expected merely because much of the circle being considered is outside the area studied. For further details see Ripley (1977), Diggle (1983) or Upton and Fingleton (1985, p. 88). Ripley (1988) provides a detailed discussion of edge correction procedures and Haase (1995) evaluates several possibilities.

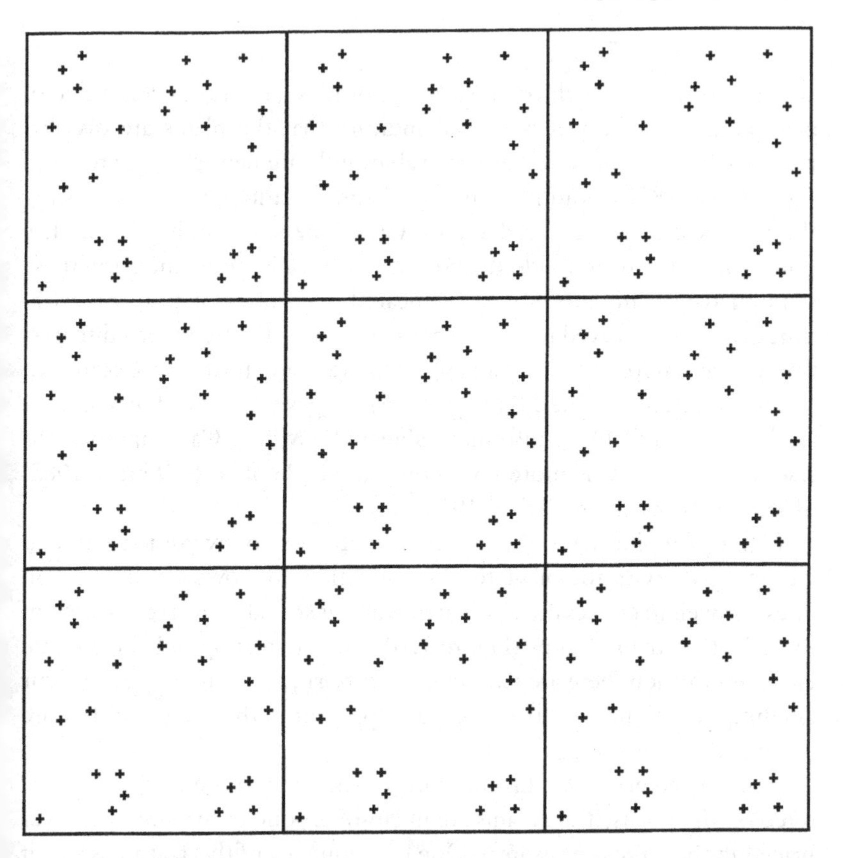

Figure 7.8 Toroidal edge correction: eight copies of the original data (central square) are placed around it. Analysis then proceeds using circles centered on the plants in the middle square only. Plants in the outer squares are only included in counting.

In addition to the weighting method described above, there are two other straightforward ways to reduce edge effects. The first is to map the plants in a buffer area around the region of interest. The second is toroidal edge correction where the map is considered to be wrapped around so that its north and south edges join and its east and west edges join. Essentially the map is used as its own buffer area (Figure 7.8). Mapping a buffer area is the most reliable method, but requires more work; the toroidal correction is not recommended because it can give biassed results for nonrandom patterns (Haase 1995).

Since $K(t) = \pi t^2$ if the plants are randomly arranged in a Poisson forest, we plot:

$$\hat{L}(t) = t - \sqrt{\hat{K}(t)/\pi} \tag{7.2}$$

as a function of t, which on the null hypothesis has an expected value of zero. Large positive values of $\hat{L}(t)$ indicate that the plants are overdispersed at scale t and large negative values indicate clumping. Upton and Fingleton (1985) recommend using a Monte Carlo approach to testing the significance of observed values. Given the study unit of area A, n points are placed at random and the analysis is performed on these random data. This procedure is repeated a number of times, say 100 times, and the observed results are compared with the frequency distribution of the random trials. As an approximate guide to the significance of the most extreme values, Ripley (1978) suggests $1.42\sqrt{A}/n$ and $1.68\sqrt{A}/n$ as 5% and 1% significance values. The Monte Carlo approach to assessing the results is more commonly used (Prentice & Werger 1985; Skarpe 1991, Zhang & Skarpe 1995).

Figure 7.9 illustrates the sensitivity of $\hat{L}(t)$ to overdispersion and clumping. As with the plant-to-all-plants method, however, the method does less well in conveying information about spatial pattern consisting of 'holes' rather than clumps (Figure 7.10a); it is detecting only large-scale clumping. When there are two scales of pattern, as when the plants occur as clumps of clumps, the two scales are apparent in the analysis if they are sufficiently distinct (Figure 7.10b).

One characteristic of this method is that, because the circles are centered on the plants, if the plants are in clusters, none of the circles are ever placed in the large empty areas. One consequence of this feature is that if the plants occur in regular strips of width d separated by gaps of width d, when $t = d/2$ even the most empty circles will have some plants in them (Figure 7.11). When $t = d$, most circles will be half in the patch and half in the gap, thus containing a number of plants close to the expected value, the overall density. The scale of the spatial pattern is d, and $\hat{L}(d)$ is about 0 (see Figure 7.11).

Getis and Franklin (1987) added an interesting refinement to the second-order analysis. They were studying the spatial pattern of *Pinus ponderosa* in the Klamath National Forest in California. In their paper, they present data from a 120 m × 120 m subsample, containing 108 trees apparently in clusters. Using a slightly different version of the statistic, they calculated, for each point i:

$$M_i(t) = \left(A \sum_{j}^{n} w_{ij} I_t(i,j) / \pi(n-1) \right)^2. \tag{7.3}$$

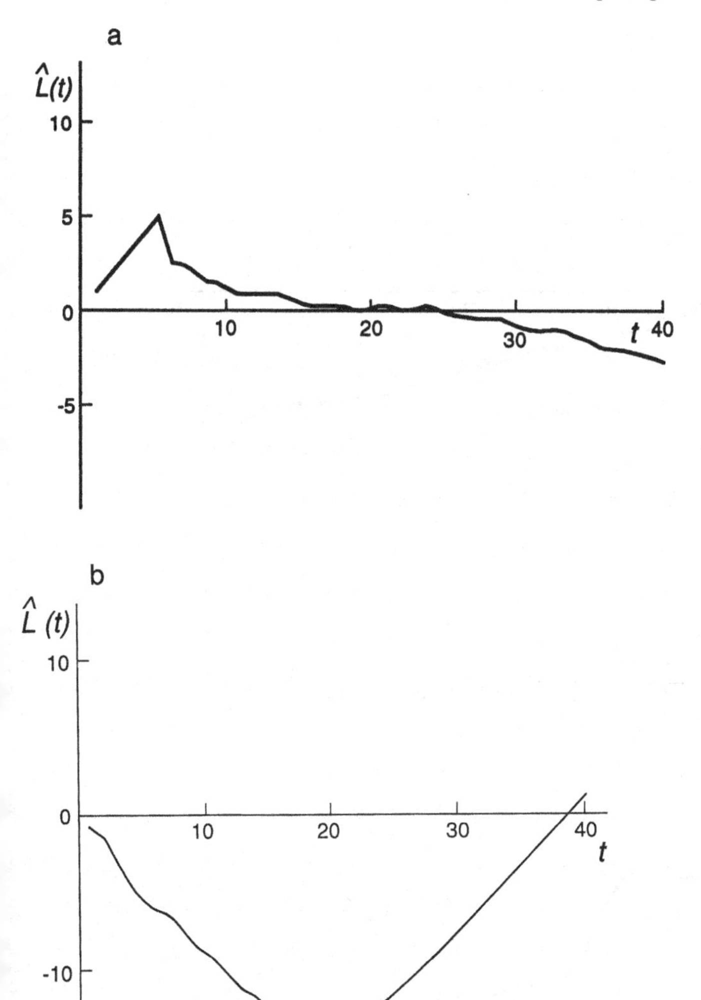

Figure 7.9 Second-order analysis: $\hat{L}(t)$ as a function of *t* when the plants are *a* overdispersed, with hard core inhibition of 5 units in which no plants are permitted within 5 units of any other and *b* clumped with clumps of radius 20.

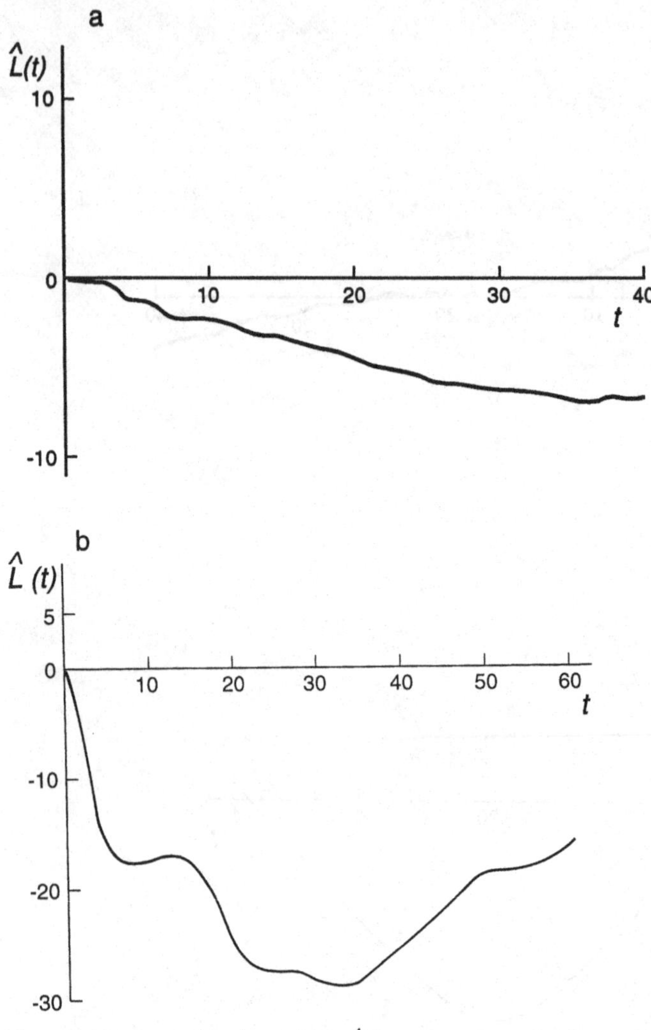

Figure 7.10 Second-order analysis: $\hat{L}(t)$ as a function of *t* when *a* the pattern consists of holes radius 30 units in a plane of otherwise randomly placed points and *b* when the points are arranged in clumps of clumps.

They then plotted contour maps based on the values of $M_i(t)$ for the individual trees, using different values of *t*. The accuracy of the contours was enhanced by calculating $M_p(t)$ for control points, *p*, in the areas where there were few trees. They then examined the contour diagrams for clusters, defined as areas in which *M* was greater than the Poisson expectation. The clusters that were identified in this way changed with the value of *t*, indicating that the spatial heterogeneity that is perceived depends on the scale of the analysis.

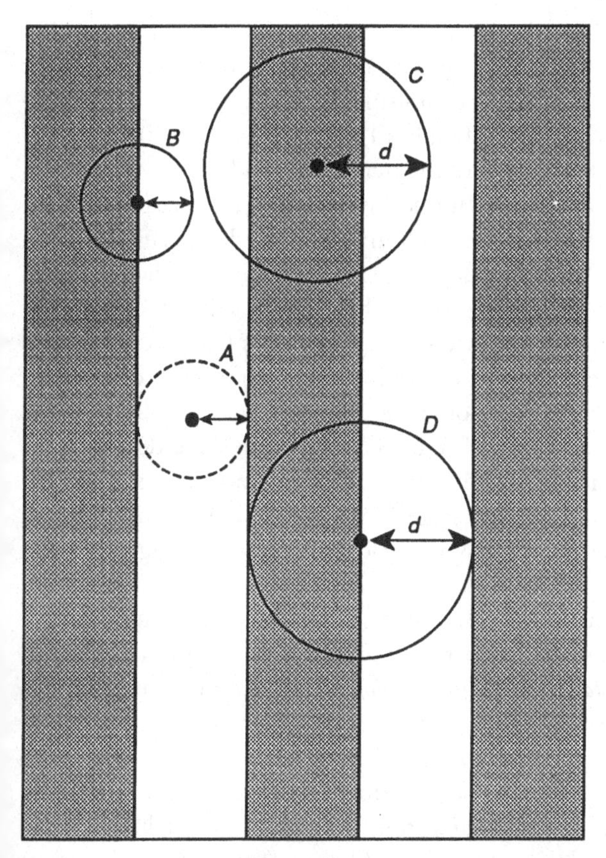

Figure 7.11 Using second-order analysis, circles centered on plants can never occur in completely empty areas; there is no plant to give circle A. Here, the plants are in strips of width d; when $t = d/2$, even relatively empty circles like B will contain some plants. When $t = d$, the majority of circles will be half in the strip and half in the gap, like C and D, thus containing a density of plants close to the overall density.

Another characteristic of this method is that because it looks at the number of plants contained in a circle of a particular radius, it does not use the information of the directions of the plants from the center. That is, it treats patterns as isotropic, whether they are or not.

The second-order analysis is an improvement over first-order methods but, as Andersen (1992) points out, a complete description of the spatial structure would require functions analogous to $K(t)$ describing the expected numbers for specific configurations of three events, four events, and so on. Such an approach would quickly become both difficult to do and difficult to interpret.

Kenkel (1988a) used the second-order analysis to investigate self-thinning in jack pine, *Pinus banksiana*, growing on a sandy plain near Elk Lake, Ontario. Self-thinning is density-dependent mortality in a plant population and it is usually accompanied by changes in the spatial pattern of the surviving plants (Hughes 1988). Kenkel (1988a) wanted to test whether the spatial pattern of the trees, living and dead, was compatible with random mortality. He found that when all trees were analyzed, their arrangement was locally random (distances less than 4 m) and clumped at intermediate distances (4–19 m). Living trees had a spatial pattern that was significantly regular, with an obvious dearth of distances between live trees in the range less than 3 m. In contrast, the dead trees were significantly clumped compared to the hypothesis of random mortality and had an excess of observed distances between dead trees at distances between 2.5 m and 17 m. He concluded that the development of strongly regular pattern in the surviving trees was the result of two phases of competition, an early stage of symmetric competition for soil resources and a later stage of asymmetric competition for light (Kenkel 1988a).

Petersen and Squiers (1995) present very different results from a similar study of the spatial pattern of tree mortality, in this case of aspen (*Populus grandidentata* and *Populus tremuloides*), growing in a mixed forest in northern Michigan. Using the same second-order method, they found that the living aspen stems in 1979 were significantly clumped at 12–16 m and that those still alive in 1989 were not just clumped, but more clumped than would be predicted from the random mortality of those alive in 1979. The scale of aggregation of living stems in 1989 (14–18 m) was comparable to that in 1979. They attribute the difference between their findings and the theoretical predictions of increasing dispersion and the contrast with Kenkel's (1988a) results to the clonal nature of aspen's growth (Petersen & Squiers 1995).

Tessellations

In the preceding sections, we have concentrated on the kinds of questions we might ask when the plants occur in clumps: how big are the patches and how far apart are they, or how big are the gaps? In this section, we will turn our attention, at least in part, to the kinds of questions we might ask having found that the plants are overdispersed: how far apart are the plants? From the earlier discussion of plant-to-plant distance techniques, it is probably clear that to answer the question of interest, it is not sufficient to examine only the distances to the first nearest neighbors. On the other hand, we probably do not need to examine the distances of each plant to all other plants, because far away plants will have little effect on growth or survival. Instead, it makes sense to think about all the plants that are somehow immediate or primary neighbors, if only because those are the plants with which interactions are probably most intense (Czárán & Bartha 1992). We need, therefore, a method to determine which plants are primary neighbors.

Models of plant competition have been based on the concept of the Dirichlet domain introduced in Chapter 1. The Dirichlet domain associated with a plant in the plane is the region of the plane that is closer to that plant than to any others. The idea is that the resources (soil nutrients, water, light) that are associated with that region of the plane are available to the closest plant before or more than they are available to others. Mithen *et al.* (1984) have shown that there is a some association between the size of a plant's Dirichlet domain and its subsequent success, although the size of the plant itself is critical. The Dirichlet domain model also provides a simple definition of which plants are primary neighbors and therefore which distances to other plants are important. Plant i and plant j are neighbors if and only if their Dirichlet domains share a boundary (*cf.* Figure 7.12). If lines are drawn joining the plants that are neighbors according to this definition, the result is the Delaunay triangulation of the plane, which technically is the dual of the Dirichlet tessellation. We can use the frequency distribution of line lengths in the Delaunay triangulation to answer some questions about the spatial pattern of the plants. For example, if the plants are overdispersed, the frequency distribution would be unimodal with lower variance indicating more regular spacing of the plants. On the other hand, if the plants occur in simple clumps, the frequency distribution will be bimodal, including short distances between neighbors in the same clump and longer distances between neighbors in adjacent clumps.

Another model that produces a tessellation that is very similar to the Delaunay tessellation starts with the distances between plants ordered from least to greatest. The two plants that are closest are joined first and then the lines joining points are added to the tessellation in order from the smallest, subject to the condition that no line can be added to the tessellation if it crosses a pre-existing line (Figure 7.13). This tessellation is referred to as the Least Diagonal Neighbor Triangulation (LDNT, *cf.* Fraser & Van den Driessche 1972).

Having joined all the plants in pairs to create a tessellation, we can examine the distribution of distances from plants to their neighbors.

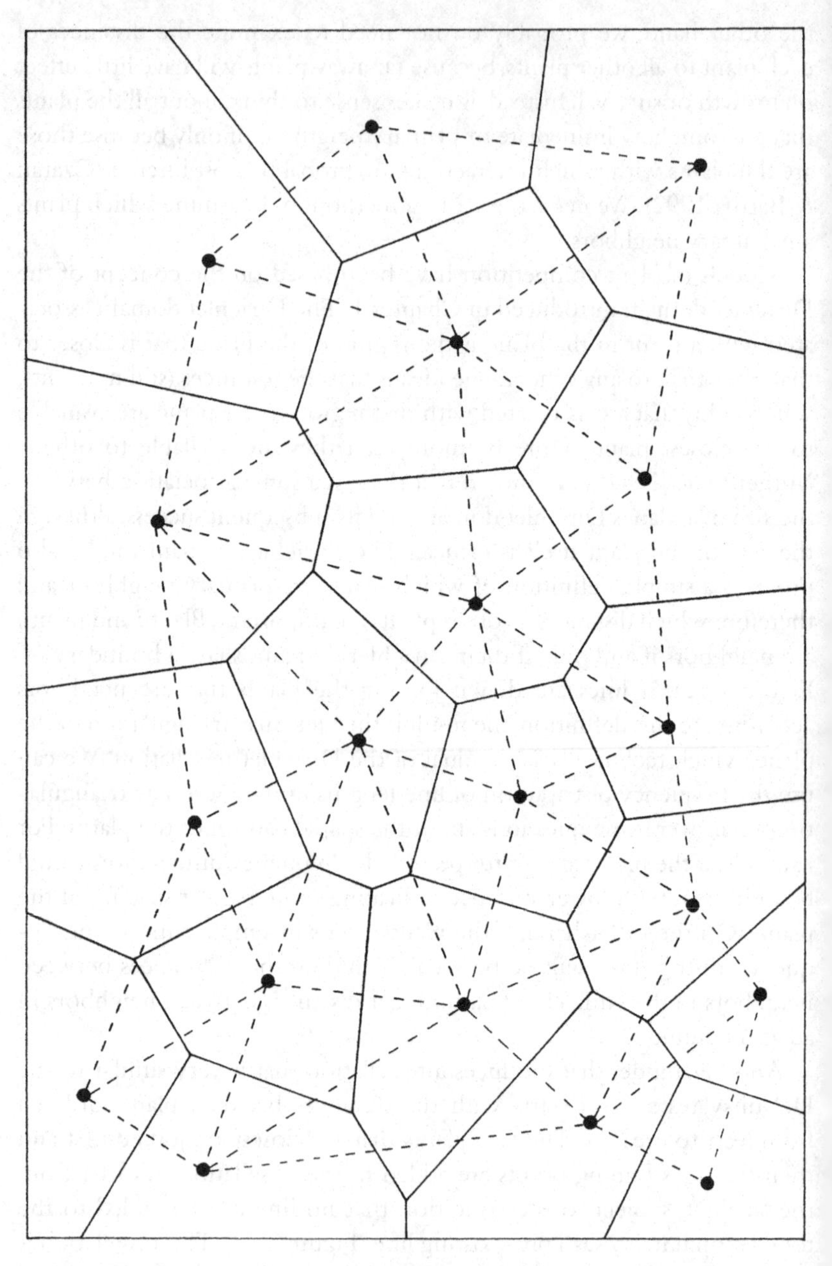

Figure 7.12 Plants i and j are neighbors where their Dirichlet domains share a boundary (solid line). The broken lines join plants that are neighbors by this definition; they form a Delaunay tessellation.

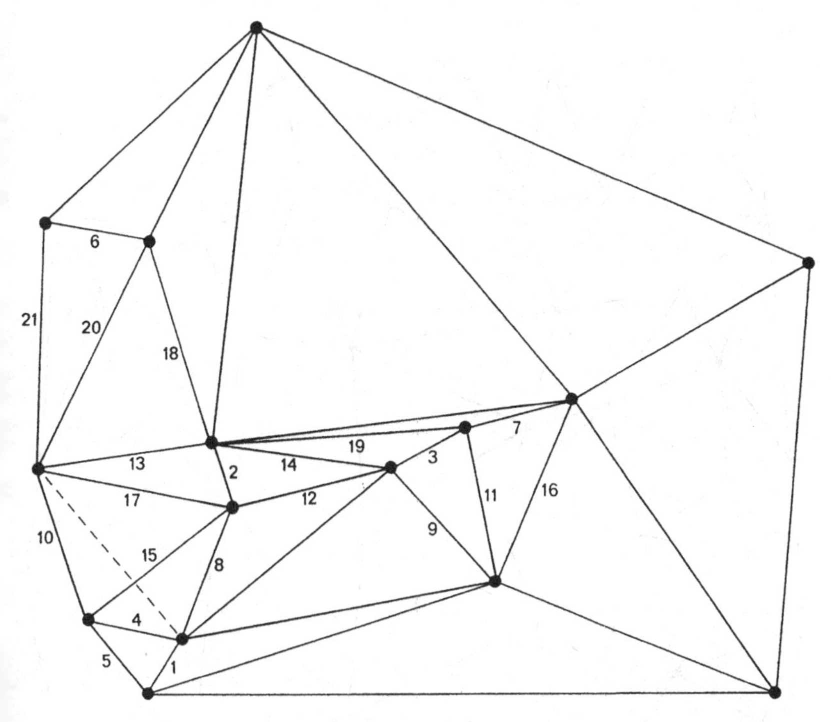

Figure 7.13 The least diagonal neighbor (LDN) tessellation joins plants in pairs based on their distance. The two closest plants are joined first; the lines joining points are then added in order from shortest to longest, provided that they do not cross any pre-existing lines. The first 21 lines are labelled in order. The dotted line is shorter than line 20 but it is not used because it crosses line 15.

Figure 7.14 shows the LDNT for points that are overdispersed by a hard-core inhibition process. (A hard-core inhibition process does not just decrease the probability of finding another point within a certain radius of an existing point; it makes the probability zero.) The associated frequency distribution of edge length reveals the underlying structure clearly. Figure 7.15 shows the frequency distribution for a clumped and for a random arrangement of points in the plane. The increase in the variance of edge length is apparent as the dispersion becomes more clumped.

Quadrat counts

In the introductory comments of this chapter, we said that we were not interested in techniques that merely distinguish among the possibilities of

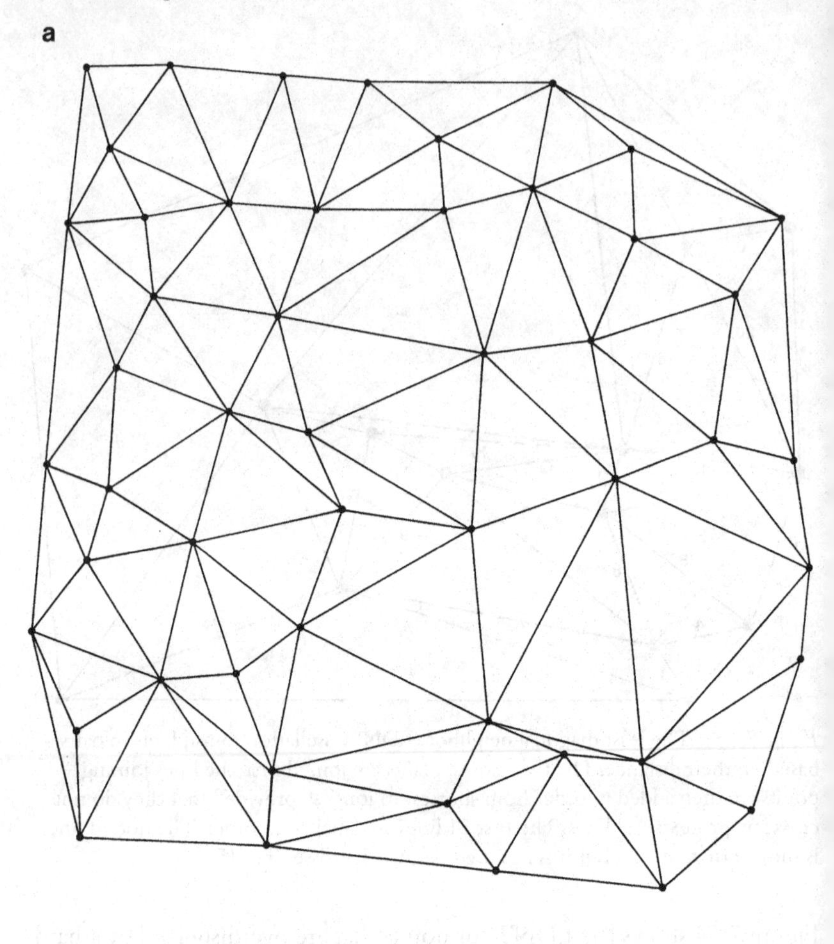

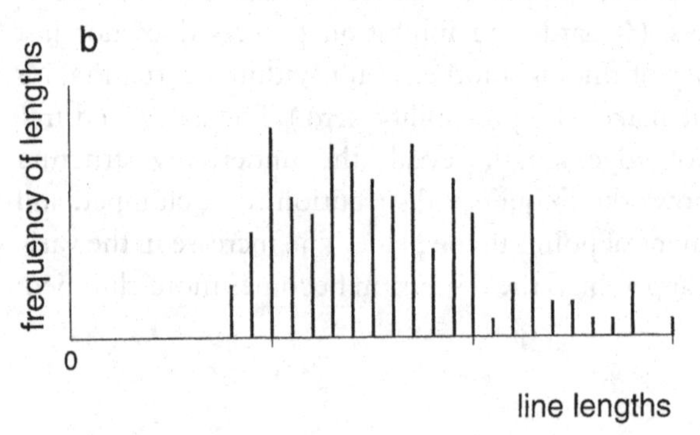

Figure 7.14 The LDN triangulation for points that are overdispersed in a hard core inhibition process (*a*); the underlying structure is revealed in the associated frequency distribution of edge length (*b*).

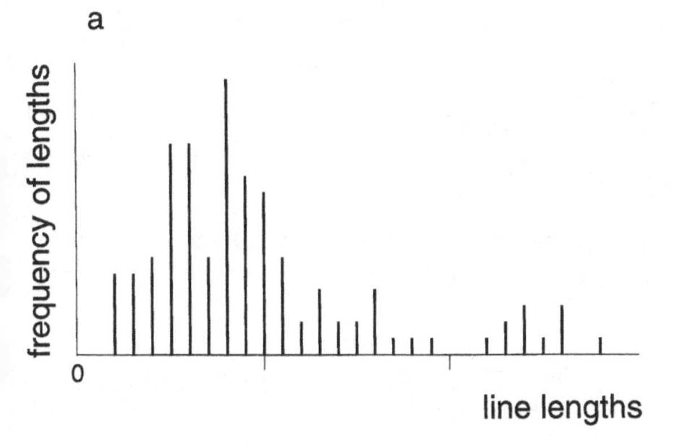

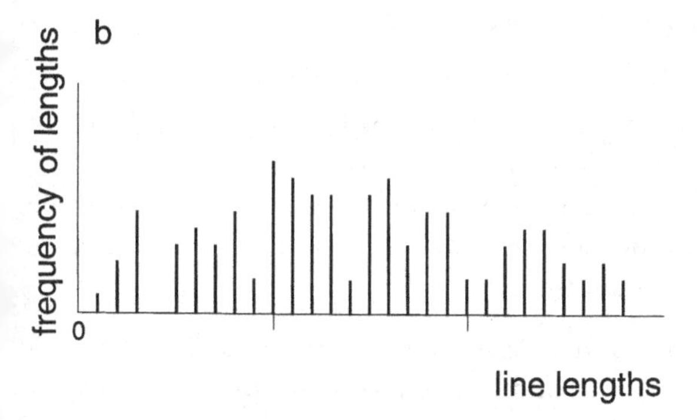

Figure 7.15 The frequency distribution of LDNT edge lengths for the clumped and random arrangements of points in a plane (*a*); the increase in the variance of edge length is apparent (*b*).

the points being clumped, random, or overdispersed. That is true, but the chapter would not be complete without a brief discussion of the commonly used technique of investigating dispersion using counts of plants in quadrats.

The basic technique is to count the plants in a set of n quadrats, letting x_i be the number of plants in the ith quadrat. The mean and sample variance are calculated and their ratio is often referred to as an index of dispersion:

$$I_d = \sum_{i=1}^{n} (x_i - \bar{x})^2 / \bar{x}. \tag{7.4}$$

This index is interpreted as indicating clumping if it is greater than 1, overdispersion if it is less than 1, and random dispersion if it is close to 1.

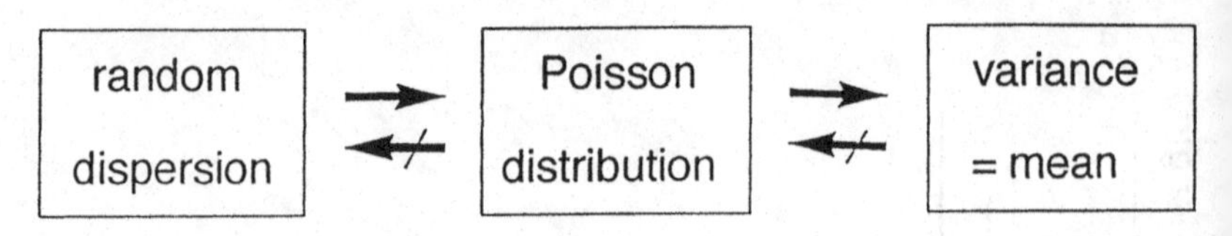

Figure 7.16 The logic of the variance:mean ratio. If the plants are randomly dispersed, then the frequency distribution in quadrats is Poisson. If the distribution is Poisson then the variance and the mean are the same. The reverse inferences are not valid.

Many texts suggest testing the significance of this index by comparing $(n-1)I_d$ with the χ^2_{n-1} distribution (see Ludwig & Reynolds 1988). The implication is that if the test result is not significant, then the plants are more-or-less random in their dispersion.

The reasoning behind the use of this method is as follows:

1. If the plants are randomly arranged then the frequency distribution of plants per quadrat will follow a Poisson distribution.
2. Since the variance and mean are equal in a Poisson distribution, if the frequency distribution is approximately Poisson the sample variance and mean should be about the same.

Both of these statements are true and if the plants are random then the index of dispersion will be around 1. The reverse reasoning, however, is not sound. Having the variance equal to the mean does not guarantee a Poisson distribution of frequencies and a Poisson distribution of frequencies does not guarantee spatial randomness (Figure 7.16).

As Hurlbert (1990) points out, there are many frequency distributions that have the variance equal to the mean; he calls this class of distribution 'unicornian' because of the well-known but surprising fact that the dispersion of all unicorn populations has this property. Because of the possibility of a nonPoisson but unicornian distribution, many authors suggest that a goodness-of-fit test comparing observed and expected frequencies is a better approach to testing for a Poisson distribution. That is a good suggestion. The next difficulty, however, is that even if the number of plants per quadrat follow a Poisson distribution, the dispersion of plants may still not be random, as Figures 7.17 and 7.18 show. The spatial relationship among the high- and low-density quadrats must be considered whether the quadrats are regularly placed as in Figure 7.17 or are themselves randomly placed as in Figure 7.18. In both examples, spatial autocorrelation is high; in the grid of quadrats the correlation coefficient of rook's move neighbors is 0.83 and using the LDNT definition of

0	0	0	1	1	2	2	1	0	0
0	1	1	2	2	3	2	2	1	0
1	1	2	3	3	4	3	3	2	1
1	1	2	3	4	5	4	3	3	2
1	2	3	4	5	6	5	4	3	2
1	2	3	4	4	5	4	3	3	2
1	1	2	3	3	4	3	3	2	2
1	1	2	2	2	3	2	2	2	1
0	1	1	2	2	2	2	1	1	0
0	0	1	1	1	1	1	0	0	0

Figure 7.17 The frequency distribution of plants in quadrats is Poisson, but their dispersion is not random: a grid of quadrats. The correlation of rook's definition neighbors is 0.83.

neighbors for the scattered quadrats, the correlation of neighboring samples is 0.88.

Quadrat counts cannot, therefore, be used alone to test for spatial randomness. The spatial autocorrelation of the quadrat counts needs to be analyzed.

Anisotropy

The methods discussed so far in this chapter have been based on the assumption of isotropic pattern; that is, the detection of pattern is averaged over all directions. In this section, we will describe some modifications of the methods in order to evaluate the directional nature of spatial point pattern.

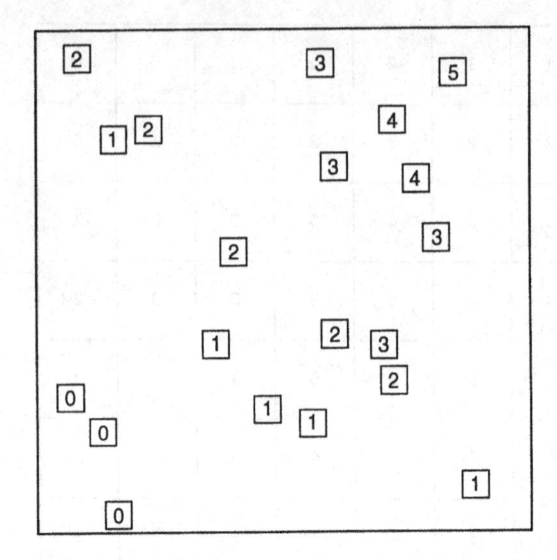

Figure 7.18 The frequency distribution of plants in quadrats is Poisson, but their dispersion is not random: random quadrats. When the samples are joined using a LDNT, the correlation among neighboring samples is 0.88.

For any of the methods that look at distances to plants or that count plants in circles, a simple modification is to divide the circle around each point into a number of sectors (12 in Figure 7.19). In this way, second-order analysis can be modified to examine anisotropy, dividing the circle around each point into N sectors and calculating $\hat{K}_s(t)$ for each sector, s:

$$\hat{K}_s(t) = A\sum_{i}^{n} \sum_{\substack{j \\ i \neq j}}^{n} w_{ij} I_t(i,j,s)/n^2, \tag{7.5}$$

where A is the area of the plot. Where d_{ij} is the distance between points i and j, $I_t(i,j,s)$ is 1 if $d_{ij} < t$ and j is in sector s of a circle around point i. If sector s centered on plant i with radius t is completely within the study plot then $w_{ij} = 1$, otherwise it is the reciprocal of the proportion of that sector's circumference that lies within the plot.

$\hat{K}_s(t) = \pi t^2/N$ if the plants are randomly arranged in a Poisson forest, and it would make sense to plot:

$$\hat{L}_s(t) = t - \sqrt{N\hat{K}_s(t)/\pi} \tag{7.6}$$

as a function of t for each sector. In most cases, however, the order of the points, i and j, is not important and so we would combine sectors on opposite sides of the circle, and thus plot:

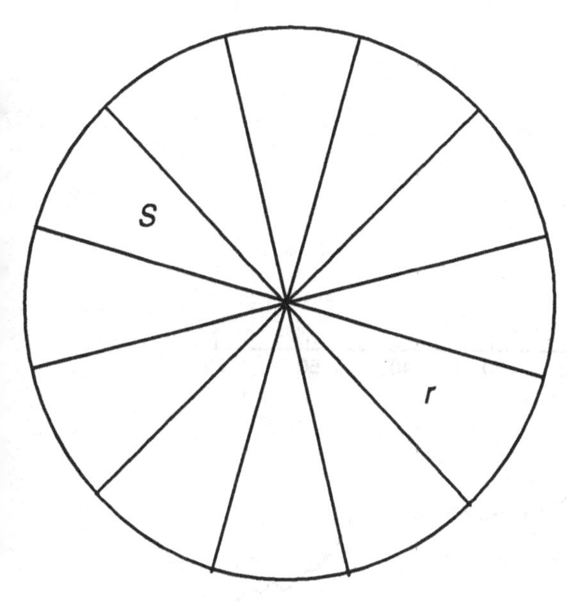

Figure 7.19 The division of the circle centered on a plant into 12 sectors. Sectors r and s are opposite each other.

$$\hat{L}_s(t) = t - \sqrt{N[\hat{K}_s(t) + \hat{K}_r(t)]/2\pi} \tag{7.7}$$

where sector r is opposite sector s (see Figure 7.19). On the null hypothesis of complete spatial randomness, $\hat{L}_s(t)$ has an expected value of zero. Large positive values of $\hat{L}(t)$ indicate that the plants are overdispersed and large negative values indicate clumping at scale t in the sector. Significance testing can use the guidelines given above in the general treatment of the second-order technique or by a Monte Carlo approach.

As an example of anisotropy, we reanalyzed part of the huon pine (*Lagarostrobos franklinii*) data presented by Gibson and Brown (1991). We used a 36 m × 32 m subsample from the top part of their Figure 2 (site D1), in which the stems greater than 10 cm in diameter appear to form three linear patches parallel to the x-axis. Figure 7.20 shows two plots of $\hat{L}_s(t)$ resulting from this analysis, using $N = 8$. Parallel to the x-axis, apart from the repulsion of individual stems at a scale of 0.5 m, the stems are clumped at a range of scales to the maximum tested, 10.8 m. In contrast, in the direction of the y-axis, the stems are clumped only to a scale of 2 m and then strongly overdispersed in the range of 7.8 m to 9 m. Gibson and Brown (1991) detected this same scale of pattern, using the standard second-order analysis that averages over directions, and attributed it to

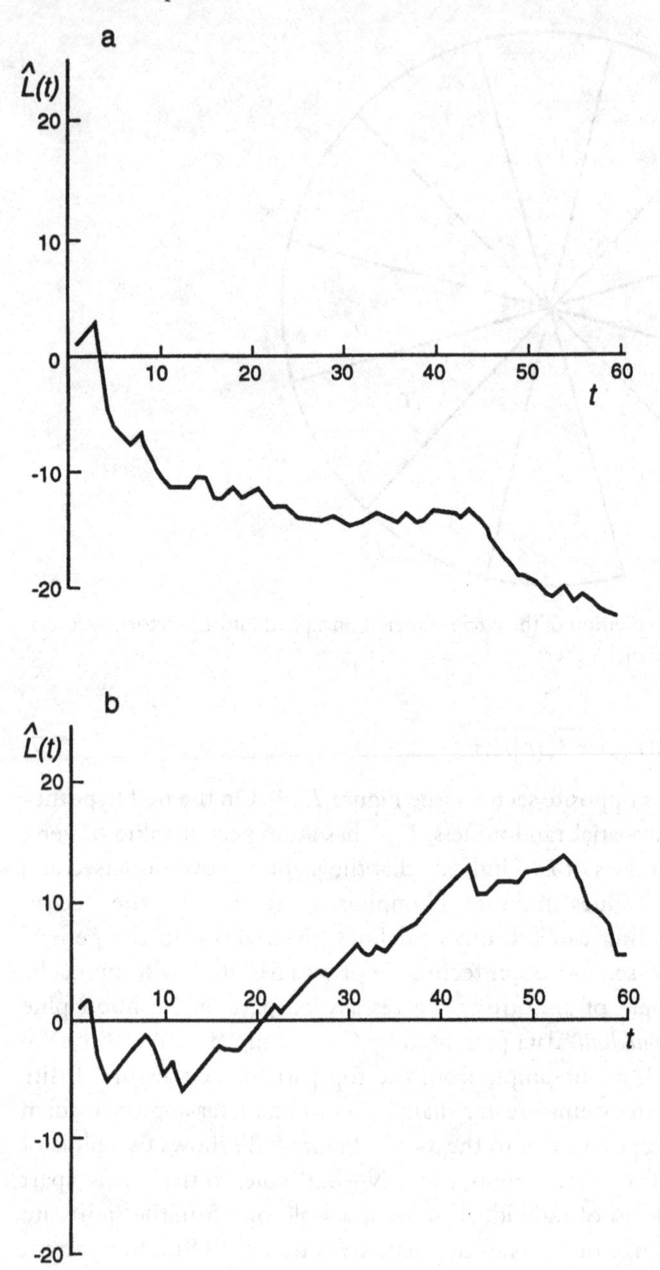

Figure 7.20 Two of four possible plots showing second-order analysis using eight sectors. The data for opposite sectors are combined. *a* Parallel to the *x*-axis, there is repulsion at small scales and clumping at larger scales. *b* Parallel to the *y*-axis, the opposite is true.

seedling dispersal or to canopy disturbance. The fact that the pattern is strongly anisotropic will affect our evaluation of which processes have given rise to it.

Bivariate point patterns

Having described methods that analyze the spatial point pattern of plants when they are treated as being of only one kind, we turn now to examine methods that consider plants of two different kinds. The different kinds may be two species, the two sexes of dioecious plants, the two forms of a dimorphic species, or plants that are attacked or not by a particular herbivore, pathogen or parasite, and so on. One of the most basic questions that can be asked about plants in natural vegetation concerns how they are arranged in space, and when the plants are of two kinds there is the further question of how the plants of the two kinds are arranged relative to each other. One way to approach the question is to describe the spatial arrangement of each kind separately and then to ask how the two arrangements are related. A second approach is to describe the spatial arrangement of all the plants, and then to consider different labellings of the plants as they are assigned to different classes. In either case, it is of interest to determine whether the plants of the different kinds are segregated from each other or whether they are aggregated.

As with many other phenomena discussed in this book, segregation and aggregation are scale dependent, as Hurlbert (1990) points out 'Degree of aggregation in nature is always strongly a function of spatial scale'. One version of this scale dependence is shown in Figure 7.21: in part b the two kinds appear to be segregated, but in the context of more empty space around them, they appear aggregated (part a). The two kinds are actually segregated within the patches; their aggregation is the result of the overall patchiness of the plants. Therefore, it may not be appropriate to ask simply whether the kinds are segregated or aggregated, but rather we should ask at what scales are they segregated and at what scales aggregated.

Nearest neighbor methods

The discussion of two-species methods can follow the structure of the above discussion of the methods used for point pattern analysis as applied to plants of one kind. The most simple approach to examining segregation and aggregation is to look at the kinds of plants that are nearest

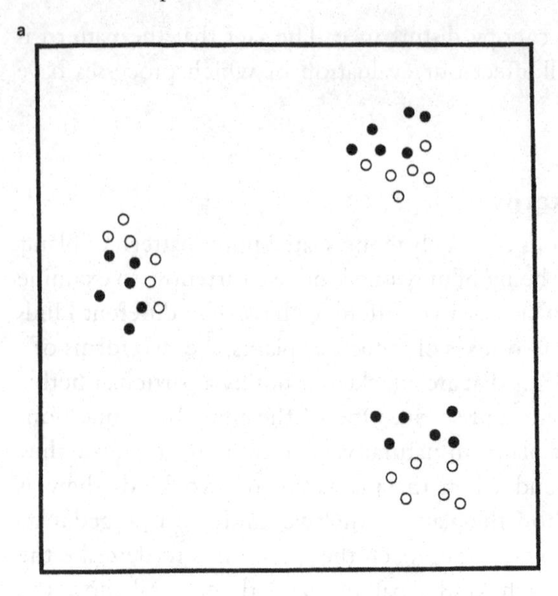

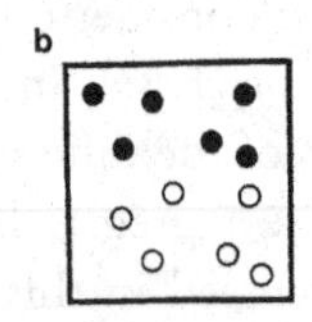

Figure 7.21 Scale of segregation and aggregation. *a* In relation to the surrounding empty space, the two kinds appear aggregated, although they are segregated within the patches. Their aggregation results from overall plant patchiness. *b* When a small area is considered, the two kinds of plants appear to be segregated.

neighbors. Whether the plants of either kind are overdispersed, clumped, or random, as a general rule (to which there may be exceptions) if the kind of plant is independent of the plant's position, then the identity of that plant's nearest neighbor will be independent of its own identity. On the other hand, if the kinds are segregated, then the frequency of nearest neighbors of the same kind will be greater than expected, and if the kinds are aggregated then the number of unlike nearest neighbors will be greater than expected.

For example, we mapped the stems of *Solidago canadensis* in 2m × 2m plots at the edge of a hay field, and recorded whether or not they had been attacked by insect herbivores, the most obvious and common result being the formation of a gall (Dale & Powell 1994). (This example is described in

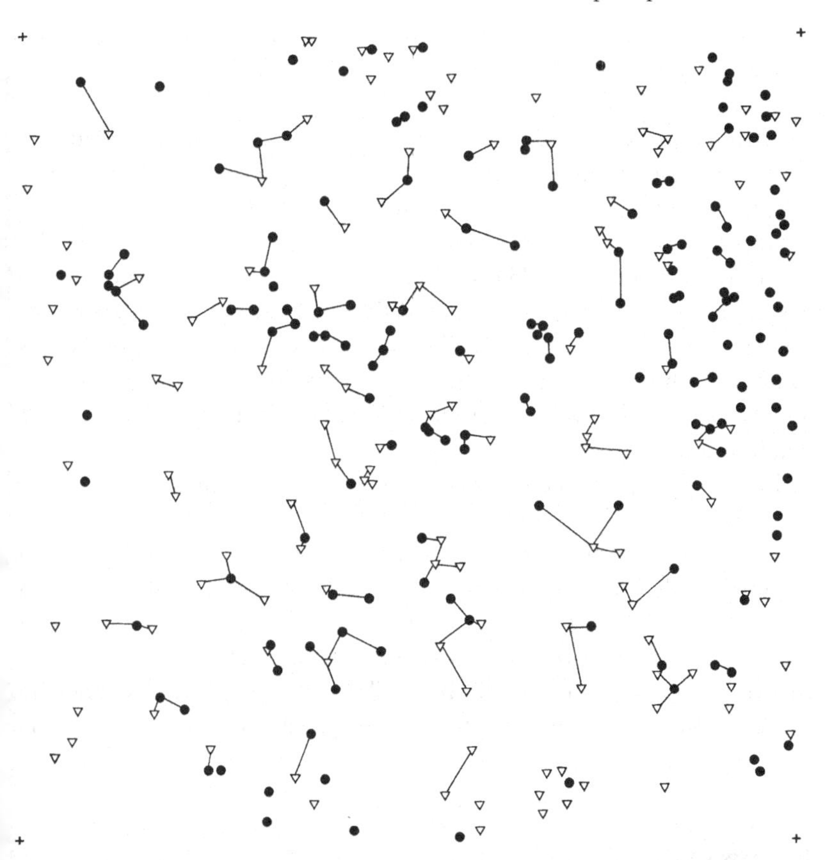

Figure 7.22 The nearest neighbors of 183 *Solidago canadensis* plants that are further than 5% of the plot's side length from the edge (plot 3 of Dale & Powell, 1994). Ninety of the nearest neighbors were of the same kind, 'attacked' or 'clean'. A randomization analysis of the plants' labels found no evidence to suggest that the two kinds were either aggregated or segregated.

greater detail in Chapter 6.) The plants were classified as obviously attacked or as 'clean'. We restricted the nearest neighbor analysis to the plants in the central 1.6m × 1.6m square to avoid edge effects; another method for avoiding edge effects is described by Kenkel *et al.* (1989). Figure 7.22 shows plot 3 and nearest neighbors of those 183 central plants. Ninety of the nearest neighbors were of the same kind. We performed 100 randomizations of all the plants' labels and found that in 38 cases there were fewer than 90 like joins and in 62 cases there were 90 or more. Based on this analysis, there is no evidence to suggest that the two kinds of plants are segregated or aggregated at the scale of nearest neighbors.

We could also test whether like nearest neighbors tend to be closer than unlike nearest neighbors. In the same data set, just analyzed, the average distance between alike nearest neighbors was 5.1 cm (s.d. 3.27 cm) and between unlike nearest neighbors was 5.8 cm (s.d. 3.10 cm). The large variance prevents the difference between them from being significant.

We can extend this kind of analysis to look at second or third nearest neighbors in the same sort of way. Interestingly enough, in analyzing the same *Solidago* plot, looking at the first and second nearest neighbors, the randomization tests finds 75 out of 100 less than the observed number of like joins; using the first, second, and third nearest neighbors 94 out of 100 randomizations have fewer like joins than observed. Clearly, there is something about the spatial relationship of the two kinds of plants that will be discovered by more extensive analysis than the simple nearest neighbor method was able to discern.

In a study of a dioecious tropical tree *Ocotea tenera* in Costa Rica, Wheelwright and Bruneau (1992) looked at the sexes of nearest neighbor trees to determine whether there was spatial segregation. In the natural population, the frequency of nearest neighbors being of different sexes was much greater than expected by chance. Because the trees of this species are able to change their sexual expression, the authors speculated that the observed pattern could be attributed to labile sexual expression modified by the neighboring trees. The spatial nonrandomness of the sexes may increase their fitness.

Frameworks

The nearest neighbor methods just described essentially provide a framework within which the plants' positions are simplified to lists of the nearest neighbors or of the first two nearest neighbors. What is done then is to examine the categories to which plants joined in this way belong and what proportion are 'like' joins, between two plants of the same kind.

Clearly, there are other and more extensive frameworks of joins between pairs of plants that can be used. A Delaunay tessellation or a least diagonal triangulation as described in the subsection entitled 'Tessellations' can obviously be used in this way. Once a framework is established, the procedure is to look at the frequency of like joins and ask whether the number of these is significantly less or greater than expected. The significance will usually be determined using a randomization procedure. It may also be informative to compare the frequency distributions of the lengths of like and unlike joins to examine the properties of the joint spatial pattern. In the *Solidago* example described in the previous section, first nearest neighbors were not significantly often of the same

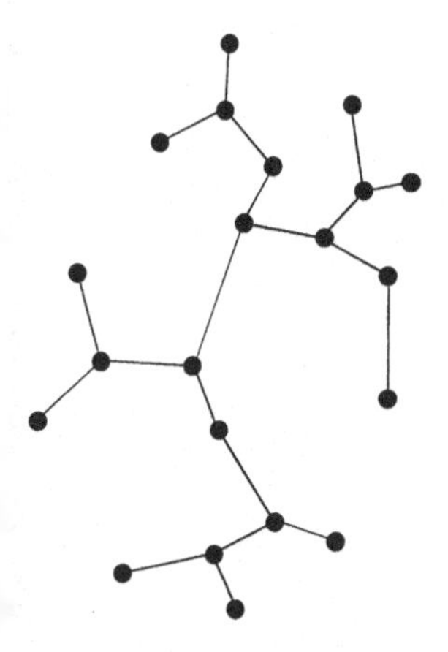

Figure 7.23 A minimum spanning tree, connecting the points using the smallest distances (artificial data).

kind but when neighbors are defined by LDNT, the result becomes significant. On the other hand, even using LDNT, there was no significant tendency for the distance to like neighbors to be shorter than the distance to unlike neighbors.

Other frameworks are available for this kind of approach, one of which is the point pattern's minimum spanning tree. A tree, in graph theory, is a set of lines that join points together without producing cycles, closed loops of lines. The minimum spanning tree is the set of lines with smallest total length that connect all the points into a single structure without cycles. It is like a chain of nearest neighbors and it includes more joins than just the nearest neighbor joins (Figure 7.23). As with the nearest neighbor analysis, the minimum spanning tree would then be used to compare the frequencies and lengths of like and unlike joins.

Another candidate is the Gabriel graph which consists of lines connecting points i and j whenever the circle of diameter d_{ij} that passes through the two points contains no other points (Gabriel & Sokal 1969, Figure 7.24). Using a framework to examine the frequency of like joins is essentially an examination of autocorrelation in the data, but the autocorrelation of categorical rather than quantitative attributes (*cf.* Sokal & Oden 1978a,b; Upton & Fingleton 1985). The disadvantage of using a minimum spanning tree or Gabriel graph as the framework is that it may be too restrictive of the plants that are considered to be neighbors. Because the minimum spanning tree allows no cycles, it tends to have fewer lines and the Gabriel graph also typically does not produce a full triangulation (see Figure 3 of Gabriel & Sokal 1969).

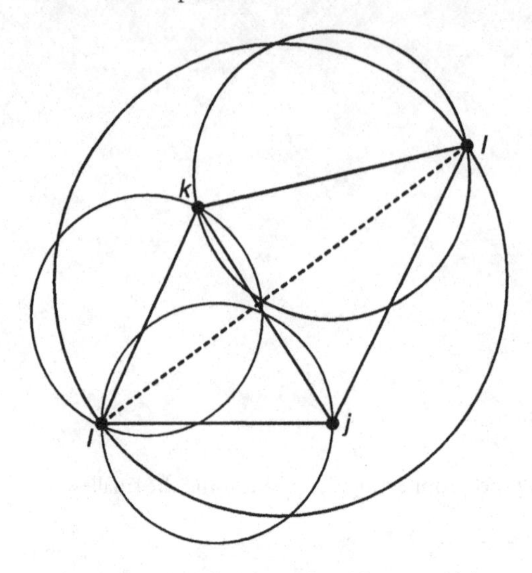

Figure 7.24 A Gabriel graph with some of the test circles shown. Points *i* and *j* are joined if the circle of diameter d_{ij} that passes through the points is empty. The dotted line is not part of the graph because the circle associated with it is not empty.

Second-order or neighborhood methods

All of the framework techniques are based on considering pairs of points joined by lines. Neighborhood methods are based on sets of plants, more than two at a time, and examining the combinations of the kinds of plants in those sets. One of these methods is an extension of the second-order method described above for single species. The procedure is to count the number of plants of type 2 within distance *t* of a plant of species 1, and so on. If the definition of $I_t(i,j)$ is changed to take the value 1 only when $d_{ij} \leq t$ and plant *i* is of species 1 and plant *j* is of species 2, then we can define the following:

$$\hat{K}_{1,2}(t) = A \sum_{i}^{n_1} \sum_{j}^{n_2} w_{ij} I_t(i,j)/n_1 n_2 \tag{7.8}$$

$$\hat{K}_{2,1}(t) = A \sum_{i}^{n_1} \sum_{j}^{n_2} w_{ji} I_t(i,j)/n_1 n_2. \tag{7.9}$$

A is the area of the plot and w_{ij} is a weighting factor defined for Equation 7.1. $\hat{K}_{1,2}(t)$ and $\hat{K}_{2,1}(t)$ are estimates of the same function, and the combined estimator is $n_2\hat{K}_{1,2}(t) + n_1\hat{K}_{2,1}(t)/(n_1 + n_2)$ (Upton & Fingleton 1985).

To investigate the joint spatial pattern of the two species, we plot:

$$\hat{L}(t) = t - \sqrt{[n_2\hat{K}_{1,2}(t) + n_1\hat{K}_{2,1}(t)]/\pi(n_1 + n_2)} \quad (7.10)$$

as a function of t, which on the null hypothesis has an expected value of zero. Large positive values of $\hat{L}(t)$ indicate that the two kinds of plants are segregation at scale t and large negative values indicate aggregation.

Figure 7.25 shows two kinds of plants, occurring together in clusters, but with the plants of one kind excluded to the outer rim of the clusters. Figure 7.26 shows the bivariate analysis of this arrangement, with strict segregation up to a distance of 5 units and aggregation at a distance of 17.

Szwagrzyk and Czerwczak (1993) used a version of this kind of analysis to study the spatial pattern of trees in old growth forests in Poland and the Czech Republic. The two most common species were *Fagus sylvatica* and *Picea abies*; the other species included *Tilia cordata*, *Carpinus betulus*, *Acer campestre*, *Fraxinus angustifolia*, *Ulmus glabra*, and *Acer pseudoplatanus*. They looked at three different kinds of bivariate analysis: pairs of species, small *versus* large trees based on diameter class, and living *versus* dead. For all three kinds of bivariate analysis the result was usually the same: the trees of the two kinds seem to occur independently of each other. The authors provide an interesting discussion of this finding, but wisely point out that the lack of significant departure from randomness and independence does not mean that the processes causing the spatial pattern are truly stochastic.

Conversion to quadrats

One approach to dealing with mapped plant positions is to convert them to quadrat form. Mapped plant locations can be converted into grids of contiguous quadrats by setting up a matrix for each kind of plant in which the elements of the matrix are the numbers of plants of that kind found in each quadrat. The smaller and more numerous the quadrats are, the less information will be lost by this conversion, but there is an obvious tradeoff between losing less information and having to deal with larger matrices. In most cases, the matrices will be 'sparse', containing mainly 0's with just a sprinkling of 1's. In Chapter 6 in the section of 4TLQC, we described a method for examining segregation and aggregation using quadrat data.

Multispecies point pattern and quantitative attributes

In investigating spatial pattern in point data where the plants are of several species, the techniques usually employed are the same as those described above for the two-species case. Nearest neighbor or framework methods

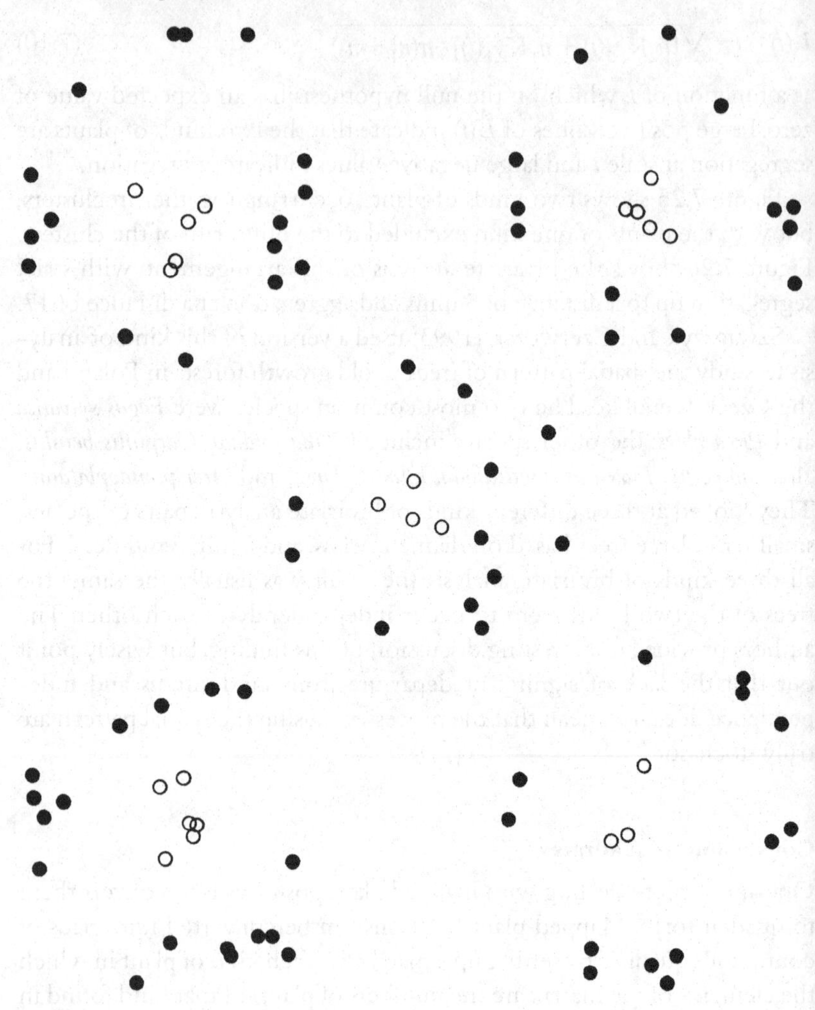

Figure 7.25 Two plant types occur together in clusters, but with one kind only occurring at the outer edge (artificial data).

can be used and the number or length frequency distributions of particular like joins or particular unlike joins can be compared. For example, Armesto *et al.* (1986) compared the spatial patterns of trees in two north temperate, two south temperate, and three tropical forests; one method they used was to examine the distance from a tree to its nearest conspecific. The south temperate forests had noticeably shorter distances to conspecifics, which the authors suggest may be related to a lower diversity of insect herbivores.

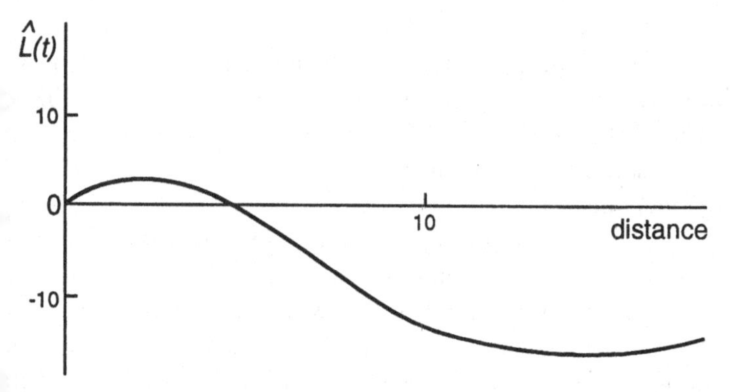

Figure 7.26 Bivariate second-order analysis of the pattern in Figure 7.25. $\hat{L}(t)$ as a function of t, showing segregation up to a distance of 5 units and aggregation up to 17.

The two-species version of second-order analysis described above has been used to analyze many-species data, as in the study by Szwagrzyk and Czerwczak (1993) discussed above. As a second example, Diggle (1983) presents an analysis of the trees in the Lansing Woods, looking at three types of tree (oaks, *Quercus* spp.; hickories, *Carya* spp.; and maples, *Acer* spp.). The three types were analyzed in pairs, using $\hat{K}(t)$. The analysis showed that maples and hickories seemed to occur as a balanced process, with maples being common where hickories are rare and hickories common where maples are rare. However, when the two were combined and analyzed as univariate data, the result did not resemble complete spatial randomness. When the oaks were included, the stems of the three kinds of trees, considered together in a univariate analysis, appeared to be randomly placed in the plane.

We have found no example of truly multispecies analysis, in the sense of looking at combinations of species simultaneously, using mapped point data. It would certainly be possible to convert point pattern data to quadrat form and then to use a two-dimensional equivalent of multi-species pattern analysis (such as multi-scale ordination) described in Chapter 5, but it has yet to be done. The area of multispecies point pattern analysis is another in which there is room for new approaches and further investigation.

Multispecies point pattern analysis can be considered as dealing with the spatial pattern of mapped categorical data, where the categories are the species. This approach can be compared with the analysis of mapped quantitative data, of which there are many examples. A simple case is a mapped stand of trees, where, in addition to the spatial coordinates of each tree, we have recorded the tree's size, whether that is height, diameter, or volume. It is then an easy matter to use a spatial framework like LDNT to define the neighbors of each tree and then investigate the spatial autocorrelation of the size of immediate neighbors, of the neighbors of neighbors, and so on.

A more complicated example would be one in which the mapped plants had several different measures associated with them, such as the height of a herbaceous plant, number of leaves, widths and lengths of leaves, number of flowers, number of flower buds, measures of herbivore damage to various parts of the plant and so on. More work needs to be done on the question of how best to analyze such multivariate spatial data, but it is an important area for our understanding of plant ecology. James and McCulloch (1990) are correct in stating that the multivariate approach to population studies is currently poorly developed but is important even at the level of description; we would add that it is even more important when the spatial pattern of that population can be included.

Quantitative attributes can be converted into categorical data by dividing the plants into classes based on their size. Gibson and Brown (1991) studied the spatial pattern in stands of Huon pine, *Lagarostrobus franklinii*, in Tasmania. When trees of all sizes were considered together, it was difficult to discern the scale of spatial pattern using second-order analysis. When the trees were divided into size classes by diameter, however, the situation was clarified, with the smallest size class consistently showing a scale of 1 m to 3.5 m, and the larger size classes either showing no significant pattern, or pattern at a larger scale, 3.5 m to 9.5 m. (Those authors used a concept of scale that is double that used in this book and therefore the values they report are 2 m to 7 m and 7 m to 19 m, respectively.) They interpreted the scale of pattern of the smaller trees as representing clumping of seedlings on fallen logs and the clumping of vegetative sprouts arising from the same fallen tree (Gibson & Brown 1991).

Categorical and quantitative data can be combined in mapped point data, and can be analyzed together. For example, we mapped the positions of all the plants in a 200 m × 300 m plot on a gravelly outwash near Kluane Lake in the Yukon (unpublished data). We also measured their heights. Three species dominated: *Picea glauca*, *Populus tremuloides*, and *Hedysarum mackenzii*. Based on an analysis of first and second nearest neighbors, we determined that there was a significant tendency for neighbors to be of the same species. There was also a significant positive correlation ($r=0.197$) of the heights of nearest neighbor pairs, which might just reflect the clumping of species if the species had different average heights. We therefore calculated the correlation of heights of nearest neighbors when they were of the same species, and it too was significant ($r=0.212$). This example shows that the vegetation is patchy not only in the tendency of neighbor plants to belong to the same species but also in the correlation of heights of conspecific neighbors. Either soil nutrients or disturbance history may be responsible for this observed pattern.

Concluding remarks

As stated at the beginning of the chapter, its purpose is not to provide a comprehensive treatment of point pattern analysis, for which there are other sources, but to discuss those methods of point pattern analysis most directly relevant to the topic. There are undoubtably other methods in this area to be explored or developed and at the time of writing it seems that there will be exciting developments in the next few years. As in the previous decades, the availability of ever faster and greater computing power will continue to increase the range of methods that we can use and thus the range of approaches that we can imagine.

Recommendations

1. For univariate point pattern, Ripley's K function, known also as second-order analysis, seems to be the best available technique. It does not do well at detecting gaps. The refinement of using counts from control points in the gaps to help draw contour lines based on the values associated with individual plants may be very informative.
2. Quadrat counts cannot be used alone to detect or quantify non-randomness.
3. For examining the characteristics of neighbors, a framework such as the least diagonal neighbor triangulation is a convenient approach. That tessellation or the Delaunay also provides a method for determining a plant's primary neighbors.
4. For bivariate point pattern analysis, the K function approach is again recommended.
5. In the area of the analysis of multispecies point pattern, there is a need for further research and the development of techniques.

8 · *Pattern on an environmental gradient*

Introduction

In this chapter, we will discuss the arrangement of plants on environmental gradients. In this context, an environmental gradient is a monotonic directional change in the intensity of an environmental factor with distance. It is the class of gradients that Keddy (1991) calls 'spatially continuous' and includes cases that may give rise to obvious zonation in the community.

Obviously the concept of spatial pattern is somewhat different in this context than in previous chapters, but it still refers to nonrandomness that has a certain predictability. As we move along a gradient, we do not expect to see the repeated alternation of different phases of a mosaic, but rather we expect species to become present and perhaps abundant where they were previously absent and then to become absent again. The predictability is in the way that species come and go along the gradient and the relationship between the ranges and densities of the species.

In Chapter 1, we discussed the importance of spatial pattern, as an area of study, pointing out that there are two facets to consider: (1) making inferences about processes based on observed pattern, and (2) the effect current spatial pattern has on future processes and interactions. The same two categories apply to the study of pattern on gradients. The potential positions of individual species are determined by their physiological responses to the gradient. Then, the arrangements of species on gradients can be used to examine questions about the forces that structure these communities, the interaction between species whether positive or negative, and the niche relations of the species in the community. Therefore, the pattern we observe arises from the interaction of the physiology of each species and the biotic effects of competition, positive association, predation and so on. On the other hand, the arrangement of species on an environmental gradient will determine which species will be able to

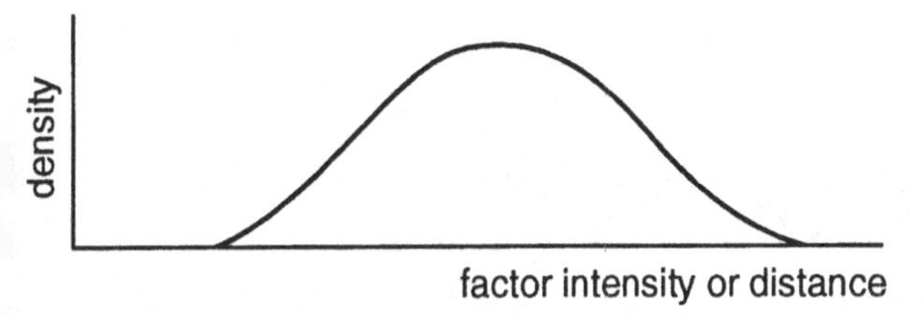

Figure 8.1 The density of a single species, *d*, as a function of distance along an environmental gradient, *x*, is often modelled by a bell-shaped curve.

interact in the future. The plants that are closest on the gradient are the ones that may be competing most strongly. Similarly, the arrangement of the plants along the gradient will determine what potential positive interactions between species may actually occur (see Bertness & Calloway 1994).

A simple model of a single species response to an environmental gradient is a symmetric unimodal response like a bell-shaped curve, when density is plotted as a function of the controlling factor on the gradient, or of distance along the gradient (Figure 8.1). These may not be the same thing: the same physical distance along a gradient may produce different degrees of change in a controlling environmental factor in different places, resulting in broader or narrower species ranges (*cf.* Figure 8.2). The symmetric unimodal response curve is seldom found, with skewed curves being more common, as we will describe later in the Chapter (Austin & Austin 1980; Minchin 1989; Collins *et al.* 1993). It is worth noting that the perceived symmetry or skewness of a unimodal response will depend in part on the scaling of the factor axis: logarithmic, arithmetic, or exponential.

Austin (1980) has pointed out that it is important to distinguish among three kinds of gradients. There are direct environmental gradients in factors that affect plant growth themselves, such as pH and temperature. There are indirect gradients like elevation or aspect that produce the observed effect on plants through factors that do act directly such as temperature or insolation. Lastly, there are resource gradients which are gradients in the amounts of nutrients. Species may respond differently to the three kinds of gradients. Having set up the classification, we should point out that some gradients may fit into several of the categories; for example, water availability may represent a gradient of water as a resource, a direct gradient and an indirect one acting through temperature. Similarly, Økland (1992) argues that in mires, the water-table gradient has aspects

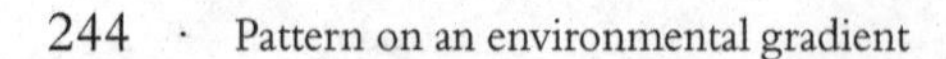

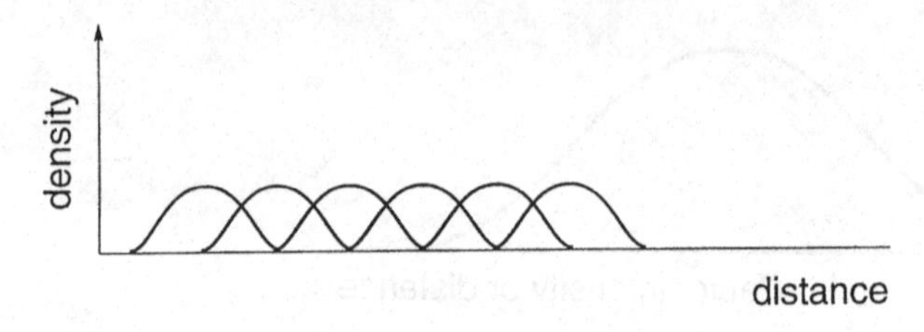

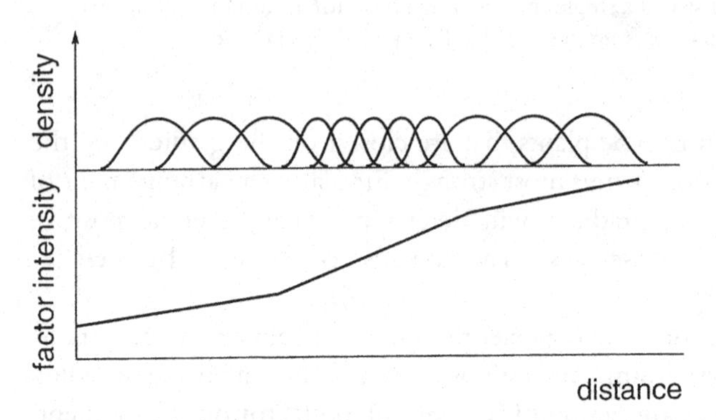

Figure 8.2 The intensity of the controlling environmental factor, i, may not change at a constant rate with distance, x, leading to a compression of species' ranges.

of all three categories. In natural vegetation, plants must respond to several gradients simultaneously and different combinations of gradients will produce differently shaped responses to the set of gradients (Austin & Smith 1989).

If the controlling factor is known and it can be measured, then a direct analysis of species' responses to the factor is possible but, in many cases, the gradient may be the result of the interactions of several environmental variables (e.g., water supply and temperature) or the actual controlling factor may be unknown, in which case we are forced to make inferences about the plants' responses to the gradient (cf. Austin *et al.* 1984; Austin 1987).

Not only may the rate of change in a controlling factor vary along the length of a gradient, it is also possible that plants may respond more or less strongly to the same amount of change in the controlling factor, depending on its intensity. One version of this thinking is called the 'critical tide level hypothesis' which suggests that in communities of intertidal algae, there are particular levels on the shore where species replacement occurs more rapidly over small differences in height, because the response of individual species to changes in the duration of immersion or exposure to

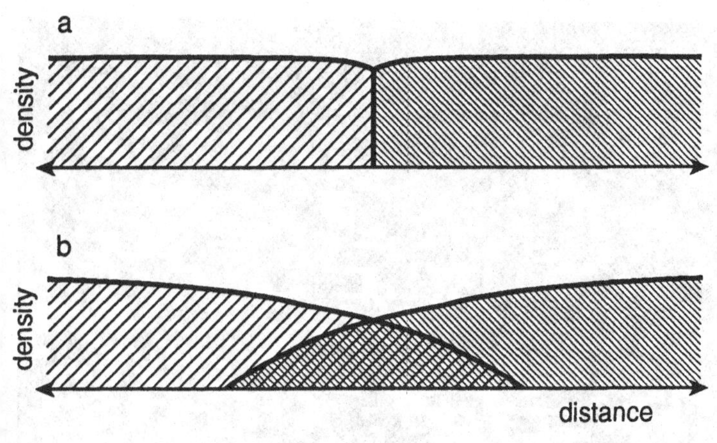

Figure 8.3 *a* Competing species replacing each other along an environmental gradient with no intermediate zone of coexistence. *b* Competing species replacing each other with an intervening zone of coexistence.

desiccation is much stronger at that level (Doty & Archer 1950). It is easy to imagine similar critical levels on other kinds of environmental gradients, such as temperature or soil moisture. For example, with increasing altitude or latitude, particular combinations of temperature and moisture regime may limit the functioning of, first, any broad-leaved evergreen tree, then any broad-leaved deciduous tree, and then any trees at all. Thus, there may be boundaries that apply to several different but functionally similar species.

There are a variety of ways in which species can be arranged on an environmental gradient, and the arrangement reveals much about the organization of plant communities. For instance we can examine their arrangement for evidence of critical levels in the controlling factor, or we can look for evidence of biological interaction among the species. It has been suggested that competition between species is an important process in the development of spatial pattern on a gradient. Depending on how competition affects the plants, for instance, if the two species cannot coexist, it could result in the beginning of one species' range following immediately after the ending of another species' range (Figure 8.3a). On the other hand, it is possible that species replacement occurs with a zone where the two species can coexist (Figure 8.3b). In that zone of coexistence, the density of one species decreases as that of the other increases. One spatial model of this kind of species replacement is Rapoport's (1982) Gruyère model in which the transition resembles Swiss cheese, going from solid white, to white with small black patches in it, the black

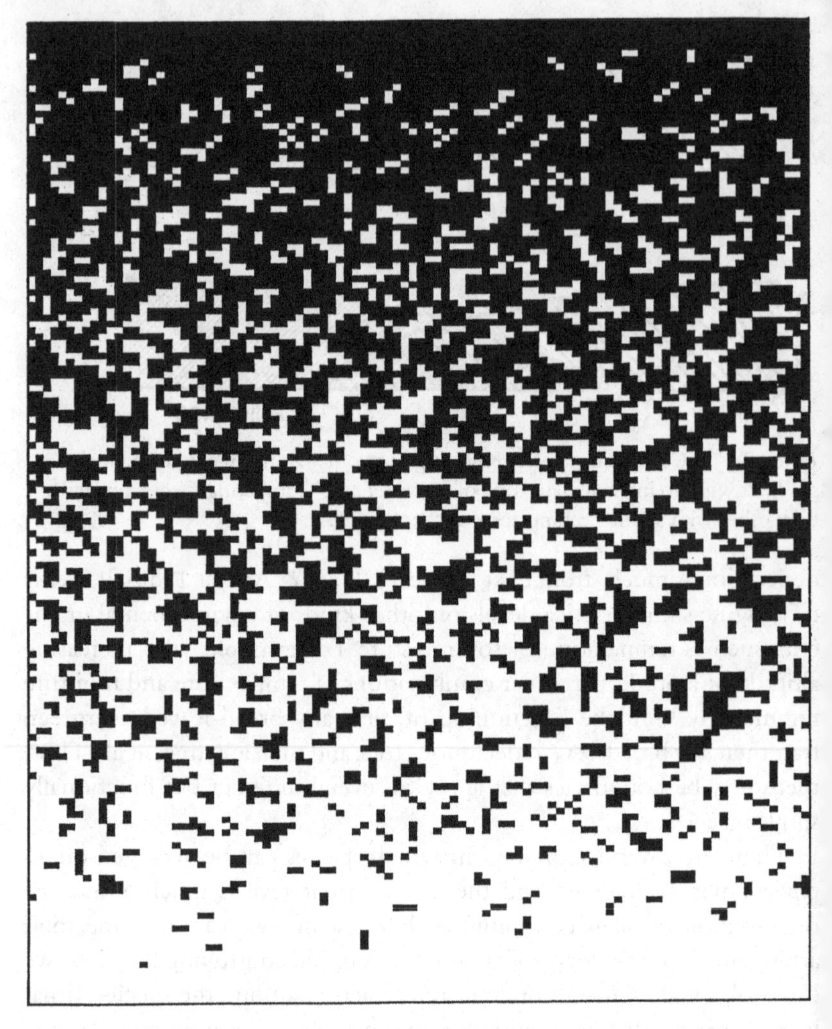

Figure 8.4 Rapoport's 'Gruyère' model of species replacement. The black represents one species or vegetation type and the white represents another.

patches then becoming larger and coalescing to the point that the end of the transition is solid black (Figure 8.4).

Shipley and Keddy (1987) discuss the arrangement of species on an enviromental gradient in terms of two opposing views of plant communities: the individualistic view and the community unit view. The community unit view suggests that groups of species will replace each other along the gradient so that clusters of upper and lower boundaries will be

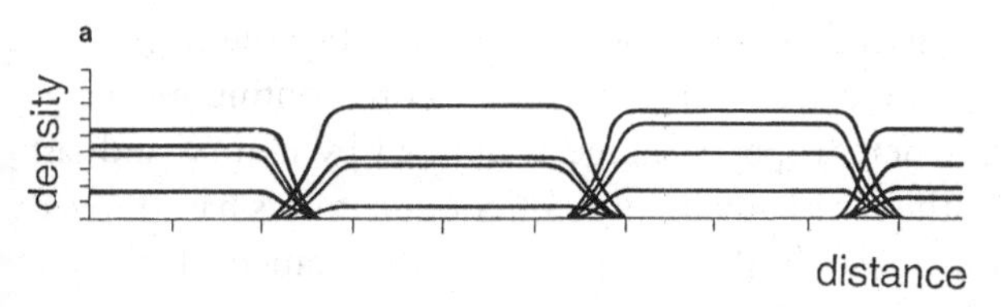

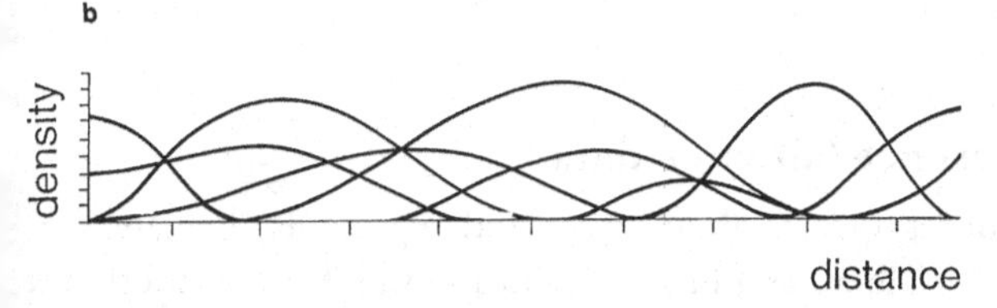

Figure 8.5 Two models of species' densities along an environmental gradient. *a* Community unit model where groups of species have similar responses and boundaries. *b* Individualistic model in which the species' responses are independent of each other.

found (Figure 8.5a). On the other hand, the individualistic view suggests that the species occur more or less independently of each other so that upper and lower boundaries occur independently and boundaries of either kind should not be clustered (Figure 8.5b). The two models illustrated in Figure 8.5 are not the only ones: Whittaker (1975) described two others and we could certainly devise more. For example, where one species replaces another, there is a choice between sharp exclusion as in Figure 8.3a and gradual replacement as in 8.3b. Distinguishing among the possible arrangements and questions related to them will be important themes in subsequent sections of this chapter.

In addition to the nature of plant communities, there are a number of other areas of plant ecology that can be studied using the patterns observed on gradients. Keddy (1991) suggests that gradients are a powerful research tool because they take some of the apparent obstacles of spatial heterogeneity for ecological studies and turn them into advantages. One particularly attractive approach is to use two different gradients such as water depth and exposure to wave action (Keddy 1991) or altitude and drainage in montane vegetation (Minchin 1989). We will not review all aspects of this fascinating area of research, but will concentrate on those topics directly related to spatial pattern on gradients.

There are two different sampling designs that need to be considered and two kinds of data. The two kinds of data are density (including cover) and presence/absence; the two designs are continuous records as a function of distance and spaced samples, such as quadrats. For the beginning sections of this chapter, we will concentrate on continuous presence/absence data, where a species range is defined by its first and last occurrence. The mathematical techniques of combinatorics have proved useful in this area of research. (Combinatorics is the branch of discrete mathematics that deals with finite problems of counting, selection and arrangement of objects.)

Continuous presence/absence data

This approach ignores species abundances and represents the range of each species on the gradient as a line segment joining its uppermost and lowermost occurrences. The whole gradient is therefore represented by a 'sheaf' of line segments, the position and lengths of which correspond to the species' ranges on the gradient. There are at least four models that we can consider, as rephrased from Whittaker (1975):

1. Distinct groups of species with sharp exclusion boundaries, comparable to Shipley and Keddy's (1987) community unit model.
2. Sharp exclusion boundaries between competing species but no natural groupings.
3. Groupings of species that are not exclusive.
4. No groupings and no exclusion, comparable to Shipley and Keddy's (1987) individualistic model.

These four models, for the kind of data under consideration, are illustrated in Figure 8.6.

We will now describe a variety of methods that can be used to detect particular features of this kind of data. These methods are illustrated using the ranges of seaweed species on rocky intertidal shores in Nova Scotia, on which the environmental gradient is the length of time of emergence from seawater each day, which acts through related desiccation and temperature effects and therefore is an indirect gradient. The data discussed here are from three sites in Yarmouth County (approximately 43°40′N, 66°W): Wedge Pt. near Tusket, St. Ann Pt. near Pubnico, and Chegoggin Pt. near Yarmouth (Dale 1986). Each site was sampled at several stations (identified by its compass point e.g. South side, East side, . . .) using three to seven line transects from above the high-tide level to the upper subtidal. The shores are gently sloping (1:23 to 1:30) and the spring tide range is 3.6 m. The most common species at these sites included *Ascophyllum nodosum*, *Fucus spiralis*, *Fucus vesiculosus*, *Fucus serratus*, *Chondrus crispus*, and *Gigartina stellata*. *Chondrus crispus*, known locally as

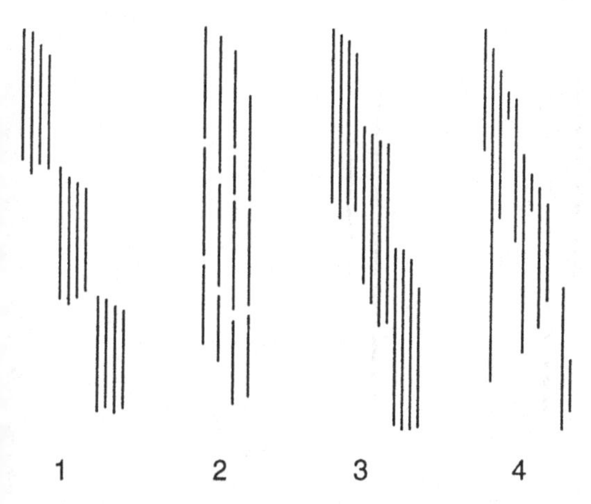

Figure 8.6 Four models of species' ranges arranged on a gradient. The presence of each species on the gradient is represented by a line. In the first model, each zone consists of a group of species that occur together and the zonal units replace each other completely over a short distance. In the second, the vegetation consists of four sequences of species; within each sequence, one species begins where another ends. In the third model, the zonal groupings of species overlap. In the fourth, the species occur more or less independently along the gradient.

'Irish moss', is harvested for commercial use. The transects used were line intercept transects, as described in Chapter 2, on which were recorded the linear positions of different species of algae as they intersected the edge of a measuring tape. Thus, for example, the first transect record might be: 0–35 cm, bare rock; 35–102 cm, *Fucus spiralis*; 102–127 cm, *Fucus spiralis* and *Ascophyllum nodosum*; and so on. From these transects, the positions in running length of the upper and lower boundaries of each species were determined. In our brief example, the upper end of the range of *Fucus spiralis* on the first transect is at 35 cm and that of *Ascophyllum nodosum* is at 102 cm. There were 29 transects in all.

In order to distinguish among the various models of how the ranges are arranged on the gradient, we will examine the overlap of ranges, the intermingling of upper and lower boundaries, the contiguity of ranges, and the clustering of boundaries. The consistency of the order of boundaries at different parts of the shore will be measured by concordance.

Overlap

The first approach to the analysis of species ranges on a gradient that we will consider was originated by Pielou (1977b) in a study of the overlap of latitudinal spans of seaweeds. The pattern of overlap is evaluated by

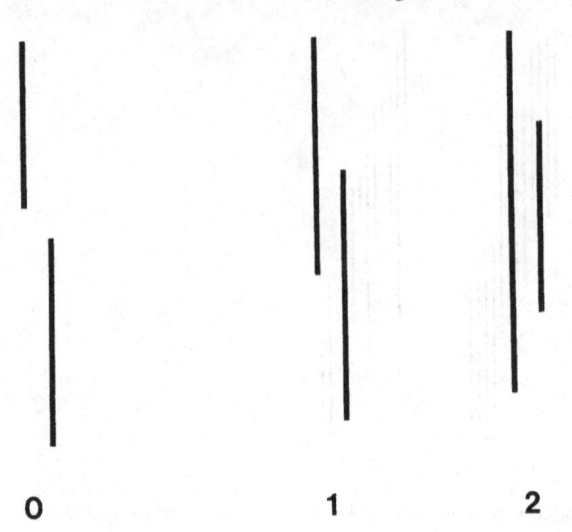

Figure 8.7 Definition of λ: $\lambda=0$, where there is no overlap; $\lambda=1$ where there is partial overlap; and $\lambda=2$ where there is complete overlap.

counting the numbers of pairs of species in each of three classes designated by the variable λ: no overlap, $\lambda=0$; partial overlap, $\lambda=1$; complete overlap, $\lambda=2$ (see Figure 8.7). The numbers of pairs of species' ranges in the classes can be compared with the expected results based on certain hypotheses, as will be illustrated below (Pielou 1977b, 1978).

Let $\mathbf{L}=(\ell_0, \ell_1, \ell_2)$ be the overlap vector in which ℓ_t is the number of pairs of lines in the sheaf for which $\lambda=t$. For n species in the sheaf, $\ell_0+\ell_1+\ell_2=n(n-1)/2=k$. For example, in Figure 8.8, $\mathbf{L}=(3, 2, 1)$ with $\Sigma\ell_i=6$.

One null hypothesis, call it H_o, is that the order of the $2n$ events, n beginnings and n endings, is fully random with the only constraint being that each ending follows its own beginning. On this simple null hypothesis, we can derive the means and variances:

$$E(\mathbf{L}|H_o)=k\,(1,1,1)/3 \quad \text{(Pielou 1977b)}, \tag{8.1}$$

and

$$\mathrm{Var}(\mathbf{L}|H_o)=2k(2n+1, n+3, n+3)/45 \quad \text{(Dale 1979)}. \tag{8.2}$$

All four patterns of overlap in Figure 8.6 have 12 species and so $E(\mathbf{L} \mid H_o)=(22, 22, 22)$ and $\mathrm{Var}(\mathbf{L}|H_o)=(73.3, 44, 44)$. Based on the assumption of convergence of the distributions of the ℓ_i to the Normal distribution, the 95% confidence limits for $\mathbf{L}$ are (5, 9, 9) and (39, 35, 35). Examining the third example in Figure 8.6, we find that $\mathbf{L}=(16, 45, 5)$. Based on H_o, we

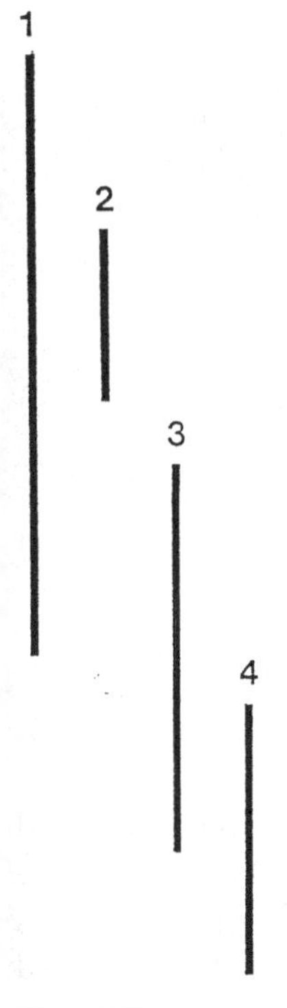

Figure 8.8 An example of overlap $\lambda_{12}=2, \lambda_{13}=1, \lambda_{14}=0, \lambda_{23}=0, \lambda_{24}=0, \lambda_{34}=1$. $\mathbf{L}=(3, 2, 1)$.

would conclude that a surprisingly large number of pairs of ranges overlap partially and a surprisingly small number overlap completely. We are not able to affirm that numbers are significantly larger or smaller than expected, with a known significance level of $\alpha=0.05$, because the three values in $\mathbf{L}$ are not independent.

A second use of H_o for a set of many transects is to examine the signs of $\mathbf{L}-E(\mathbf{L}|H_o)$ for general tendencies. For example, in the 29 transects from the three Nova Scotia sites described above, there was a tendency for $\mathbf{L}-E(\mathbf{L}|H_o)$ to be of the form of $(+,-,-)$. This form matches examples one and two in Figure 8.6 best.

H_o is a useful null hypothesis to examine $\mathbf{L}$, but to examine the amount of overlap of species' ranges, we can use a second null hypothesis, call it H_d. This hypothesis assumes that the length of each species measured along the gradient is fixed, but that the boundaries are placed randomly and independently, apart from the constraints imposed by the lengths. Given the range of length a and a total length of gradient w, being considered, the position of the upper boundary can be treated by a rectangular probability distribution for the distance from the top of the gradient running from 0 to $w-a$. In considering a pair of species' ranges, let a be the longer and b the shorter. It is obvious that if $a+b>w$ then $\lambda \neq 0$, so that those pairs need to be treated separately. Let A be the set of pairs for which $a+b>w$ and let B be the set of all other pairs. We can show that:

$$E(\ell_0 | \mathrm{H_d}) = \sum_A \frac{w-a-b^2}{(w-a)(w-b)} \tag{8.3}$$

$$E(\ell_1 | \mathrm{H_d}) = \sum_A \frac{b(2w-2a-b)}{(w-a)(w-b)} + \sum_B \frac{w-a}{w-b} \tag{8.4}$$

and

$$E(\ell_2 | \mathrm{H_d}) = \sum_A \frac{a-b}{w-b} + \sum_B \frac{a-b}{w-b} \quad \text{(Dale 1986).} \tag{8.5}$$

The observed overlap vectors for the seaweed data were compared with the expected values based on the null hypothesis $\mathrm{H_d}$, as was done above for $\mathrm{H_o}$. As with $\mathrm{H_o}$, the general finding was an excess of ℓ_0 values and a deficit for both ℓ_1 and ℓ_2 (Dale 1986); that is, $\mathbf{L} - E(\mathbf{L} | \mathrm{H_d})$ tended to be of the form $(+,-,-)$ as in examples one and two of Figure 8.6. The fact that this was found for $\mathrm{H_d}$ as well as for $\mathrm{H_o}$ means that fewer pairs of ranges than expected overlapped at all, even when the lengths of the ranges were considered.

Intermingling of boundary types

A related characteristic of communities on gradients is the extent to which the two kinds of events, upper and lower boundaries, are intermingled along the gradient. To measure the extent to which they are intermingled, we can use ℓ_0, already introduced, but it is useful to define it somewhat differently. Let q_i be the number of lower boundaries that are above the ith upper boundary; then:

$$\ell_0 = \sum_{i=1}^{n} q_i. \tag{8.6}$$

This is the same variable that counts the number of nonoverlapping ranges, but it also measures the intermingling of the two types of boundaries. For example, where X represents an upper boundary, and o repre-

Table 8.1. *Upper one-sided critical values for* ℓ_0*, calculated from the exact distribution ('exact' in table) and from the Normal approximation using a continuity correction of 0.5 ('approx.')*

	ℓ_0					
	$p=0.05$		$p=0.025$		$p=0.01$	
n	exact	approx.	exact	approx.	exact	approx.
5	8	7.5	9	8.2	9	9.0
6	11	10.3	12	11.3	13	12.3
7	15	13.7	16	14.8	17	16.2
8	19	17.4	20	19.3	21	20.5
9	22	21.6	24	23.3	26	25.3
10	27	26.2	29	28.2	31	30.6
11	32	31.2	34	33.5	37	36.3
12	37	36.6	40	39.3	43	42.4
13	43	42.4	46	45.5	50	49.0
14	50	48.6	53	52.1	57	56.0
15	56	55.3	60	59.1	64	63.5
20	–	94.4	–	95.3	–	107.1
30	–	202.0	–	212.8	–	225.4
40	–	347.7	–	364.4	–	383.8
50	–	530.8	–	554.2	–	581.3

Notes:
The null hypothesis is rejected if ℓ_0 is greater than or equal to the value in the table: p is the significance level. (From Dale 1988a.)

sent a lower boundary, in the sequence XXXXoooo, the two kinds are separated and $\ell_0=0$; in the sequence XoXoXoXo, they are highly intermingled and $\ell_0=6$. We commented above that because ℓ_0, ℓ_1, and ℓ_2 are not independent we cannot legitimately test all three. We can, however, derive a statistical test for one of them, here ℓ_0. We have derived the mean and variance of ℓ_0, based on H_o, but there is no direct derivation of its frequency distribution which we need in order to derive or confirm approximations of critical values. Dale (1988a) uses a bivariate distribution of ℓ_0 and T, the number of lower boundaries at the end of the sequence, to solve this problem.

Based on that approach, we find that the frequency distribution of ℓ_0 seems to be asymptotically Normal. (Remember, we assumed asymptotic normality for all three components of **L** in our discussion above.) Table 8.1 gives the critical values for ℓ_0 derived from the frequency distributions described by Dale (1988a), and also from the Normal approximation using a continuity correction of 0.5. The two sets of critical values are similar. When we apply this test to the Nova Scotia seaweed data, 26 of the 29 transects had an ℓ_0 greater than the expected value based on H_0, but only 11 were significant at the 5% level. These results confirm a tendency toward greater intermingling of upper and lower boundaries than predicted by H_0.

Contiguity

The next method investigates a particular kind of boundary intermingling in order to test the contiguity hypothesis mentioned above. In describing the zonation of marine algae, Chapman (1979) stated that, 'on a shore, species zones are commonly contiguous, so that the lower limit of one species marks the upper limit of another.' (See also Chapman 1973.) Two versions of this hypothesis can be tested: a strong form, that it is a general rule that the number of such contiguities (upslope boundaries immediately following a downslope boundary) is significantly greater than the number expected; and a weak form, that the number of contiguities merely tends to be greater than the number expected. The boundaries are considered contiguous even if there is some physical distance between them.

To test the weak form of the hypothesis, we need to know only the expected number of contiguities. We can then use the sign test to see whether the observed values exceed the expected value significantly often. To test the strong form of the hypothesis, we need to derive the frequency distribution of the number of contiguities so that we can evaluate the statistical significance of the observed numbers of contiguities (Dale 1984).

As we did in evaluating the intermingling of boundaries, we will treat the transect data as sequences of paired events, n beginnings and n endings. Let c be the number of contiguities, the number of times an ending immediately precedes a beginning. For example, in the sequence $X_1\ X_2\ o_2\ X_3\ o_3\ o_1\ X_4\ o_4$, $c=2$ since o_2 immediately precedes X_3 and o_1 precedes X_4. If the events are labelled, as in this example, and either the beginnings or the endings are constrained to be in fixed order, the number of such sequences that can be produced with n pairs of events is $(2n)!/2^n n!$ or $(2n-1)!!$ The frequency distribution of c in these sequences is described by the Münch numbers, Table 8.2, where $h(n,j)$ is the frequency with which $c=j$ in rankings of n pairs of (labelled) events when the order of the beginnings is fixed. We can prove that $E(c\,|\,n)=(n-1)/3$ and we can generate frequency distributions of c (Dale 1984). The critical

Table 8.2. *The frequency distribution of* c, *the number of contiguities, as a function of* n, *the number of species or of pairs of events. These are the Münch numbers (Dale 1984)*

	c							
n	0	1	2	3	4	5	6	Σ
1	1							1
2	2	1						3
3	6	8	1					15
4	24	58	22	1				105
5	120	444	328	52	1			945
6	720	3708	4400	1452	114	1		10395
7	5040	33984	58140	32120	5610	240	1	135135
8	40320	341136	785304	644020	195800	19950	494	2027025

values of *c* can be calculated from these distributions and the 1%, 5%, 95% and 99% critical values of *c* for $n=4$ to $n=45$ are given in Table 8.3. McGregor (1988) has shown that the distribution of *c* is asymptotically Normal with increasing *n*; for large values of *n*, therefore, the Normal approximation can be used.

The relative positions of the upper and lower boundaries of species of marine algae in zoned communities on rocky shores were recorded on line transects at various sites in Nova Scotia and at one site on the South Coast of England. In only 35 of the 139 transects studied was the observed number of contiguities significantly greater than expected at the 5% significance level. In most of the transects (110 of 139), however, the observed number of contiguities was greater than the expected (see Dale 1984) and, in fact, the observed number of contiguities exceeded the expected number significantly often. It appears, therefore, that the upper and lower boundaries of species do not occur in a truly independent fashion and so the individualistic view of the nature of plant communities is not supported by this kind of data from a spatially continuous gradient (see Shipley & Keddy 1987).

It is obvious that there is a relationship between the number of contiguities and the degree of intermingling between the two kinds of boundaries because, for instance, a sequence such as XXXXXXoooooo has no contiguities and no intermingling and the sequence XoXoXoXoXo has many contiguities and much intermingling. However, the two are not measuring the same thing. For example, for $n=10$, the sequence XXXXXoooooXoXoXoXo has a value of

Table 8.3. *Critical values for c, the number of contiguities (Dale 1984)*

n	1%	5%	95%	99%
4	–	–	3	3
5	–	–	4	4
6	–	0	4	4
7	–	1	4	5
8	–	1	5	6
9	–	1	5	6
10	1	1	6	6
11	1	2	6	7
12	1	2	7	7
13	1	2	7	8
14	2	2	7	8
15	2	3	8	9
16	2	3	8	9
17	2	3	9	10
18	2	3	9	10
19	3	4	9	10
20	3	4	10	11
21	3	4	10	11
22	3	4	11	12
23	4	5	11	12
24	4	5	11	12
25	4	5	12	13
26	4	6	12	13
27	5	6	13	14
28	5	6	13	14
29	5	6	13	15
30	6	7	14	15
31	6	7	14	15
32	6	7	14	16
33	6	8	15	16
34	7	8	15	17
35	7	8	16	17
36	7	8	16	17
37	7	9	16	18
38	8	9	17	18
39	8	9	17	19
40	8	10	17	19
41	8	10	18	19
42	9	10	18	20
43	9	10	19	20
44	9	11	19	20
45	10	11	19	20

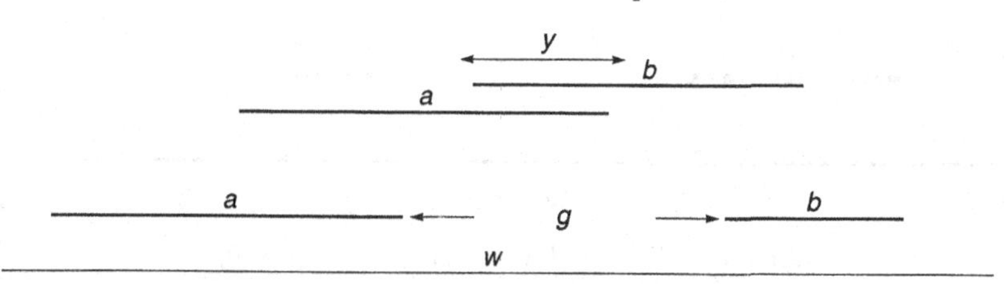

Figure 8.9 The gap size is g and y measures the amount of partial overlap. w is the length of the entire transect; b is the length of the shorter line and a that of the longer.

ℓ_0 that is significantly high (35) but the number of contiguities (5) is not significant. On the other hand, the sequence XXXXoXoXoXoXoXoXoooo has significantly more contiguities than expected (6), but ℓ_0 (21) is not significantly high. The difference is that contiguities are a very local feature of the pattern, since they are a characteristic of adjacent boundaries; whereas the amount of intermingling is a characteristic of the overall pattern. The same is true ecologically. Contiguities may reflect the interaction between pairs of species, where one replaces another along the gradient, but the intermingling of boundaries reflects the response of all the species in the community to that gradient.

Gap and partial overlap

The next step in our study of transect data giving the ranges of species along an environmental gradient is to examine the sizes of gaps between nonoverlapping ranges and the sizes of partial overlaps.

When a pair of lines do not overlap, the gap between them can be measured by variable g and when they overlap partially, variable y can be used to measure that overlap (see Figure 8.9). We do not need to define a similar sort of measure for cases of complete overlap, since it will always be the length of the shorter line.

Since g and y are measured lengths of gap and overlap, in deriving their expected values we need to use the null hypothesis H_d that the lines are placed independently of each other but keep their measured length. The expected values of g and y are:

$$E(g\,|\,H_d) = (w-a-b)/3, \qquad \text{if } a+b<w. \tag{8.7}$$

$$E(y\,|\,H_d) = b/2 + b^2/6(w-a) \qquad \text{where } a+b<w \text{ and} \tag{8.8}$$

$$= b-(w-a)/3 \qquad \text{where } a+b>w \text{ (Dale 1986).} \tag{8.9}$$

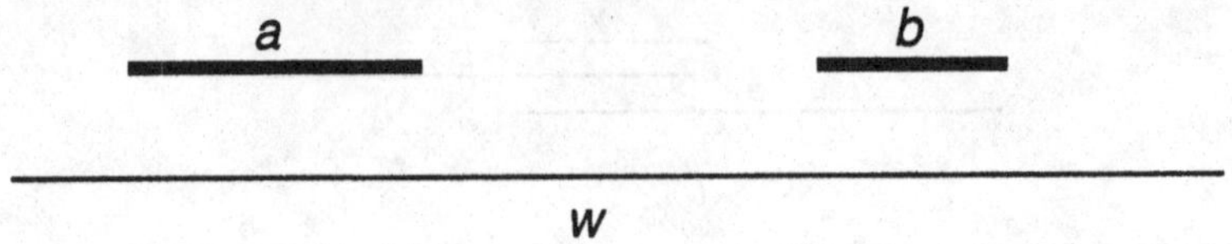

a b

w

Figure 8.10 Illustration of $E(g)$: $w=100, a=25, b=15; E(g)=20$.

For example, in Figure 8.10 $w=100$, $a=25$ and $b=15$; $E(g|H_d)=(100-25-15)/3=20$. In the upper part of Figure 8.11, $w=100$, $a=25$, and $b=20$; $E(y|H_d)=10+[400/(6\times 75)]=10.89$. In the lower part of Figure 8.11, $w=100$, $a=73$, and $b=50$; $E(y|H_d)=50-(27/3)=41$.

In devising statistical tests for these two variables we can consider testing individual observed values or compare the overall distributions of g or y with the expected distributions. The fact that the set of values of g or y are not independent makes the second kind of test difficult, so we will examine only the first kind of tests.

We can show that g is significantly different from its expected value if:

$$g>(1-\sqrt{\alpha/2})\,(w-a-b) \tag{8.10}$$

or

$$g<(1-\sqrt{1-\alpha/2})\,(w-a-b) \quad \text{(Dale 1986)}. \tag{8.11}$$

Using the usual 5% level, the critical values are $0.01258(w-a-b)$ and $0.8419(w-a-b)$ (Dale 1986). We can also show that the critical values for y are:

$\sqrt{\alpha/2})\,b$ and $\sqrt{1-\alpha/2})\,b$ if $a+b<w$,

$\sqrt{(1-\alpha/2)(a+b-w)^2+\alpha/2b^2}$, and

$\sqrt{\alpha/2(a+b+w)^2+(1-\alpha/2)b^2}$ if $a+b>w$ (Dale 1986).

For example, based on Figure 8.10, even if g is as low as 2, it is not significantly small because the critical value is 0.755, but if g is 55, it is significantly large because the critical value is 50.5. Based on the lower part of Figure 8.11, the critical values for y are 24.05 and 49.5, which are very close to its minimum (23) and maximum (50).

The same seaweed data set was analyzed using these techniques and the results show that not all contiguities have significantly small gaps. There are, however, a large number of partial overlaps that are significantly small. This result means that if species replacement along the

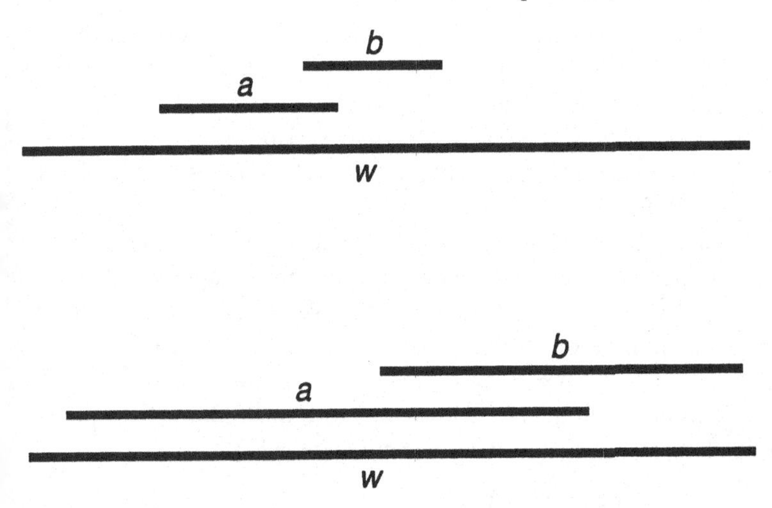

Figure 8.11 Illustration of $E(y)$: in the upper example, $w=100$, $a=25$, $b=20$; $E(y)=10.89$. In the lower example $w=100$, $a=73$, $b=50$; $E(y)=41$.

gradient is mediated by competition, there may be a small zone in which the competitors overlap, a phenomenon which is theoretically possible (Pielou 1974; see also Czárán 1991).

Event measure

The use of H_d, which measures the positions of the species' boundaries by the length along the transect, can be criticized because the same distance at one part of the gradient may produce a greater change in the environment experienced by the plants than at some other part. One version of this phenomenon for intertidal communities is the critical tide level hypothesis (Doty & Archer 1950). Therefore running length may not be the best measure of position. To avoid this problem we could measure position using the boundaries or events themselves, and redefining the length of a range as the number of events it includes. For example, using capital letters to represent upper boundaries, in ABCbac, the length of the range of the first species is the number of events from A to a, which is 5. We can re-examine overlap using a third null hypothesis H_e, that any two ranges occur independently of each other, but keeping their own length as measured by events. For enumeration, we can assume that the order of the upper boundaries is fixed. If there are n species we know that their arrangement can be described by a sequence of n pairs of labelled events and that there are $(2n-1)!!$ such sequences (Dale and Narayana 1984).

If the ranges of two species do not overlap, the gap between them is called g. For example, in the sequence ABCbDEaFdecGfg, the gap between Bb and Ff is $g=3$, occupied by D, E, and a.

If the ranges overlap partially, the partial overlap is y.

In the same example, the overlap of Aa and Cc is $y=5$, occupied by events C, b, D, E, and a.

If the ranges overlap completely, the overlap is the length of the shorter range, z. In the sequence we have been using as an example, the overlap of the first two species, Aa and Bb, is $z=3$, the events B, C, and b. For n pairs of events, g can have values from 0 to $2n-4$, and y and z can have values from 2 to $2n-2$.

Let $f_g(n,j)$ be the frequency with which $g=j$ in all $(2n-1)!!$ sequences of n labelled pairs of events. We can show that for $n>2$:

$$f_g(n,j) = (2n-5)!!\binom{2n-j-1}{3}. \quad (8.12)$$

$E(g\,|\,H_e) = (2n-4)/5$ and $\mathrm{Var}(g\,|\,H_e) = 4(2n+1)(n-2)/75$ (Dale 1988b). The distribution is skewed and does not approach the Normal distribution with increasing n.

Let $f_y(n,j)$ be the frequency with which $y=j$ in the $(2n-1)!!$ sequences of n pairs of events and define $f_z(n,j)$ similarly for z. We can show that:

$$f_y(n,j) = f_z(n,j) = f_g(n,j-2) \quad \text{(Dale 1988b)}. \quad (8.13)$$

In devising a one-sided statistical test for small values of g, we know that $\Pr(g=0) = 2/n$, so that $g=0$ can be significantly small only in cases where $n>40$, which may not be very useful. One-sided tests for significantly large values of g are more practical. Table 8.4 gives the upper 5% critical values for g, y, and z. In the sequence ABaCDEcFdebGgf, the gap between Aa and Gg is 8 which is greater than the critical value of 7 and is therefore significantly large.

These methods allow us to examine gaps and overlaps of species' ranges from the 'plant's eye view', by using the upper and lower boundaries of other species as measures of the intensity of change in the environmental factor. The approach also provides a test for values of z, which is not possible if the species' ranges are measured by length along a transect. The main disadvantage of this approach is that to test for significantly small values of the variables, large values of n are required. We should point out also that all pairs of species cannot be tested because of the lack of independence among such tests. We would therefore

Table 8.4. *Critical values of* g, y *and* z *(upper 5%) for* $n = 4$ *to* $n = 100$

n	g	y and z
4–15	n	$n+2$
16–33	$n+1$	$n+3$
34–51	$n+2$	$n+4$
52–69	$n+3$	$n+5$
70–88	$n+4$	$n+6$
89–>100	$n+5$	$n+7$

Notes:
These measures of gap and overlap use events as units of length (Dale 1988b).

reserve this approach for special cases in which particular pairs of species are of interest.

Clumping and spacing of boundaries

So far we have examined a variety of properties of the arrangements of species in communities on environmental gradients. We now examine the extent to which the boundaries of the species (either upper or lower) are clumped together (or widely spaced out) along the gradient. This is a useful characteristic for distinguishing among the different models of the arrangement of species on a gradient. For example, the clumping of boundaries is one of the characteristics predicted by the community unit model of plant communities (Shipley & Keddy 1987). It is also useful for testing hypotheses such as the critical tide level hypothesis (Doty and Archer 1950), described above, which would cause the boundaries of species to be clumped at the critical levels of the gradient where the effect of the gradient is most intense.

To evaluate the spacing and clumping of boundaries, it is usual to standardize the gradient to a length of 1 between the upslope boundary of the first species at 0 and the last species' downslope boundary at 1. The other boundaries of the species ranges are then treated as points in the unit interval, $x_1, x_2, \ldots x_m$, which break the interval into $m+1$ parts, $u_1, u_2, \ldots u_{m+1}$ (see Figure 8.12). Where there are n species and the whole gradient is considered, $m = 2n - 2$. If the boundaries are clumped, there should

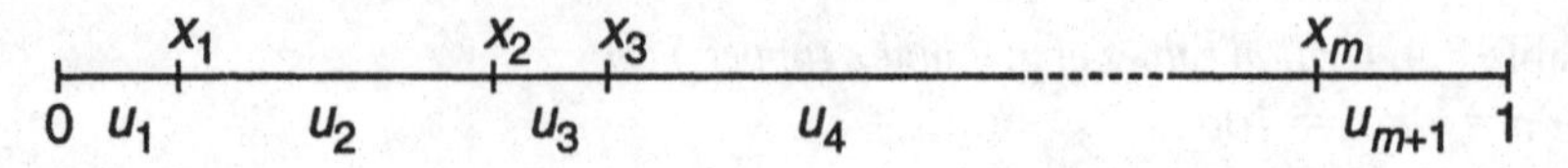

Figure 8.12 Random numbers in the interval 0 to 1, $x_1, x_2 \ldots x_m$, break the interval into $m+1$ segments, $u_1, u_2, \ldots u_{m+1}$

be a large number of small segments and a number of large segments separating the clumps, and the first step in looking for clumping would be to measure the variability of the u_1 values.

Several authors have described the measure

$$W_m = \sum_{i=1}^{m+1} u_i^2 \tag{8.14}$$

under a variety of names (Greenwood 1946; Kimball 1947; Darling 1953). On the null hypothesis that the x_i's are randomly and independently placed in the unit interval, each governed by a rectangular distribution on (0,1), the mean and variance of the distribution of W_m are:

$$E(W_m) = 2/(m+2) \tag{8.15}$$

and

$$V(W_m) = 4m/[(m+2)^2(m+3)(m+4)]. \tag{8.16}$$

For example, for $m=20$, $E(W_m)=0.091$ and $V(W_m)=0.000299$.

The distribution of W_m is asymptotically Normal with increasing m, but the convergence is very slow (Moran 1947). The slow convergence makes it impossible to use the Normal approximation for significance testing. Instead, we used a Monte Carlo approach to derive critical values for W_m. Table 8.5 presents the approximate critical values for $m=4, 5, \ldots 50$, based on 100000 randomly broken unit intervals.

The measure W_m really examines the pairwise clumping of boundaries because the small values of the u_i's are the result of two boundaries being close together. That measure cannot detect whether the boundaries occur in clumps of more than two and therefore we need a second statistic that measures the serial autocorrelation of the sizes of adjacent segments:

$$h_m = \sum_{i=1}^{m} u_i u_{i+1}. \tag{8.17}$$

Based on the same null hypothesis, the mean and variance of h_m are:

Table 8.5. *Approximate critical values of* W_m *from 100000 simulations.* p *is the cumulative probability*

	p					
m	0.001	0.025	0.05	0.95	0.975	0.999
4	0.214	0.222	0.231	0.511	0.566	0.639
5	0.184	0.192	0.200	0.434	0.487	0.551
6	0.162	0.169	0.176	0.373	0.415	0.477
7	0.146	0.151	0.157	0.330	0.370	0.417
8	0.132	0.137	0.143	0.293	0.327	0.369
9	0.121	0.126	0.131	0.265	0.294	0.334
10	0.112	0.116	0.121	0.239	0.265	0.302
11	0.104	0.108	0.113	0.221	0.245	0.276
12	0.097	0.101	0.105	0.203	0.225	0.254
13	0.092	0.095	0.099	0.189	0.208	0.234
14	0.086	0.090	0.093	0.175	0.193	0.214
15	0.082	0.085	0.088	0.163	0.180	0.203
16	0.077	0.081	0.084	0.154	0.169	0.189
17	0.074	0.077	0.080	0.145	0.158	0.178
18	0.071	0.073	0.076	0.136	0.148	0.165
19	0.068	0.070	0.073	0.130	0.141	0.156
20	0.065	0.067	0.070	0.123	0.135	0.150
25	0.054	0.056	0.058	0.098	0.105	0.117
30	0.046	0.048	0.050	0.081	0.088	0.097
35	0.041	0.042	0.043	0.070	0.074	0.081
40	0.036	0.037	0.039	0.069	0.064	0.070
50	0.030	0.031	0.032	0.048	0.051	0.054

$$E(h_m) = m/[(m+1)\ (m+2)] \qquad (8.18)$$

and

$$V(h_m) = (m^3 + 3m^2 + 4m - 4)/[(m+1)^2(m+2)^2(m+3)(m+4)] \quad \text{(Dale 1988a).} \qquad (8.19)$$

For example, for $m=20$, $E(h_m)=0.0433$ and $V(h_m)=0.0000787$, which would give the confidence interval of (0.0259, 0.0607) if we could reliably use a Normal approximation.

We have not investigated the asymptotic distribution of this statistic in a formal way, and so we give the approximate critical values for h_m based on 100000 trials in Table 8.6. The distribution of h_m seems to approach the Normal distribution much more rapidly than that of W_m. For

Table 8.6. *Approximate critical values of* h_m *from 100000 simulations.* p *is the cumulative probability*

	p					
m	0.001	0.025	0.050	0.0950	0.975	0.999
4	0.022	0.033	0.046	0.211	0.223	0.232
5	0.026	0.037	0.048	0.188	0.199	0.211
6	0.028	0.038	0.048	0.165	0.177	0.191
7	0.030	0.039	0.047	0.147	0.159	0.173
8	0.031	0.038	0.046	0.133	0.143	0.156
9	0.030	0.037	0.044	0.120	0.130	0.143
10	0.030	0.036	0.043	0.110	0.118	0.130
11	0.029	0.035	0.041	0.101	0.109	0.119
12	0.029	0.034	0.039	0.094	0.101	0.110
13	0.028	0.033	0.038	0.087	0.094	0.102
14	0.028	0.032	0.036	0.082	0.088	0.096
15	0.027	0.031	0.035	0.077	0.082	0.090
16	0.026	0.030	0.034	0.072	0.077	0.084
17	0.026	0.029	0.033	0.068	0.073	0.079
18	0.025	0.028	0.031	0.064	0.069	0.075
19	0.025	0.028	0.030	0.061	0.065	0.071
20	0.024	0.027	0.029	0.058	0.062	0.068
25	0.021	0.024	0.025	0.047	0.050	0.054
30	0.019	0.021	0.022	0.039	0.041	0.044
35	0.017	0.019	0.020	0.033	0.035	0.038
40	0.016	0.017	0.018	0.029	0.031	0.033
50	0.013	0.014	0.015	0.023	0.024	0.026

instance, when we compare the approximate critical values for $m=40$ and $p=0.025$ and $p=0.975$ which are 0.017 and 0.031 in Table 8.6 with those from the Normal approximation, 0.0164 and 0.0307, they are much closer than the equivalent values for W_m.

As a numerical illustration of these two indices, consider the sequence of values 0.001, 0.001, 0.001, 0.001, 0.001, 0.147, 0.15, 0.45, 0.25. These data give $W_m=0.309$ and $h_m=0.202$, both of which are significantly high for $m=8$. When the same data are rearranged to give alternating high and low values, 0.001, 0.147, 0.001, 0.15, 0.001, 0.45, 0.001, 0.25, 0.001, W_m is unchanged, but $h_m=0.002$ which is significantly low. Figure 8.13 illustrates the usefulness of the two statistics in detecting the high variability of interboundary distances and their clumping. Almost all combinations of test results for W_m and h_m seem to be possible, except for the combina-

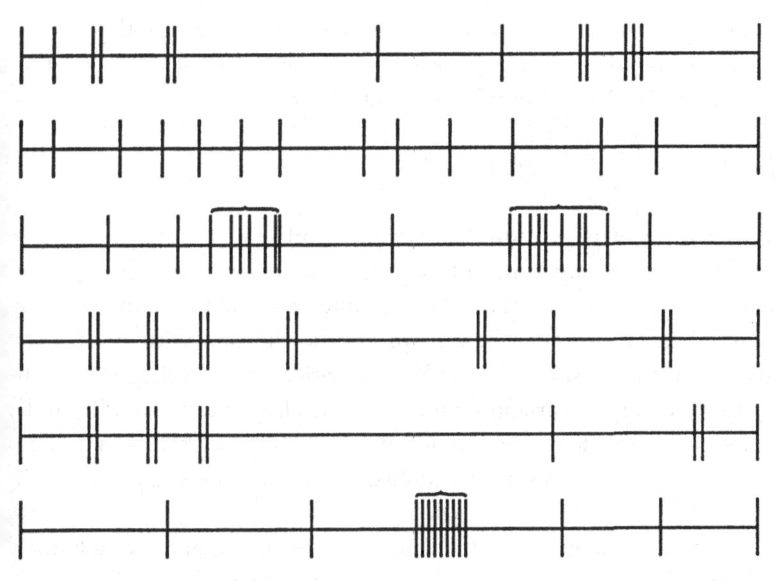

Figure 8.13 Different test results for w_m and h_m arising from the arrangement of boundaries. The long lines represent environmental gradients and the short bars are the species' boundaries. The curly brackets indicate regions of clumping of boundaries. From the top down, the results are as follows: w_m significantly high, h_m not significant; w_m low, h_m not significant; w_m not significant, h_m high; w_m not significant, h_m low; w_m high, h_m low; w_m high, h_m high. (Redrawn from Dale 1988a.)

tion of W_m being significantly low and h_m being significantly high. This exception occurs because when the segments are significantly more equal than expected, the autocorrelation of adjacent segments cannot be significantly high.

When large values of h_m indicate that the boundaries are clumped, we need to identify the parts of the gradient where the boundaries are most clumped together. Those parts, if any, will indicate the critical levels of the controlling factor. What we do is to identify the sections of the gradient where the observed concentrations of boundaries have the lowest probabilities based on the null hypothesis of independent rectangularly random placement.

For each pair of boundaries x_i and x_j, where $i<j$, we calculate the probability that the interval between them contains at least as many boundaries as it does; call it $p(i,j)$. That interval contains exactly $j-i-1$ boundaries. On the null hypothesis, any boundary, except x_i and x_j and those that define the end points, falls between them with a probability of

$x_i - x_j$. Let q be the number of other boundaries available; if both x_i and x_j are endpoints, $q = m$ and if only one of them is an endpoint, $q = m - 1$. For most pairs if x_i and x_j, $q = m - 2$. Then (Dale 1988a):

$$p(i,j) = 1 - \sum_{k=0}^{j-i-1} \binom{q}{k} (x_j - x_i)^k (1 - x_j + x_i)^{q-k}. \tag{8.20}$$

Small values of $p(i,j)$ indicate that the number of boundaries in the interval between x_i and x_j has a low probability of occurring by random processes. If there are several overlapping intervals containing improbably large numbers of boundaries, we can choose the one with the lowest probability as the most likely to indicate a critical level on the gradient. If there are several nonoverlapping intervals with large numbers of boundaries, we can look for a preset number of clumps of boundaries and choose those with the lowest probability, or we may choose all those with $p(i,j)$ less than some threshhold value such as 5%. The $p(i,j)$ are not independent of each other, and so this is not a proper statistical test of whether the boundaries clumped, with a known level of significance. Combined with a statistical test for h_m, however, which examines the whole gradient, calculating $p(i,j)$, for all i and j provides an objective method for detecting critical levels.

These methods can be illustrated using the Nova Scotia seaweed data which we have already analyzed in other ways. For these data, we started at the top of the shoe and the transect ended in the upper subtidal. Because the range of some species extended into deeper water that we did not sample, the last records of several species do not necessarily represent the end of their range. The run of T endings at the end of the transect, the terminal run of the sequence, may be broken by unperceived beginnings, the upper boundaries of species growing in water deeper than we could sample. Therefore, the terminal run of endings should be omitted from the analysis, and we should set $m = 2n - 2 - T$ and have the end of the unit interval defined by the last species' upper boundary.

The values of W_m and h_m calculated from these data were tested using a two-sided test at the 5% significance level. For all transects, even if h_m was not significantly large, we identified the regions of greatest boundary clumping using $p(i,j)$ and a threshold probability of 5%. All 29 transects had values of W_m that were greater than expected, and 13 were significantly greater. In 22 transects, h_m was greater than expected and in 11 of them was significantly high. In two transects, h_m was significantly low and in those cases W_m was significantly high. We found that most of the transects have one possible critical level, even if h_m was not significantly high; 5 of the 29 have two. Most of the critical levels were detected in the bottom quarter of the transect and most had probabilities less than 0.1%.

To summarize the seaweed data analysis:

1. The observed values of ℓ_0 tended to be greater than expected and 11 of the 29 were significantly high.
2. There was a significant tendency to have more contiguities than expected but not all contiguities had significantly small gaps.
3. A large number of partial overlaps were significantly small.
4. There is some evidence for the existence of clumps of boundaries in the lower parts of the transects.

When comparing these results with the various models of the arrangements on gradients, it must be remembered that we have not looked at densities, only the ranges of species defined by their highest and lowest occurrences. The evidence we have, however, does not support the individualistic model. In a study of marsh vegetation, Shipley and Keddy (1987) examined the upper and lower boundaries separately and also found reason to reject the individualistic model. The relationship of evidence to models will be discussed further in more general terms at the end of the chapter.

Concordance

The last topic that needs to be discussed with regard to the analysis of the ranges of species on a spatially continuous environmental gradient is a measure of the similarity of the orders of the upper and lower boundaries. This topic is an important one in evaluating the community unit model, because if such units exist the order of boundaries should be very similar in all transects through the same vegetation. On the other hand, if the individualistic model holds and the endpoints of each species' range is strongly and precisely controlled by its physiology, the order of the boundaries on the gradient will show little variability.

The order of the boundaries is essentially a ranking of events and the similarity of such rankings is generally referred to as concordance. There are several measures available for the evaluation of the concordance of several rankings (Conover 1980). Let n be the number of objects in each ranking and let m be the number of rankings. Let r_{ij} be the rank of the ith object in the jth ranking and R_i be its rank sum: $R_i = \Sigma r_{ij}$. Let μ be the mean of the R_i and V their variance; $\mu = m(n+1)/2$; $V = m(n^3 - n)/12$. The usual form of the test is to calculate:

$$T = (n-1)\sum_{i=1}^{n}(R_i - \mu)^2/nV, \tag{8.21}$$

which then can be compared to a table or for larger values of m and n, the χ^2 distribution with $n-1$ degrees of freedom (*cf.* Conover 1980).

This technique can be used to examine either all the upper boundaries or all the lower boundaries separately. In studying seaweed zonation, we might hypothesize that upper limits are probably set by the plants' physiology but that lower limits are set by biological interactions such as competition and predation. The prediction would then be that the order of the lower boundaries would be much more variable than that of the upper boundaries. The Nova Scotia seaweed data we have been using to illustrate these methods give no clear evidence that the prediction is true.

In dealing with both the upper and lower boundaries of species' ranges, we are dealing with rankings of paired events; the lower boundary must be below the upper boundary of any species, so the $2n$ events are not all independent. Clearly then, we cannot use the procedure for testing rankings of simple events unmodified; the lack of independence must be incorporated into the test.

In devising a test of concordance for rankings of paired events, there are at least two possible approaches. The first is to use standard techniques to evaluate the concordance of the orders of the upper boundaries considered separately, and of the lower boundaries considered separately, to create a new measure of the concordance of order of upper and lower boundaries (now ignoring the species) and then to combine the three into a single measure. An alternative is to modify an existing measure of concordance to deal with paired data and determine its limiting distribution. We shall explore the second approach.

There are m rankings of n pairs of events. Let R_{x_i} be the rank sum for the ith beginning and R_{o_i} be the rank sum for the ith ending. The overall mean of the rank sums is $\mu = m(2n+1)/2$, but the expected value of the R_{x_i} is $\mu_x = m(2n+1)/3$ and $\mu_o = 2m(2n+1)/3$. It is important to realize, however, that the actual means of the R_x or the R_o may not be those expected values in a particular set of rankings; the observed means depend on the intermingling of the two kinds of events. The variances of the rank sums about their own expected values are both the same: $V_x =$

$V_o = m(2n+1)(n-1)/9$. From this fact we can derive their variance about the common mean:

$$V_\mu = m(2n+1)[m(2n+1)+4(n-1)]/36. \quad (8.22)$$

The test statistic to consider is then:

$$T_p = S/V_\mu \quad (8.23)$$

where:

$$S = \sum_{i=1}^{n} [(R_{x_i} - \mu)^2 + (R_{o_i} - \mu)^2]. \quad (8.24)$$

Unfortunately, the limiting distribution of the statistic is unknown, but it is probably a noncentral χ^2 distribution (*cf.* Kendall & Stuart 1973). The best we can offer the reader with our present knowledge is to suggest randomization procedures to evaluate the statistic and to provide an index for comparison.

The randomization procedure shuffles the labels of the beginnings and endings in the rankings, but leaves the 'template', the pattern of beginnings and endings, of each ranking intact. The number of times that the observed value of S is exceeded in 1000 such relabellings is counted to test whether the observed level of concordance can be considered to be statistically significant. A Monte Carlo procedure can be used to complement this randomization test: S is calculated for each of 1000 randomly generated sets of m rankings for n species and the statistics again compared.

For example, one of the Nova Scotia seaweed study sites gives the following rankings of the upper and lower boundaries of the five species common to the three transects:

1,2; 3,5; 4,9; 6,8; 7,10
1,2; 3,6; 4,5; 7,10; 8,9
1,2; 4,6; 3,5; 7,10; 8,9.

Calculations give $S = 690.5$, $V_\mu = 44.9$ and thus $T_p = 15.3$. Both the randomization and Monte Carlo tests show that the rankings are highly significantly concordant.

Another way in which to evaluate and compare the degree of concordance is to create an index that runs between 0 and 1:

$$W = (S_{obs} - S_{min})/(S_{max} - S_{min}). \quad (8.25)$$

The subscripts refer to the observed value, the minimum and maximum values. Dale (1979) shows that:

$$S_{max} = m^2 n(2n-1)(2n+1)/6 \tag{8.26}$$

and

$$S_{min} = m^2 n/2. \tag{8.27}$$

The minimum value is actually somewhat larger if m is odd and n is even, but the value given above is a convenient generalization. For the data just described:

$$W = (690.5 - 44.9)/(742.5 - 44.9) = 0.925,$$

clearly indicating a high degree of concordance.

When we use this technique on the Nova Scotia seaweed data, while the concordance of transects at the same site is usually significantly high, the values of W are usually between 0.8 and 0.9, showing a high degree of similarity, but also a certain amount of variability.

Quadrats: presence/absence data

Many studies of the pattern of species on environmental gradients have used contiguous or spaced quadrats and recorded density or presence/absence in them. For some kinds of analysis, density data can be converted to the presence/absence form. Quadrat data can be used for many of the kinds of analyses described above, such as overlap, gap size, intermingling of boundary types, and clustering of boundaries. If the quadrats are small enough that there is never more than one boundary in a quadrat, then the methods described above for continuous ranges can be transferred over directly and used, with quadrat position in the sequence being the equivalent of distance along the transect. The problem comes when the quadrats are larger and there are many ties in the ranking of boundary order. For instance, Wulff and Webb (1969) used quadrats 4 cm in height to study seaweed zonation on pilings and the data they present have many ties, with as many as six boundaries in a single height interval. These ties obviously represent a loss of information. If the relationships among the ranges and boundaries of species are of interest and several quadrats contain more than one boundary, the scale of sampling does not match the scale of the pattern: the quadrats are too small.

For data in which the positions and orders of boundaries are not all known, we need tests for the clustering of species boundaries that are different from the ones already described. A number of workers have examined the detection of clustering of upslope boundaries or of downslope boundaries using data from transects of quadrats (Pielou 1975, 1979; Pielou and Routledge 1976; Underwood 1978a,b; Shipley & Keddy 1987). There are several different methods for testing for clustered boundaries based on two different null hypotheses.

Suppose there are Q quadrats of which q contain at least one of the K boundaries being considered. Pielou's approach (Pielou & Routledge 1976) was to calculate p_q:

$$p_q = \binom{Q}{q}\binom{K-1}{q-1} \Big/ \binom{Q+K+1}{K}. \tag{8.28}$$

Having observed that exactly z quadrats contain boundaries, we can test whether z is sufficiently small by calculating the probability, P, and seeing whether it is less than our chosen significance level:

$$P = \sum_{q=1}^{z} p_q. \tag{8.29}$$

The approach of Underwood (1978b) was to consider a set of transects. Using the same symbols, the expected value of q is

$$E(q) = Q[1-(1-1/Q)^K]. \tag{8.30}$$

For each transect of the set, the deviation of the observed value from the expected is calculated, d_i for the ith transect. The mean and sample variance of the d_i are then calculated and the standard t-test is used to see whether the mean is significantly different from zero.

The second method is less appealing than the first since its requires several transects for the test and since its resolving power will increase the more transects are used in a particular area. There might be some concern about spatial autocorrelation in the data. Neither method provides an objective method for determining where the critical levels (if any) are on the shore. It is interesting that Underwood (1978a,b), using his method, found no evidence for the clustering of boundaries and thus no critical levels on rocky shores in England. On the other hand, using the autocorrelation statistic, h_m, on the Nova Scotia data, we found that there was significant clustering of boundaries. Underwood's quadrats were 0.5 m by 0.5 m, and it is not entirely clear what the effect of different quadrat sizes might have on this kind of test. Shipley and Keddy (1987) comment that Underwood's test is conservative in that it has a tendency to miss nonrandom patterns.

In their own study, Shipley and Keddy (1987) used spaced quadrats to examine the spatial pattern in a marsh near Breckenridge, Quebec. The data were presence/absence and the first and last occurrence of each species was considered a boundary. They wanted to evaluate whether the data supported the community unit or the individualistic model of community structure described above by testing for the clustering of both types of boundary and the coincidence of upper and lower boundaries. They used a technique called the analysis of deviance to test the hypotheses, which tests the significance of the improvement of goodness of fit as successive terms are added to a model (the residual deviance is analogous to the residual sum of squares in analysis of variance). Their results showed that both kinds of boundaries were clustered in relative height classes, but the two kinds were clustered in different height classes. The conclusion was that both the individualistic and the community unit models being tested should be rejected.

Working at a different scale, Auerbach and Shmida (1993) analyzed the distribution of vascular plants along an altitudinal gradient on Mt. Hermon in Israel. They used 50 m and 100 m altitude bands and looked at the number of upper and lower boundaries in each. Expected values were calculated using a randomization method in which the altitudinal ranges of the species were preserved, as in our H_d described earlier in this chapter (p. 251). They found that the downslope boundaries were not different from expected but that the upslope boundaries were. They also found a clustering of the two kinds of boundaries at about 1200 m a.s.l. indicating the transition from maquis (shrubland) to open deciduous forest. They concluded that while there were vegetation discontinuities along the gradient, there was no evidence of discrete communities.

Density data

Up to this point of the chapter, we have been discussing methods for analyzing presence/absence data, recorded in a continuous fashion, so that the endpoints of the species' ranges are known. It is easy to imagine density data continuously recorded along an environmental gradient to give a picture of the species' response as in Figure 8.1. Such figures are frequently presented in papers and textbooks, but they are usually based on data from spaced quadrats, not contiguous quadrats. In using the species' ranges in the first part of this chapter, we have represented them as solid lines, although in any single transect a more realistic representation of the species occurrence would be an irregularly broken line with lots of small gaps. In the same way, a continuous record of density would be extremely ragged which would then have to be smoothed to produce a curve (*cf.* Gauch 1982, Figure 3.8). We have found no studies that have used continuous density data, probably because in most vegetation types, collecting such data would be extremely difficult.

Density data from quadrats, either contiguously or regularly placed along an environmental gradient, can, of course, be used to draw density response curves such those in Figure 8.5. The uses to which such data can be put get us into the broad area of multivariate analysis, including direct gradient analysis, niche overlap measurement, and so on. Those topics are covered with varying degrees of detail and sophistication in other places (Gauch 1982; Ludwig & Reynolds; Kent & Coker 1992), and we will therefore not deal with all of them here. Many of the kinds of analysis already presented in this chapter can be used with such data.

Keddy (1991) distinguished between spatially continuous and spatially discontinuous gradients. In the first case, the result will be some kind of zonation but in the second the gradient will have to be (re)constructed by the researcher. Since we are interested in spatial pattern on environmental gradients, it is the first kind that is more straightforward to deal with.

In the second kind of gradient, there are three components that need to be dealt with: the spatial relationships among samples in which the controlling environmental factor(s) have the same and different values, the relationship of the vegetation to the factor(s), and the spatial relationships among the plants in different samples. Keddy (1991) suggests that in these cases direct, rather than indirect, gradient analysis is to be preferred for looking at the relationship between the plants and the environment. Even using direct analysis, such a study would be complex, and given natural variability it could be very difficult to interpret, especially if more than one gradient is active.

One kind of discontinuous gradient that can be dealt with in spatial pattern analysis is the case where the gradient is actually piecewise continuous. That is what gives the spatial pattern in hummock-hollow systems for instance, where there are moisture gradients for wet to dry and dry to wet repeated through the community. We have discussed this kind of situation at several places in this book (for example in Chapters 1,

3, and 5). The pattern of the gradients and the pattern of the plants match up because the pattern in both is the result of interactions between the biotic and abiotic factors.

Perhaps the most important feature of density data is that it allows us to examine the response of a species to the environmental gradient within its range in a quantitative way. As we said at the beginning of this chapter, one underlying model of a species' response is the Gaussian or bell-shaped curve illustrated in Figure 8.1. The overall finding is that few species follow such an idealized response, but it is worth examining some of the details.

Austin and Austin (1980) studied the behavior of 13 grass species along an experimental nutrient gradient. They found no evidence to support the assumption of a Gaussian response curve, the physiological responses being asymmetric. Werger *et al.* (1983) examined a transect through a grassland in the Netherlands. Of the species, 21% showed a Gaussian response, 23% had strongly skewed unimodal responses, 10% were bimodal and 15% had complex irregular curves. Austin (1987) examined the distribution of canopy species of a sclerophyll forest in Australia with respect to mean annual temperature and found that positively skewed curves rather than bell-shaped curves were characteristic of the major canopy species. Minchin (1989) studied the distribution of species in the montane vegetation of Tasmania with respect to elevation and drainage. Of the species 45% had responses that were unimodal and symmetric, 33% were skewed, and 22% were complex. Gignac *et al.* (1991) examined the response surfaces of six species of mire bryophytes with respect to ecological and climatic gradients; most are skewed or bimodal. Austin *et al.* (1994) found that all of nine eucalypt species that they examined had skewed responses to temperature and that the direction of skew was a function of the position on the gradient. Austin and Gaywood (1994) extended this research, suggesting that the tail of the skewed distribution is toward the more mesic portion of the gradient; they found that 21 of 24 species of eucalypt conformed to the model.

Density data also allow us to look at among-species pattern along a gradient, by examining the relative positions of the modal densities of the species. For instance, Minchin (1989) tested some of the propositions of Gauch and Whittaker (1972) concerning the organization of species' responses. He tested whether the modes of the major species were evenly distributed along the gradient, perhaps because of competition and resource partitioning, and whether the modes of the minor species were randomly distributed. Using the same data on Tasmanian montane vegetation used to look at individual species' responses he found that the modes of the major species were in fact randomly distributed. He divided the minor species into structural groups and all but one supported the hypothesis of random dispersion. A somewhat similar result was obtained by Austin (1987) for the eucalypt forest trees: the proposition that the modes of the major species were evenly distributed was not supported, but in this instance there was some concern that changes in species' richness along the gradient might be a confounding factor.

Concluding remarks

It is clear from the material discussed in this chapter that spatial pattern on environmental gradients is an important aspect of the spatial organization of vegetation. The spatial pattern here takes into account several different kinds of characteristics: the upper and lower boundaries of species, the ranges of their presence, the way in which the densities of individual species respond to the gradient and the way in which the maxima of different species are arranged with respect to each other. All these characteristics can be used to help ecologists generate ideas and to test hypotheses. Based on the studies cited here, it seems that while there is little evidence for the existence of vegetation units on gradients, neither are the arrangements of species random. More work waits to be carried out to explain this kind of spatial pattern and the processes that give rise to it.

Recommendations

1. It is probably best to choose the sampling method that gives the most detail on the species' distributions along the gradient because that will increase the number of different characteristics that can be examined, as well as the precision with which they can be quantified. Depending on the spatial scale of the gradient, the measured line intercept method is preferred. If quadrats are used, they should be small.
2. It is also best to use several sampling transects, in order to evaluate the variability in the patterns.
3. The continuous presence/absence data derived from line transects give the measured positions of the upper and lower boundaries of each species. These data can be used to investigate several properties

that will help distinguish among models of how species are arranged on the gradient: overlap of ranges, intermingling of boundary types, contiguity of upper and lower boundaries, gaps between ranges, and the clustering of boundaries.

4. The degree of agreement among the transects should be measured by a concordance statistic.
5. For quadrat data, randomization methods are probably the safest statistical procedures for testing their properties.
6. In using the density of species along spatially continuous environmental gradients, it is unsafe to rely on the assumption that the response is a symetric bell-shaped curve.

9 · *Conclusions and future directions*

Summary of recommendations

In order to study spatial pattern and to answer questions about the relationship between the pattern and the processes that either give rise to it or are affected by it, we need to be able to detect pattern reliably and to quantify its characteristics. The highest quality data for pattern analysis come from strings or grids of relatively small contiguous quadrats in which some quantitative measure of species' densities has been recorded, or from the direct mapping of individual plant units such as stems. Even with the best data, no single method of analysis can quantify all the important characteristics of pattern.

For contiguous quadrat data, three-term local quadrat variance (3TLQV) and new local variance (NLV) form a good combination of methods to evaluate single-species pattern (Chapter 3). These methods detect the scales of pattern and the size of the smaller phase; the intensity associated with a 3TLQV peak can be used to evaluate the consistency of the pattern. For two species, three-term local quadrat covariance (3TLQC) is recommended; paired quadrat covariance (PQC) cannot be used because of the effects of resonance peaks, and the correlation coefficient cannot be used by itself to detect scale (Chapter 4). Multispecies pattern is best investigated using the modified multiscale ordination (MSO) technique based on 3TLQV; among its advantages are an evaluation of the evenness of the species' contributions to the multispecies pattern (Chapter 5).

In studying the associations of species, it is an important consideration that the results of pairwise tests are not independent of each other. It is also important to realize that the associations may not themselves be pairwise, and the choice of sampling and analysis methods should reflect this possibility (Chapter 5). Studying multispecies association of k species using the 2^k contingency table method is not practical for large numbers

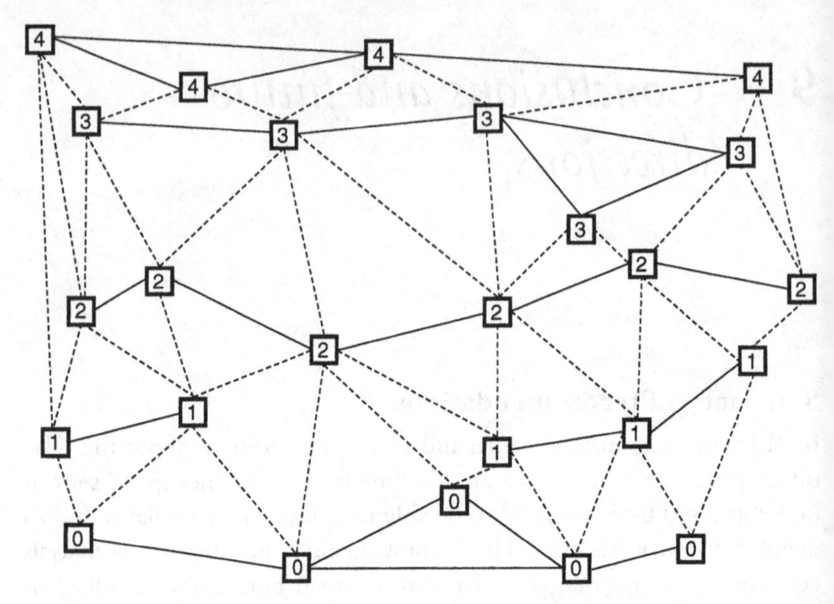

Figure 9.1 A triangulation of spaced samples as a framework for further analysis. The numbers in the quadrats are some quantitative characteristics. Solid lines join samples with the same value; broken lines for pairs that are different.

of species. It may, however, provide valuable insights into the structure of a community when applied to a selected group of important species.

In most applications, spaced samples are not recommended for pattern analysis, but in some instances the collection of data in contiguous quadrats may be impractical or impossible. This situation will arise, for example, when the substrate is itself discontinuous or when the size of the sample that can practically be collected or processed is at a much smaller scale than the scale of pattern that is of interest. In any of these cases, the most important thing is to keep the spatial relationship among the samples and to use it in the analysis (Chapter 6). The samples and their values can then be treated as objects in space and their autocorrelation evaluated. For example, the samples can be joined in a least diagonal neighbor triangulation (Figure 9.1) and the autocorrelation of first-order neighbors, second-order neighbors and so on can be plotted. Another approach is to evaluate the amount of correlation between pairs of samples in defined distance classes.

For mapped point data, the group of methods based on Ripley's K function are the most highly recommended, but there is room for the development of new methods for point pattern analysis to complement that approach, particularly for the evaluation of gaps in the point pattern (Chapter 7). A major strength of the method is that it can be applied to univariate or bivariate data, to answer different questions about scales of overall pattern or of segregation or aggregation of two kinds of plants.

To study the response of species' densities to a continuous environmental gradient, there are advantages to sampling methods that give detailed information on how density varies with distance and that gives the order of the end points of species' ranges with some precision. Such data can then be analyzed by several different techniques to answer questions about the relationships between different species ranges, such as the patterns of overlap. The same data can be used to evaluate hypotheses about the spatial distributions of the upper and lower boundaries of species; for example, whether they occur in clusters or whether the two types of boundary are intermingled (Chapter 8).

A general theme throughout this book is the lack of independence, whether in the form of associations that are not pairwise by nature, tests that are not independent, or spatial autocorrelation. Even statistical tests for analysis methods that we wish to recommend, such as 3TLQV, are made difficult because of lack of independence arising from the way in which the data are used. In many instances, the covariation in the data we wish to analyze may be so difficult to quantify or partial out that it will make randomization procedures the best approach to evaluating the results.

What next?

Having worked through the preceding eight chapters that represent the body of the book, it will have become clear that we are not yet at the point where final conclusions can be drawn; the field of spatial pattern analysis in plant ecology is really just beginning. Several of the technical chapters have ended with concluding remarks and recommendations, but it seems appropriate to end the book with a look ahead. We will end, therefore, with a sample of problems to be solved, techniques to be developed, and questions to be answered.

Three dimensions

Chapters 4–6 described extensions of the basic methods of Chapter 3 to two species, to many species and to two dimensions. Many communities of plants are three dimensional and in some, at least, the vertical dimension is very important to an understanding of their physical structure. Epiphytes in a rain forest, corticolous lichens, and phytoplankton in an aquatic community are just three examples. Much of the influence of a plant community on the animals that live within it arises from the three-dimensional structure. Quantitative description of the three-dimensional pattern of vegetation is therefore an important component of defining habitat structure. Without working out full details, we can speculate about new methods that could be developed to deal with three-dimensional data.

What will the data look like? As discussed in earlier chapters, the data may consist of points arranged in an otherwise empty volume. For instance, we could make a three-dimensional map of the positions of fruit or aphids on a tree, or of the positions of a particular species of orchid in a patch of rain forest. The alternative is to have sampling units from which we obtain presence/absence data, counts or density estimates. An obvious arrangement would be to use the equivalent of a lattice of cubic sampling units and, for instance, record the counts of various species of phytoplankton in each cube of a volume of water.

For point pattern data, many of the methods described in Chapter 7 for two-dimensional analysis can be transferred into a three-dimensional form without difficulty. For instance, nearest neighbor methods and point-to-all-points can be used without modification. On the other hand, Ripley's second-order method will require some changes.

As in the two-dimension case, the number of pairs of points with distance t or less between them will be counted. This is now the equivalent of counting points in spheres of radius t centered on the data points. As in Chapter 7, d_{ij} is the distance between points i and j and $I_t(i,j)$ is 1 if $d_{ij} \leq t$ and 0 otherwise. Where λ is the density of points per unit volume, the expected number of other plants within radius t of a randomly chosen plant is just λ multiplied by the function $K(t)$. We estimate $K(t)$ by $\hat{K}(t)$:

$$\hat{K}(t) = V \sum_{i \neq j}^{n} \sum^{n} w_{ij} I_t(i,j)/n^2, \tag{9.1}$$

where V is the volume of the study plot, and w_{ij} is a weighting factor used to reduce edge effect. If the circle centered on point i with radius t lies totally within the study plot then $w_{ij} = 1$, otherwise it is the reciprocal of the proportion of that sphere's surface that lies within the plot.

If the points are randomly arranged in space, $K(t) = 4\pi t^3/3$, and we therefore plot:

$$\hat{L}(t) = t - \sqrt[3]{3\hat{K}(t)/4\pi} \tag{9.2}$$

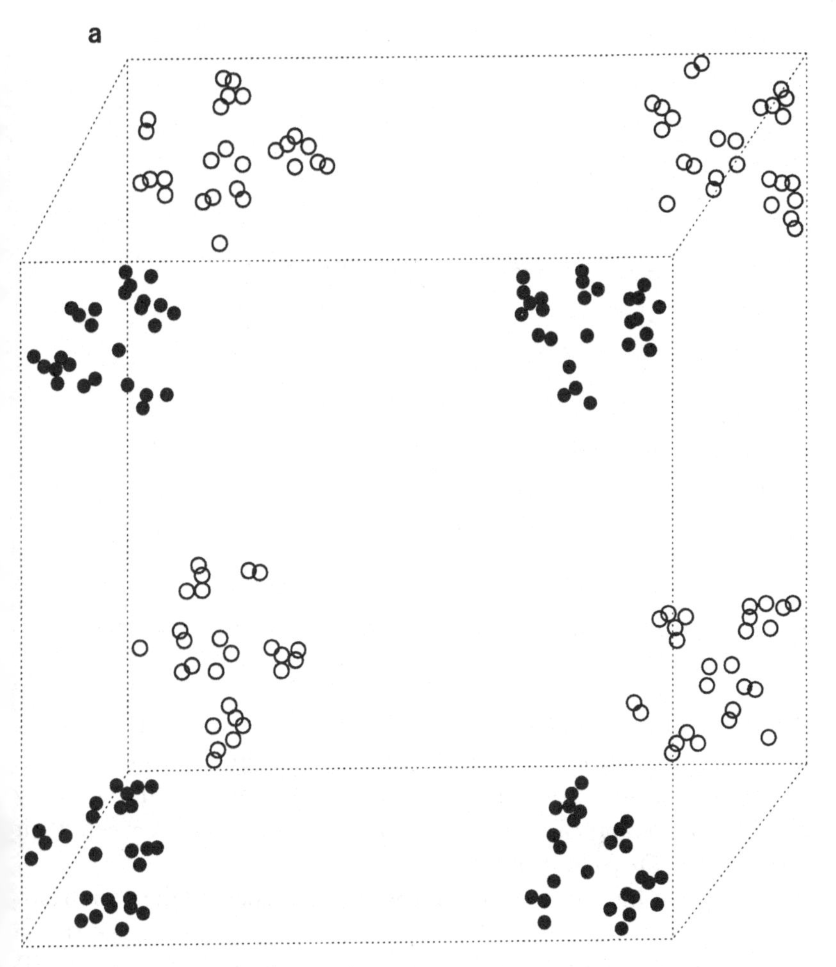

Figure 9.2 *a* Three-dimensional data for second-order analysis. The 64 points are in clumps of clumps.

as a function of *t*. On the null hypothesis of complete spatial randomness, the expected value is zero. Large positive values indicate overdispersion and large negative values indicate clumping. Figure 9.2 illustrates the method with data consisting of clumps of clusters of points (a) and the detection of the two scales of underdispersion (b).

It is possible to create the three-dimensional equivalent of a tesselation based on the positions of points. For instance, Okabe *et al.* (1992) describe the division of three-space into domains that are the parts of space closest to each of the data points than to any other. The walls of the domains are made up of parts of the planes that are right bisectors of the lines joining

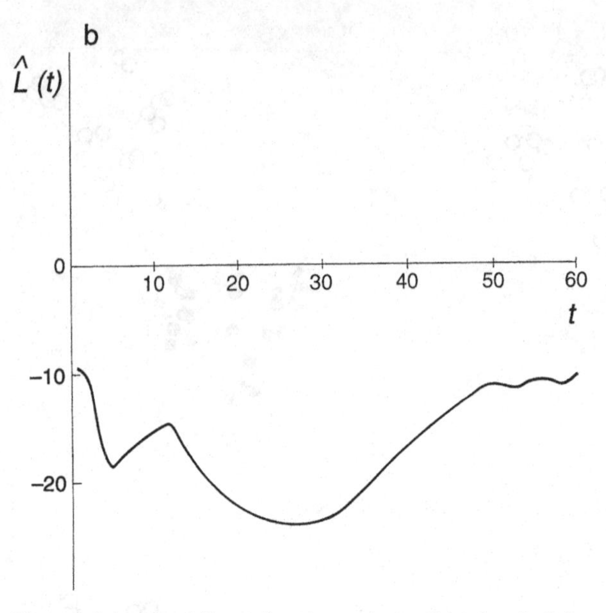

Figure 9.2 (cont.) *b* Second-order analysis of the data in 9.2*a*; two scales of pattern are obvious in the plot.

the points. This division of space into domains is the same as the Dirichlet or Voronoi tessellation described previously (Chapters 1 and 7). That division has a dual that is the three-space equivalent of the Delaunay tessellation, joining all points that share a boundary between their domains (see Okabe *et al.* 1992).

We can also imagine a three-dimensional equivalent of the least diagonal neighbor triangulation (LDNT). The two points that are closest join first. Thereafter, the lines joining pairs of points are added to the triangulation in order from shortest to longest on condition that they do not pass through any triangle (not the plane of a triangle) already in existence. As in two dimensions, the LDNT will be similar to the Delaunay, but we have found no theoretical work on this subject.

In three dimensions, the concept of anisotropy becomes more complex. We can treat the three dimensions using spherical systems and, as Upton and Fingleton (1989) point out, there are a number of different notations available for their description. Plant ecologists would be best to follow the lead of geographers and use the horizontal x–y plane as 0° vertical. We can then measure the angle to any point from that plane with angle ϕ and the angle in that plane with θ, breaking with the convention used by Upton and Fingleton but consistent with the summary of two-

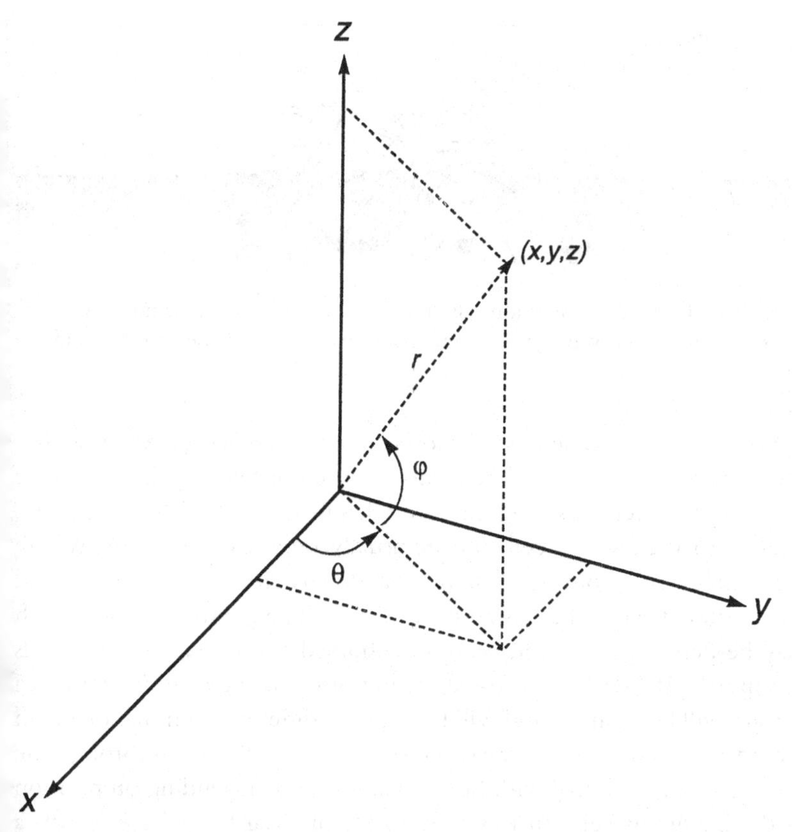

Figure 9.3 The relationship between Cartesian coordinates, x, y, and z, and polar coordinates using r, θ, and ϕ.

dimensional spectral analysis (Figure 9.3). Any (x,y,z) coordinate in three-space can be related to polar coordinates using ϕ and θ as follows:

$$x = r\cos\phi\cos\theta,\ y = r\cos\phi\sin\theta, \text{and } z = r\sin\phi. \tag{9.3}$$

(See Figure 9.3.) Anisotropy can now be thought of as having a ϕ component related to elevation and a θ component. In thinking about anisotropy in point pattern, we will have to keep that distinction in mind. In addition, it is possible that the pattern will be isotropic in one plane but not in others. For example, for motile phytoplankon in a water body, the cells might be isotropically dispersed in the x–y plane, but highly clumped vertically at the same range of depth throughout. On the other hand, with a steady wind blowing, the spatial pattern may be very different in the east-west direction compared to the north-south direction depend-

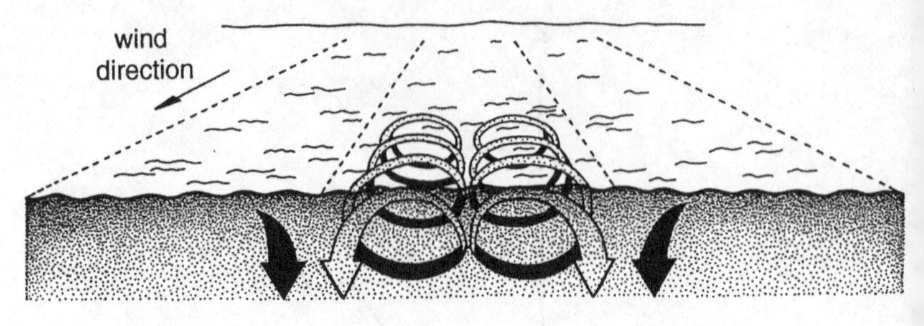

Figure 9.4 Langmuir convection cells develop in water when the wind blows across the surface. They are typically a metre or two in width (based on Reynolds 1984).

ing on the wind. When wind blows over the surface of water with a velocity greater than $3\,\mathrm{m\,s^{-1}}$, it creates convection cells in pairs that rotate in opposite directions (Figure 9.4; Chapman 1987). These cells are referred to as Langmuir cells and are usually a metre or two wide. Where their circulation converges there is a downward flow produced and where they diverge, there is an upward flow. The phytoplankton, which may be denser than water, become entrained in these convection cells (Chapman 1987). Under these conditions, the vertical variation in density will be changed and will have a scale different from that of any in the horizontal directions. What is more, the vertical pattern through the water column will vary with horizontal location, depending on position in the Langmuir cells. In this way, simple physical forces give rise to a complex three-dimensional pattern in the phytoplankton.

Any analysis of spatial pattern in three-space could be summarized in three diagrams related to the three axes, x, y, and z. It could equally well be summarized with respect to radius r and the two angles, ϕ and θ, or by r, θ, and z, depending on the data set under discussion.

As in the discussion of two-dimensional point patterns in Chapter 7, if there are two or more kinds of object in the three-dimensional data, we can examine the scales of their segregation and aggregation using frameworks of lines joining pairs of points and looking at the autocorrelation in that structure. The second-order method outlined above can also be modified to investigate the joint spatial pattern of two species in three-space, in an obvious way. Lastly, mapped point data can be converted to sample unit data and dealt with in that form, which we will describe next.

The following discussion of sampling unit data will be based on the assumption that the units are cubes. Other shapes are possible, but probably not practical. The data collected from the sampling units may be presence/absence, counts (as with mapped data converted to this form), or densities. For example, the data for a single cube might be: 20% spruce foliage, 10% spruce twigs, 10% aspen foliage, 5% aspen twigs, 3% aspen trunk. Many of the methods that we can imagine for analyzing this sort of data will just be modifications of methods already described elsewhere in this book.

Three-dimensional versions of PQV and random paired quadrat frequencies (RPQF) are obvious extensions of existing methods that could be used to analyze this kind of data. If PQV is used the results can be summarized as an *R*-spectrum, looking at variance as a function of distance between units, without regard for direction, or looking at variance as a function of angle, either ϕ or θ. In displaying the results of the RPQF analysis in two dimensions, to evaluate anisotropy, the results were presented as a two-dimensional array of circles or squares representing the Freeman–Tukey standardized residuals (as in Figures 6.14–6.16). A similar approach would be more difficult in three dimensions, but computer graphics could provide us with an apparent three-dimensional lattice of spheres the size and color of which represent the magnitude and sign of the Freeman–Tukey standardized residual for each combination of displacement in the x, y and z directions.

In Chapter 6, we saw how spectral analysis could be modified for use in two dimensions and there is nothing to prevent us from extending its application to three. Let the data be in a lattice $m \times n \times w$ units in size, with species density x_{ijk} at grid position (i,j,k). The periodogram for frequencies p, q, and g is I_{pqg}:

$$I_{pqg} = mnw(c^2_{pqg} + s^2_{pqg}), \tag{9.4}$$

where

$$c_{pqg} = \sum_{i=1}^{m} \sum_{j=1}^{n} \sum_{k=1}^{w} x_{ijk}[\cos 2\pi(ip/m + jq/n + kg/w)] \tag{9.5}$$

$$s_{pqg} = \sum_{i=1}^{m} \sum_{j=1}^{n} \sum_{k=1}^{w} x_{ijk}[\sin 2\pi(ip/m + jq/n + kg/w)]. \tag{9.6}$$

The results of this analysis can be presented as a three-dimensional plot with three axes representing the values of p, q, and g, with a sphere at each lattice point representing the magnitude of I_{pqg}. The results can be collapsed into summary figures in three ways, the R-spectrum, the Θ-spectrum, and the Φ-spectrum. The R-spectrum combines all I_{pq} for which the values of $r = \sqrt{p^2 + q^2 + g^2}$ are the same and plots I as a function of r. The Θ-spectrum combines all I for which the values of θ, as defined above, are the same and plots I as a function of θ; the Φ-spectrum does the same for angle ϕ, as defined above.

One method that seems to hold promise for the assessment of spatial structure is the measurement of fractal dimension. The method involves counting the number of cubes of size δ that the object(s) in question occupies; call the number C (Kenkel & Walker 1993). The fractal dimension, $\mathfrak{D}$, is then the slope of the line in a plot of $\log(C)$ as a function of $\log(\delta)$. Because of the practical difficulties of counting cubes of different sizes, what is actually done is to count squares in photographs and determine the fractional dimension in a single plane through the habitat, $\mathfrak{D}_2$. The actual fractal dimension, $\mathfrak{D}_3$, then lies between $\mathfrak{D}_2 + 1$ and $2\mathfrak{D}_2$ (Morse *et al.* 1985). Shorrocks *et al.* (1991) used this method to relate the fractal dimensions of lichens to arthropod body lengths and found a clear relationship: because of the fractal spatial structure, there is more usable space for smaller organisms and, therefore, they are more abundant. While determining the fractal dimension does not tell us everything about the spatial structure, it may be an important characteristic to which other organisms respond. Kenkel and Walker (1993) add an important cautionary note that the orientation of the grid or lattice can affect the outcome, so that several orientations should be used.

Most of the approaches to three-dimensional pattern analysis that we have sketched here have not been used, mainly because the kind of data appropriate for these analyses is difficult to collect. Three-dimensional pattern analysis is the key to the quantification of habitat structure; it is just not clear at the moment how it is best done.

Relation to spatial structure of physical factors

In Chapter 1, we discussed various causes of spatial pattern in vegetation, and stated that one cause is spatial pattern in some environmental factor which affects the plants; topography is an obvious example (Matérn 1986). It seems clear that, for this system to work, there must be pattern in the environmental factor; the difference in the factor's intensity at two places should at first increase with distance and then decrease. In contrast to this picture, the underlying model of geostatistics is a semivariogram that increases from some minimum value, the nugget, at distance 0 to a maximum, the sill, at some critical distance, the range (Figure 9.5). As distance increases beyond R, the semivariance remains more or less constant

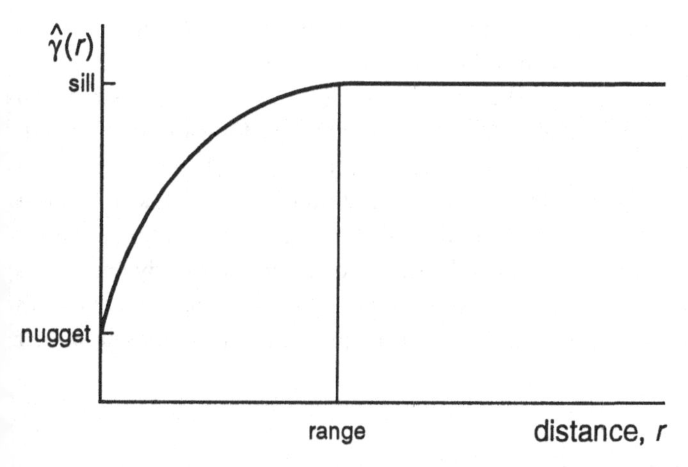

Figure 9.5 The standard form of a variogram from geostatistics, after Palmer (1990a). The nugget is related to the inherent variability in the material at a very small scale. The sill is the variance among spatially independent samples. The range is the distance at which the sill is reached, and the distance beyond which samples can be considered to be spatially independent.

Remember that it is the variogram that PQV estimates (Chapter 3) so that the geostatistical model is predicting a variance plot that looks like Figure 9.5, rather than one with a peak. If analysis of an environmental factor does indeed give a variance curve that increases to a constant value, we might be concerned that it indicates the absence of repeating spatial pattern.

Palmer (1990b) studied the variation in 16 soil characteristics using samples taken at 10 cm intervals along a transect. The semivariograms for some of the characteristics seem to peak and then decrease, indicating repeating pattern, but others seem to level off, and yet others continue to increase over the range of distances studied (to 40 m). Oliver and Webster (1986) also examined the semivariograms of soil properties; in their study all the variance plots flattened out at a maximum value, with the range being about 30 m.

Bell *et al.* (1993) investigated the spatial structure of physical variables at much larger scales, up to 10^6 m. They examined forest soil properties, water chemistry of lakes, and temperature and precipitation in Northeastern North America. What they found was no clear indication that variance reaches a maximum value with increasing distance. For example, there was a clear linear increase in log variance of soil calcium, nitrogen, and phosphorous with log distance. This finding has interesting implications for the interpretation of large-scale pattern in vegetation, an area of study that has received little attention to date.

One interpretation of the increasing variogram result is that spatial patterns may be locally stationary but exhibit 'long-distance' trends (Matérn 1986). The variograms given by Bell *et al.* (1993) deal with a range of distances from about 1 km to 250 km or even 1500 km. In most of the examples we have used in this book, the scale of sampling has been an order of magnitude less than the smallest scale in their study, and so it is perhaps not surprising that different kinds of pattern are detected. This comparison may demonstrate that natural phenomena are not self-similar at all scales.

Obvious extensions

There are a number of methods that have been described in this book that have obvious extensions that are yet to be developed. One method that we can consider, although it has yet to be fully developed, is a multi-scale version of the popular canonical correspondence analysis. That technique produces an ordination that includes both the vegetation data and the environmental variables simultaneously. One of the results of the analysis is a biplot of species and environmental variables in ordination space. The strength of the relationship between a species and a variable is interpreted by the species' position with respect to the variable's arrow in the ordination plot (see Figure 9.6). Most published examples of this analysis have used quadrats that are not transect strings. We have two data sets where the quadrats are contiguous and part of transects, from Ellesmere Island sedge meadows and Yukon shrub communities. For these data sets, the analysis could be carried out on individual quadrats, blocks of two quadrats, blocks of three and so on. This approach is yet to be explored and evaluated, but it will permit us to examine how the relationships among species and environmental descriptors change with scale without using a size hierarchy of sampling units

Temporal aspects of spatial pattern analysis

In previous chapters, especially Chapter 1, we discussed the relationship between temporal processes and spatial pattern. One approach to studying the change of spatial pattern through time is just to consider time as one more dimension, but with the understanding that any pattern in the system will be anisotropic. Figure 9.7 shows an example of a one-dimen-

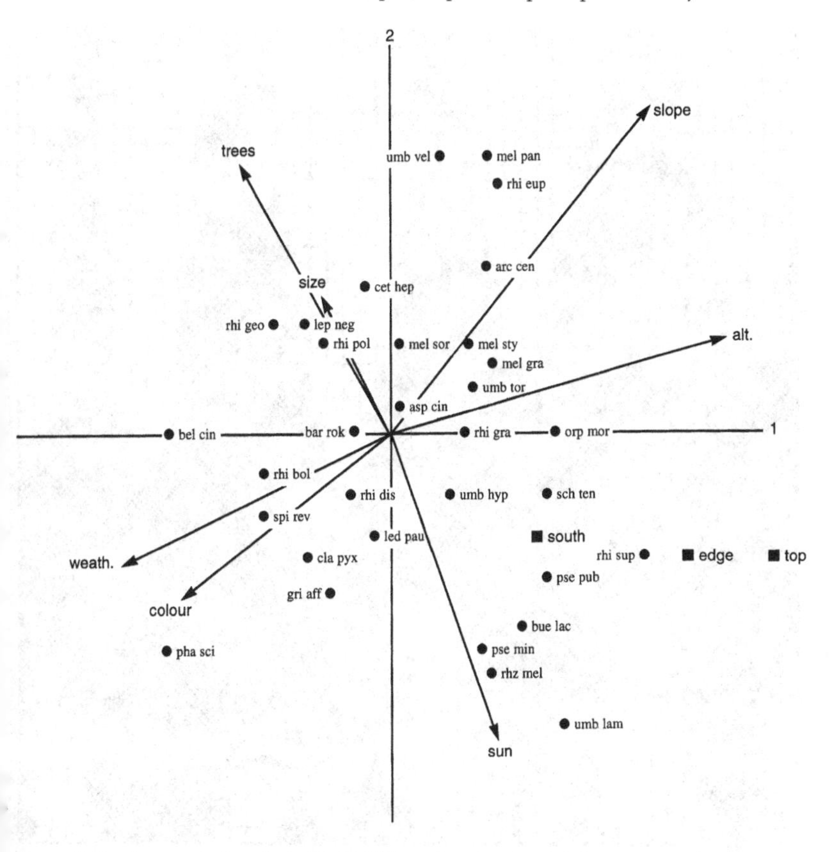

Figure 9.6 An example of a CANOCO ordination in which species and environmental variables are ordinated together (redrawn from John & Dale 1990). The arrows represent the quantitative environmental variables, the squares are the categorical variables and the dots are individual species. The relationship between the species and the variables can be interpreted from their relative positions. For example, *Rhizocarpon superficiale* ('rhi sup') is closely associated with the edge and top of rock faces ('edge' and 'top'); *Phaeophyscia sciastra* ('pha sci'), on the other hand, is associated with rock faces that are weathered, dark in color, not steeply sloping and low in altitude ('weath.', 'colour', 'slope', and 'alt.').

sional transect (as in the bottom part of Figure 1.1) observed through time as colonization, mortality, coalescence, breakup, and re-establishment occur. In its initial stages, to the beginning of breakup, it is somewhat similar to the Rapoport Gruyère model. In a similar way the growth and mortality of patches of lichens or the projections of forest trees' canopies can be pictured and analyzed as solid objects, the cross-section of which represent the two-dimensional extent at a particular point in time.

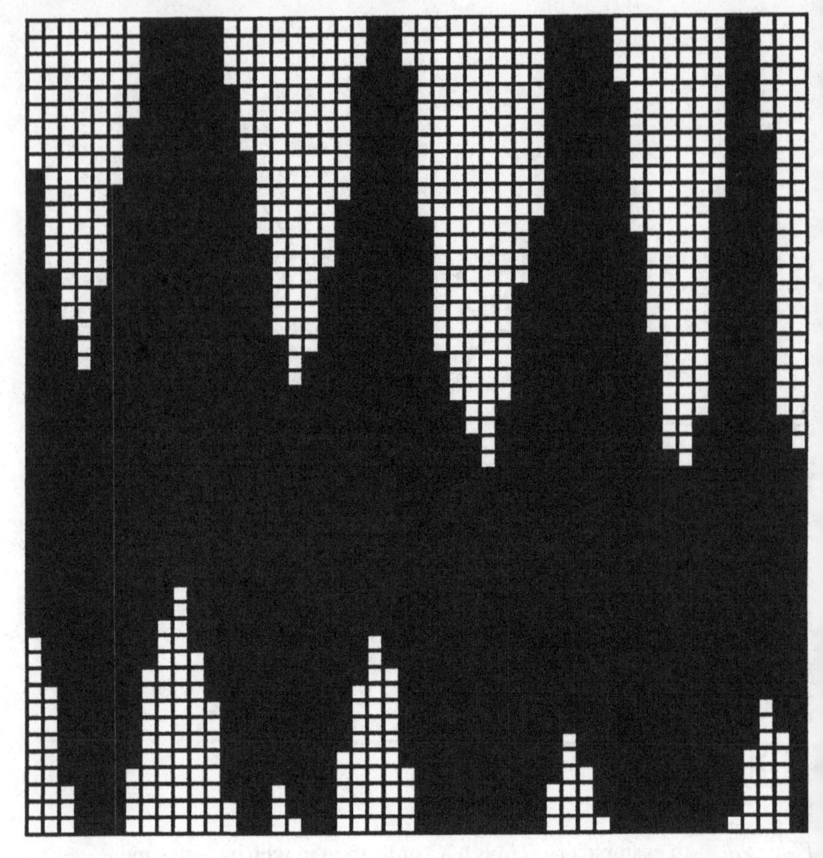

Figure 9.7 A one-dimensional transect, such as the one depicted in Figure 1.1, is recorded through time with the horizontal lines representing presence (black) or absence (white) in the quadrats at each of several times. It produces the equivalent of an anisotropic pattern in two dimensions.

Wavelets

A method known as wavelet analysis has been enjoying increasing popularity in a variety of applications (Chui 1992; Daubechies 1993). The method resembles spectral analysis or Fourier analysis to a certain extent but does not assume stationarity of the underlying process. Therefore, it can examine a sequence both with regard to frequency and to position in the sequence rather than just frequency. Bradshaw and Spies (1992) suggest that one way of viewing the technique is to visualize a particular wave form and a moving window along the sequence. The

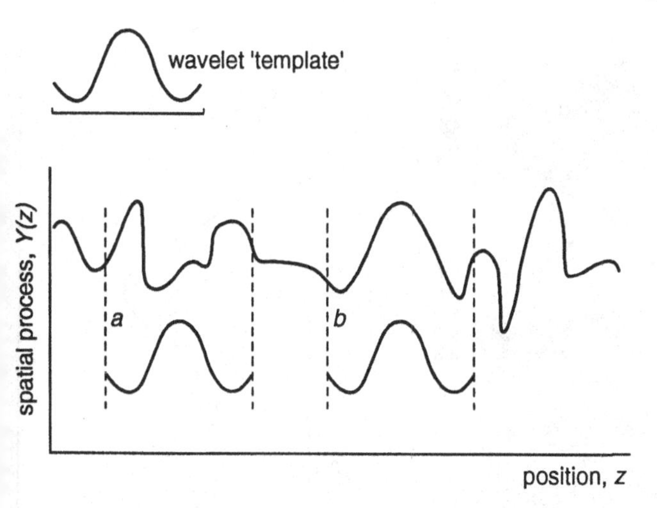

Figure 9.8 A wavelet template (upper left) is compared to the data sequence at a range of positions and scales. At point (a) the match is poor; at point (b) it is good. Y is a spatial process that is a function of position, z.

transform gives a low value when the data in the window do not match the wave template and a high value when the data match the wavelet in shape and dimension (Figure 9.8).

It is possible that this approach will be able to provide insights in the area of spatial pattern analysis. For example, Antoine *et al.* (1992) used two-dimensional wavelet transform in image analysis and found that it could be used to detect the position, orientation and visual contrast of simple objects. Slezak *et al.* (1992) used wavelet analysis to detect clusters of galaxies and voids between them at a range of scales. The similarity of these studies to spatial pattern analysis in ecology suggests that this approach may prove very useful in our future work. The only application of this kind we have found is Bradshaw and Spies (1992), who used wavelets to characterize canopy gap structure in forests.

They used the wavelet transform

$$W(a,x_i) = \frac{1}{a}\sum_{j=1}^{n} f(x_j)\, g[(x_i - x_j)/a], \tag{9.7}$$

where x_i measures distance along a transect of n units, $f(x)$ is the measure in unit x, such as density, a is scale and g is some windowing function or wavelet. Different functions can be used but one function they used was the 'Mexican hat' template, which for $a=1$ is:

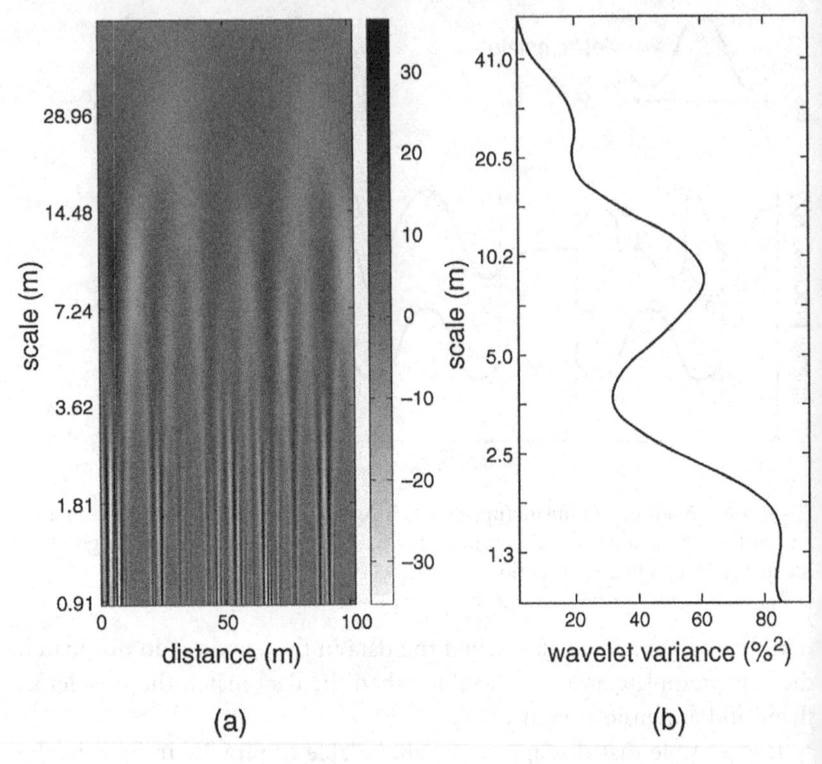

Figure 9.9 *a* Gray-scale output from wavelet analysis for *Betula glandulosa* at the Yukon site called Flint. The Mexican hat wavelet was used. The darker the color the better is the match of the wavelet template to the data. *b* Wavelet variance as a function of scale. This analysis indicates strongest pattern at 1.5 m, with a second scale of pattern at 9.0 m.

$$g(x) = \frac{1}{3^{0.5}} \pi^{-0.25} (1 - 4x^2) e^{-2x^2}. \tag{9.8}$$

The results are present in gray-scale diagrams with distance along the transect on the X-axis, scale on the Y-axis and the intensity of the gray scale representing the magnitude of the wavelet coefficient (Figure 9.9). The results were also summarized by wavelet variance, which, for scale a, is the squares of the wavelet coefficients averaged over positions:

$$V(a) = \sum_{i=1}^{n} W^2(a, x_i)/n. \tag{9.9}$$

The plot of this variance as a function of scale is very much like a plot of one of the quadrat variance techniques (Figure 9.9). With artificial

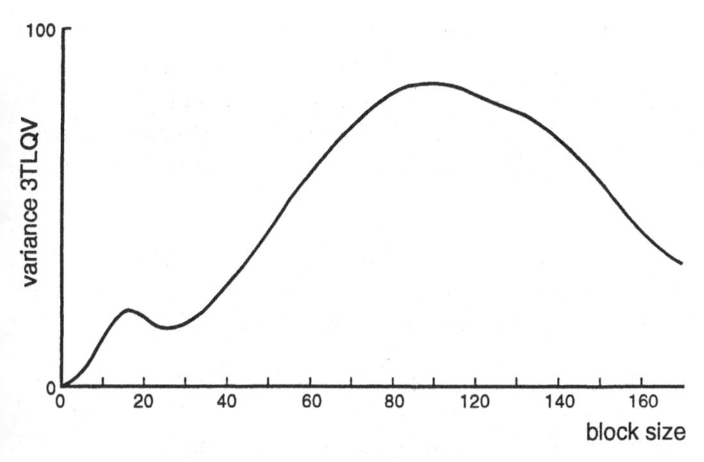

Figure 9.10 3TLQV analysis of the same data as in Figure 9.9. The same scales of pattern are detected, but the variance is greater for the larger scale.

data, input scales of 5 and 25 produce peaks in the wavelet variance at 5 and about 20, just as in TTLQV. Figure 9.9 shows the gray-scale and wavelet variance plots for a sample of data from the Yukon shrub study (*Betula glandulosa* at Flint, transect 1). The variance plot indicates spatial pattern at 1.5m and 9m; the accompanying 3TLQV analysis (Figure 9.10) shows the same scales. The wavelet method reduces the apparent importance of the larger scale pattern. Using the same method, Bradshaw and Spies (1992) found that in the Douglas Fir-Hemlock forests small to intermediate disturbances significantly influence canopy structure.

The advantage of the wavelet approach is that different wavelet templates can examine the data for different shapes of spatial pattern, thus providing fuller evaluation of its characteristics. This flexibility may make the technique very popular in the future.

Questions and hypotheses

Throughout this book, we have alluded to a variety of ecological questions and hypotheses related to the interplay between pattern and process in plant communities. We can organize those questions into groups.

The first group is a set of questions about the relationship between spatial pattern in vegetation and the environment.

1. 'Patch-gap' models of vegetation dynamics deal with community heterogeneity and dynamics based on cycles of disturbance and recovery. What is the relative importance of disturbances of different sizes and how

do the dynamics of disturbance and recovery depend on patch size? The cycle-mosaic hypothesis, described in Chapter 1, is closely related, suggesting that different patches of vegetation are at different phases of the same temporal cycle. How is the mosaic structure of vegetation (particularly in forests) related to local spatial effects on the regeneration of dominant species?

2. What is the relative importance of biotic compared to abiotic factors in the spatial distribution of species? Do abiotic factors act at different scales than biotic factors? Do they tend to act at larger scales?

3. The community's 'memory' is the extent to which the past distribution and pattern of species determines the future distribution and pattern. How strong is this effect and what factors determine its strength? Is spatial pattern a long-lasting equilibrium characteristic or is it, by nature, transitory?

4. To what extent do the spatial patterns of environmental factors, such as soil nutrients, determine the spatial patterns of the plants? To what extent are they themselves determined by the past plant pattern and how long does the effect persist?

5. What factors determine the observed densities of species as composition changes along an environmental gradient?

The second group of questions are about interactions among the plants.

1. Are there natural groupings of species in vegetation and do vegetation units exist? If there are groupings, how tightly are the members of a group linked? How sharp are spatial boundaries between groupings? If there are assembly rules for plant communities, how important are spatial effects in those rules?

2. Do some species have stronger spatial pattern because of life-history characteristics such as dispersal or regeneration niche? To what extent is the scale of a species' spatial pattern dependent on its abundance?

3. Interactions between plants can be negative or positive. Can we see the effects of competition in spatial pattern? Can we detect indirect effects such as those of second-order neighbors? How common are positive interactions? How important are spatial effects such as nucleation for the facilitation of vegetation change?

4. In some communities, the vegetation consists of several layers. In layered vegetation, what is the relationship between spatial pattern in the different layers?

5. Spatial pattern may affect the coexistence of species in natural vegetation. How is species' coexistence affected by spatial aspects of plant growth, for example phalanx *versus* guerilla growth of clones? In Chapter 5, we introduced Pacala's 'spatial segregation hypothesis' which suggests that ecological stability is enhanced by limited dispersal and local interactions. How can we use spatial pattern analysis to evaluate the importance of this effect? At what scales does it act in different kinds of vegetation?

6. In looking at communities, when do we expect true multispecies pattern? What hypotheses predict that it exists? In multispecies pattern, does the evenness of species' contributions to the pattern increase through succession?

The third group of questions are about the nature of spatial pattern itself.

1. How can spatial data be used to test whether orderly spatial pattern exists only at relatively small scales? It is possible that repeating pattern occurs only at smaller scales, with trends or greater disorder at larger scales. How is the perception of large-scale spatial pattern or large-scale spatial disorder affected by temporal scale?

2. Over what range of scales are spatial patterns self-similar? What can we learn about the biological processes of spatial organization from comparing the fractal dimensions of different kinds of plant communities?

3. What is the general and fundamental relationship between spatial pattern and temporal cycles?

4. How does the strength of the relationship between horizontal pattern and vertical structure change with vegetation type and successional stage?

5. What is the importance of repeating spatial pattern as an aspect of habitat structure?

This list of questions is not intended to be exhaustive, but only to give an idea of the kinds of questions that we can associate with the study of spatial pattern in plant ecology. While some in this list of questions can be answered using methods described in this book, for others we may need new methods or further extensions of the ones we already have.

Concluding remarks

A common thread in the concluding remarks of preceding chapters is that there is more work to be done on the development and evaluation of methods to be used for spatial pattern analysis. In effect, this chapter has followed this theme by pointing out at least some of the directions in which this area of endeavour might or should move in the next few years. Of course, there is no way of knowing what new and different approaches will appear and which of the yet to be explored approaches will prove to be the most fruitful. Fractal geometry is an attractive approach that is receiving a fair amount of attention and it will be interesting to see whether it proves to be as useful as is hoped. Similarly, some of the speculations of this chapter will prove useful but others may not.

The writing of this book has been a fascinating journey of discovery and rediscovery; that journey will continue but the book must end here.

Bibliography

Amarasekare, P. (1994). Spatial population structure in the banner-tailed kangaroo rat, *Dipodomys spectabilis*. *Oecologia*, **100**, 166–176.

Andersen, M. (1992). Spatial analysis of two species interactions. *Oecologia*, **91**, 134–140.

Anderson, A. J. B. (1971). Ordination methods in ecology. *Journal of Ecology*, **59**, 713–726.

Anderson, D. J. (1967). Studies on structure in plant communities. *Journal of Ecology*, **55**, 397–404.

Antoine, J. P., Nurenzi, R., Piette, B. & Duval-Destin, M. (1992). Image analysis with 2D continuous wavelet transform: detection of position, orientation, and visual contrast of simple objects. In *Wavelets and Applications*, ed. Y. Meyer, pp. 144–59. Masson: Springer-Verlag.

Armesto, J. J., Mitchell, J. D., & Villagran, C. (1986). A comparison of spatial pattern of trees in some tropical and temperate forests. *Biotropica*, **18**, 1–11.

Arriaga, L., Maya, Y., Diaz, S., & Cancino, J. (1993). Association between cacti and nurse perennials in a heterogeneous tropical dry forest in northwestern Mexico. *Journal of Vegetation Science*, **4**, 349–356.

Auerbach, M. & Shmida A. (1993). Vegetation change along an altitudinal gradient on Mount Herman, Israel – no evidence for discrete communities. *Journal of Ecology*, **81**, 25–33.

Austin, M. P. (1980). Searching for a model for use in vegetation analysis. *Vegetatio*, **42**, 11–21.

Austin, M. P. (1987). Models for the analysis of species' response to environmental gradients. *Vegetatio*, **69**, 35–45.

Austin, M. P., & Austin, B. O. (1980). Behaviour of experimental plant communities along a nutrient gradient. *Journal of Ecology*, **68**, 891–918.

Austin, M. P., Cunningham, R. B., & Fleming, P. M. (1984). New approaches to direct gradient analysis using statistical scalars and statistical curve-fitting procedures. *Vegetatio*, **55**, 11–27.

Austin, M. P., & Gaywood, M. J. (1994). Current problems of environmental gradients and species response curves in relation to continuum theory. *Journal of Vegetation Science*, **5**, 473–482.

Austin, M. P., Nicholls, A. O., Doherty, M. D., & Meyers, J. A. (1994). Determining species response functions to an environmental gradient by means of a β-function. *Journal of Vegetation Science*, **5**, 215–228.

Austin, M. P., & Smith, T. M. (1989). A new model for the continuum concept. *Vegetatio*, **83**, 35–47.

Bach, C. E. (1984). Plant spatial pattern and herbivore population dynamics: plant factors affecting the movement patterns of a tropical cucurbit specialist (*Acalymma innubum*). *Ecology*, **65**, 175–190.

Bach, C. E. (1988a). Effects of host plant patch size on herbivore density: patterns. *Ecology*, **69**, 1090–1102.

Bach, C. E. (1988b). Effects of host plant patch size on herbivore density: underlying mechanisms. *Ecology*, **69**, 1103–1117.

Bailey, T. C., & Gatrell, A. C. (1995). *Interactive Spatial Data Analysis*. Essex, UK: Longman Scientific & Technical.

Ballantyne, C. K., & Matthews, J. A. (1983). Desiccation cracking and sorted polygon development, Jotunheimen, Norway. *Arctic and Alpine Research*, **15**, 339–349.

Bartlett, M. S. (1964). The spectral analysis of two-dimensional point processes. *Biometrika*, **51**, 299–311.

Beals, E. (1960). Forest bird communities in the Apostle Islands of Wisconsin. *The Wisconsin Bulletin*, **72**, 156–181.

Beals, E. W. (1969). Vegetational change along altitudinal gradients. *Science*, **165**, 981–985.

Bell, G., Lechowicz, M. J., Appenzeller, A., Chandler, M., DeBlois, E., Jackson, L., MacKenzie, B., Preziosi, R., Schallenberg, M., & Tinker, N. (1993). The spatial structure of the physical environment. *Oecologia*, **96**, 114–121.

Bell, S. S., McCoy, E. D., & Mushinsky, H. R. (1991). *Habitat Structure*. London: Chapman and Hall.

Bertness, M. D., & Calloway, R. (1994) Positive interactions in communities. *Trends in Ecology and Evolution*, **9**, 191–193.

Besag, J., & Diggle, P. J. (1977). Simple randomization tests for spatial pattern. *Applied Statistics*, **26**, 327–333.

Bishop, Y. M. M., Fienberg, S. E., & Holland, P. W. (1975). *Discrete Multivariate Analysis*. Cambridge, MA: MIT Press.

Blundon, D. J., MacIsaac, D. A., & Dale M. R. T. (1993). Nucleation during primary succession in the Canadian Rockies. *Canadian Journal of Botany*, **71**, 1093–1096.

Boots, B. N., & Getis, A. eds. (1988). *Point Pattern Analysis*. Beverly Hills: Sage Publications Inc.

Borcard, D., Legendre, P., & Drapeau, P. (1992). Partialling out the spatial component of ecological variation. *Ecology*, **73**, 1045–1055.

Bouxin, G., & Gautier, N. (1982). Pattern analysis in Belgian limestone grasslands. *Vegetatio*, **49**, 65–83.

Bradshaw, G. A., & Spies, T. A. (1992). Characterizing canopy gap structure in forests using wavelet analysis. *Journal of Ecology*, **80**, 205–215.

Burnham, K. P., Anderson, D. R., & Laake, J. L. (1980). Estimation of density from line transect sampling of biological populations. Wildlife Monograph 72, supplement to *Journal of Wildlife Management*, **44**.

Burton, P. J., & Bazzaz, F. A. (1995). Ecophysiological responses of tree seedlings to invading different patches of old-field vegetation. *Journal of Ecology*, **83**, 99–112.

Cale, W. G., Henebry, G. M., & Yeakley, J. A. (1989). Inferring process from pattern in natural communities. *BioScience*, **39**, 600–605.

Cappuccino, N. (1988). Spatial patterns of goldenrod aphids and the response of enemies to patch density. *Oecologia*, **76**, 607–610.

Carlile, D. W., Skalski, J. R., Batker, J. E., Thomas, J. M., & Cullinan, V. I. (1989). Determination of ecological scale. *Landscape Ecology*, **2**, 203–213.

Carpenter, S. R., & Chaney, J. E. (1983). Scale of spatial pattern: four methods compared. *Vegetatio*, **53**, 153–160.

Carter, A. J., & O'Connor, T. G. (1991). A two-phase mosaic in a Savanna grassland. *Journal of Vegetation Science*, **2**, 231–236.

Castro, I., Sterling, A., & Galiano, E. F. (1986). Multi-species pattern analysis of Mediterranean pastures in three stages of ecological succession. *Vegetatio*, **68**, 37–42.

Chapman, A. R. O. (1973). A critique of prevailing attitudes towards the control of seaweed zonation on the sea shore. *Botanica Marina*, **15**, 80–82.

Chapman, A. R. O. (1979). *Biology of Seaweeds*. London: Edward Arnold.

Chapman, A. R. O. (1987). *Functional Diversity of Plants in the Sea and on Land*. Portola Valley Boston: Jones & Bartlett.

Chui, C. K. (1992). *An Introduction to Wavelets. Wavelet Analysis and its Applications*, Vol. 1. San Diego: Academic Press.

Clark, P. J., & Evans, F. C. (1954). Distance to nearest neighbor as a measure of spatial relationships in populations. *Ecology*, **35**, 445–453.

Cohen, W. B., & Spies, T. A. (1990). Semivariograms of digital imagery for analysis of conifer canopy structure. *Remote Sensing of the Environment*, **34**, 167–178.

Collins, S. L., Glenn, S. M., & Roberts, D. W. (1993). The hierarchical continuum concept. *Journal of Vegetation Science*, **4**, 149–156.

Conover, W. J. (1980). *Practical Nonparametric Statistics*, 2nd edn. New York: Wiley.

Cooper, C. F. (1961). The ecology of fire. *Scientific American*, **204**, 150–160.

Coxson, D. S., & Kershaw, K. A. (1983). The ecology of *Rhizocarpon superficiale*. II. The seasonal response of net photosynthesis and respiration to temperature, moisture and light. *Canadian Journal of Botany*, **61**, 3019–3030.

Crawley, M. J. ed. (1986). The structure of plant communities. In *Plant Ecology*. pp. 1–50. London: Blackwell Scientific Publications.

Cressie, N. A. (1991). *Statistics for Spatial Data*. New York: Wiley.

Cullinan, V. I., & Thomas, J. M. (1993). A comparison of quantitative methods for examining landscape pattern and scale. *Landscape Ecology*, **7**, 211–227.

Czárán, T. (1989). Coexistence of competing populations along an environmental gradient: a simulation study. *Coenoses*, **2**, 113–120.

Czárán, T. (1991). Coexistence of competing populations along and environmental gradient: a simulation study. In *Computer Assisted Vegetation Analysis*. eds. E. Fioli & L. Orlócu, pp. 317–324. The Netherlands: Kluwer Academic Publishers.

Czárán, T., & Bartha, S. (1992). Spatiotemporal dynamic models of plant populations and communities. *Trends in Ecology and Evolution*, **7**, 38–42.

Dale, M. R. T. (1979). *The Analysis of Patterns of Zonation and Phytosociological Structure of Seaweed Communities*. Ph.D. Thesis. Halifax, Nova Scotia: Dalhousie University.

Dale, M. R. T. (1984). The contiguity of upslope and downslope boundaries of species in a zoned community. *Oikos*, **42**, 92–96.

Dale, M. R. T. (1985). A graph theoretical technique for comparing phytosociological structures. *Vegetatio*, **63**, 79–88.

Dale, M. R. T. (1986). Overlap and spacing of species' ranges on an environmental gradient. *Oikos*, **47**, 303–308.

Dale, M. R. T. (1988a). The spacing and intermingling of species boundaries on an environmental gradient. *Oikos*, **53**, 351–356.

Dale, M. R. T. (1988b). The distribution of variables measuring gap and overlap in sheaves of line segments: measured by events. *Utilitas Mathematica*, **33**, 163–172.

Dale, M. R. T. (1990). Two-dimensional analysis of spatial pattern in vegetation for site comparison. *Canadian Journal of Botany*, **68**, 149–158.

Dale, M. R. T. (1995). Spatial pattern in communities of crustose saxicolous lichens. *Lichenologist*, **27**, 495–503.

Dale, M. R. T., & Blundon, D. J. (1990). Quadrat variance analysis and pattern development during primary succession. *Journal of Vegetation Science*, **1**, 153–164.

Dale, M. R. T., & Blundon, D. J. (1991). Quadrat covariance analysis and the scales of interspecific association during primary succession. *Journal of Vegetation Science*, **2**, 103–12.

Dale, M. R. T., Blundon, D. J., MacIsaac, D. A., & Thomas, A. G. (1991). Multiple species effects and spatial autocorrelation in detecting species associations. *Journal of Vegetation Science*, **2**, 635–642.

Dale, M. R. T., Henry, G. H. R., & Young, C. (1993). Markov models of spatial dependence in vegetation. *Coenoses*, **8**, 21–24.

Dale, M. R. T., & MacIsaac, D. A. (1989). New methods for the analysis of spatial pattern in vegetation. *Journal of Ecology*, **77**, 78–91.

Dale, M. R. T., & Moon, J. W. (1988). Statistical tests on two characteristics of the shapes of cluster diagrams. *Journal of Classification*, **5**, 21–38.

Dale, M. R. T., & Narayana, T. V. (1984). Tables of some properties of sequences of paired events. *Utilitas Mathematica*, **26**, 319–326.

Dale, M. R. T., & Powell, R. D. (1994). Scales of segregation and aggregation of plants of different kinds. *Canadian Journal of Botany*, **72**, 448–453.

Dale, M. R. T., & Zbigniewicz, M. W. (1995). The evaluation of multi-species pattern. *Journal of Vegetation Science*, **6**, 391–398.

Darling, D. A. (1953). On a class of problems related to the random division of an interval. *Annals of Mathematical Statistics*, **24**, 239–253.

Daubechies, I. (1993). *Different Perspectives on Wavelets*. Providence, RI: American Mathematical Society.

David, M. (1977). *Geostatistical Ore Reserve Estimation*. Amsterdam: Elsevier.

Day, T. A., & Wright, R. G. (1989). Positive plant spatial association with *Erigonum ovalifolium* in primary succession on cinder cones: seed-trapping nurse plants. *Vegetatio*, **80**, 37–45.

de Jong, P., Aarssen, L. W., & Turkington, R. (1980). The analysis of contact sampling data. *Oecologia*, **45**, 322–324.

de Jong, P., Aarssen, L. W., & Turkington, R. (1983). The analysis of contact sampling data *Journal of Ecology*, **71**, 545–559.

Dietz, H., & Steinlein, T. (1996). Determination of plant species cover by means of image analysis. *Journal of Vegetation Science*, **7**, 131–136.

Diggle, P. J. (1981). Binary mosaics and the spatial pattern of heather. *Biometrics*, **37** 531–539.

Diggle, P. J. (1983). *Statistical Analysis of Spatial Point Patterns*. London: Academic Press.

Doak, D. F., Marino, P. C., & Kareiva, P. M. (1992). Spatial scale mediates the influence of habitat fragmentation on dispersal success: implications for conservation. *Theoretical Population Biology*, **41**, 315–336.

Doty, M. S., & Archer, J. G. (1950). An experimental test of the tide factor hypothesis. *American Journal of Botany*, **37**, 458–464.

Eberhardt, L. L. (1978). Transect methods for population studies. *Journal of Wildlife Management*, **42**, 1–31.

Egan, R. S. (1987). A fifth checklist of the lichen-forming, lichenicolous and allied fungi of the continental United States and Canada. *The Bryologist*, **90**, 77–173.

Embleton, C., & King, C. A. M. (1975). *Periglacial Geomorphology*. London: Edward Arnold.

Errington, J. C. (1973). The effect of regular and random distributions on the analysis of pattern. *Journal of Ecology*, **61**, 99–105.

Felt, E. P. (1940). *Plant Galls and Gall Makers*. Ithaca, NY: Comstock Co.

Ford, E. D., & Renshaw, E. (1984). The interpretation of process from pattern using two-dimensional spectral analysis: modelling single species patterns in vegetation. *Vegetatio*, **56**, 113–123.

Forman, R. T. T., & Godron, M. (1986). *Landscape Ecology*. New York: Wiley.

Fortin, M. J. (1994). Edge detection algorithms for two-dimensional ecological data. *Ecology*, **75**, 956–965.

Franco, M., & Harper, J. (1988). Competition and the formation of spatial pattern in spacing gradients: an example using *Kochia scoparia*. *Journal of Ecology*, **76**, 959–974.

Franco, A. C., & Nobel, P. S. (1989). Effect of nurse plants on the microhabitat and growth of cacti. *Journal of Ecology*, **77**, 870–886.

Fraser, A. R., & Van den Driessche, P. (1972). Triangles, density, and pattern in point populations, pp. 277–286 In *Proceedings of the 3rd Conference of the Advisory Group of Forest Statisticians*. Jouy-en-Josas, France: International Union for Research Organization, Institut National de la Recherche Agronomique.

Gabriel, K. R., & Sokal, R. R. (1969). A new statistical approach to geographic variation analysis. *Systematic Ecology*, **18**, 259–270.

Galiano, E. F. (1982a). Détection et mesure de l'hétérogénéité spatiale des espèces dans les pâturages. *Acta Oecologica/Oecologia Plantarum*, **3**, 269–278.

Galiano, E. F. (1982b). Pattern detection in plant populations through the analysis of plant-to-all-plants distances. *Vegetatio*, **49**, 39–43.

Galiano, E. F. (1983). Detection of multi-species patterns in plant populations. *Vegetatio*, **53**, 129–138.

Galiano, E. F. (1985). The small-scale pattern of *Cynodon dactylon* in Mediterranean pastures. *Vegetatio*, **63**, 121–127.

Galiano, E. F. (1986). The use of conditional probability spectra in the detection of segregation between plant species. *Oikos*, **46**, 132–138.

Gauch, H. G. (1982). *Multivariate Analysis in Community Ecology*. Cambridge: Cambridge University Press.

Gauch, H. G., & Whittaker, R. H. (1972). Coenocline simulation. *Ecology*, **53**, 446–451.

Gellert, W., Küstner, H., Hellwich, M., & Kästner, H. (1977). *The VNR Concise Encyclopedia of Mathematics*. New York: Van Nostrand Reinhold.

Getis, A., & Franklin, J. (1987). Second-order neighborhood analysis of mapped point patterns. *Ecology*, **68**, 473–477.

Gibson, D. J., & Greig-Smith, P. (1986). Community pattern analysis: a method for quantifying community mosaic structure. *Vegetatio*, **66**, 41–47.

Gibson, N., & Brown, M. J. (1991). The ecology of *Lagarostrobos franklinii* (Hook.f.) Quinn (Podocarpaceae) in Tasmania. 2. Population structure and spatial pattern. *Australian Journal of Ecology*, **6**, 223–229.

Gignac, L. D., & Vitt, D. H. (1990) Habitat limitations of *Sphagnum* along climatic, chemical and physical gradients. *The Bryologist*, **93**, 7–22.

Gignac, L. D., Vitt, D. H., & Bayley, S. E. (1991). Bryophyte response surfaces along ecological and climatic gradients. *Vegetatio*, **93**, 29–45.

Gill, D. E. (1975). Spatial patterning of pines and oaks in the New Jersey pine barrens. *Journal of Ecology*, **63**, 291–298.

Glaser, P. H., Wheeler, G. A., Gorham, E., & Wright, H. E. (1981). The patterned mires of Red Lake peatland, northern Minesota: Vegetation, water chemistry, and landforms. *Journal of Ecology*, **69**, 575–599.

Goldsborough, L. G. (1989). Examination of two dimensional spatial pattern of periphytic diatoms using an adhesive surficial peel technique. *Journal of Phycology*, **25**, 133–143.

Goodall, D. W. (1963). Pattern analysis and minimal area – some further comments. *Journal of Ecology*, **51**, 705–710.

Goodall, D. W. (1974). A new method for the analysis of spatial pattern by the random pairing of quadrats. *Vegetatio*, **29**, 135–146.

Goodall, D. W. (1978). Numerical classification. In *Classification of Plant Communities*, ed. R. H. Whittaker. The Hague: Junk.

Goodchild, M. F. (1994). Integrating GIS and remote sensing for vegetation analysis and modelling: methodological issues. *Journal of Vegetation Science*, **5**, 615–626.

Grace, J. B. (1987). The impact of preemption on the zonation of two *Typha* species along lakeshores. *Ecological Monographs*, **57**, 283–303.

Grace, J. B., & Wetzel, R. G. (1981). Habitat partitioning and competitive displacment in cattails (*Typha*): experimental field studies. *The American Naturalist*, **118**, 463–474.

Green, D. G. (1990). Cellular automata models in biology. *Mathematical Computer Modelling*, **13**, 69–74.

Greenwood, M. (1946). The statistical study of infectious diseases. *Journal of the Royal Statistical Society*, **109**, 85– 110.

Greig-Smith, P. (1952). The use of random and contiguous quadrats in the study of structure in plant communities. *Annals of Botany*, **16**, 293–316.

Greig-Smith, P. (1961a). Data on pattern within plant communities. I. The analysis of pattern. *Journal of Ecology*, **49**, 695–702.

Greig-Smith, P. (1961b). Data on pattern within plant communities. II. *Ammophila arenaria* (L.) Link. *Journal of Ecology*, **49**, 703–708.

Greig-Smith, P. (1979). Pattern in vegetation. *Journal of Ecology*, **67**, 755–779.

Greig-Smith, P. (1983). *Quantitative Plant Ecology*, 3rd edn. Berkeley: University of California Press.

Greig-Smith, P., & Chadwick, M. J. (1965). Data on pattern within plant communities. III Acacia-Capparis semi-desert scrub in the Sudan. *Journal of Ecology*, **53**, 465–474.

Haase, P. (1995). Spatial pattern analysis in ecology based on Ripley's K-function: introduction and methods of edge correction. *Journal of Vegetation Science*, **6**, 575–582.

Haase, P., Pugnaire, F., Clark, S. C., & Incoll, L. D. (1996). Spatial pattern in a two-tiered semi-arid shrubland in southeastern Spain. *Journal of Vegetation Science*, **7**, 527– 534.

Harris, S. A. (1990). Dynamics and origin of saline soils on the Slims River delta, Kluane National Park, Yukon Territory. *Arctic*, **43**, 159–175.

Hartnett, D. C., & Abrahamson, W. G. (1979). The effects of stem gall insects on life history patterns in *Solidago canadensis*. *Ecology*, **60**, 910–907.

Heilbronn, T. D., & Walton, W. H. (1984). Plant colonization of actively sorted stone stripes in the subantarctic. *Arctic and Alpine Research*, **16**, 161–172.

Heusser, C. J. (1956). Post glacial environments in the Canadian Rocky Mountains. *Ecological Monographs*, **26**, 263–302.

Hill, M. O. (1973). The intensity of spatial pattern in plant communities. *Journal of Ecology*, **61**, 225–235.

Hnatiuk, R. J. (1969). *The Pinus contorta vegetation of Banff and Jasper National Parks*. M.Sc. Thesis. Edmonton, Canada: University of Alberta.

Hughes, G. (1988). Spatial dynamics of self-thinning. *Nature*, **336**, 521.

Hurlbert, S. H. (1990). Spatial distribution of the montane unicorn. *Oikos*, **58**, 257–271.

Hutchings, M. J. (1979). Standing crop and pattern in pure stands of *Mercurialis perennis* and *Rubus fruticosus* in mixed deciduous woodland. *Oikos*, **31**, 351–357.

Ireland, R. R., Bird, C. D., Brassard, G. R., Schofield, W. B., & Vitt, D. H. (1980). *Checklist of the Mosses of Canada*. Publications in Botany, No. 8., National Museum of Canada.

Jackson, R. B., & Caldwell, M. M. (1993). Geostatistical patterns of soil heterogeneity around individual perennial plants. *Journal of Ecology*, **81**, 683–692.

James, F. C., & McCulloch, C. E. (1990). Multivariate analysis in ecology and systematics: panacea or Pandora's box? *Annual Review of Ecology and Systematics*, **21**, 129–166.

Jenkins, G. M., & Watts, D. G. (1969). *Spectral Analysis and its Applications*. San Fransisco: Holden-Day.

John, E. A. (1989). An assessment of the role of biotic interactions and dynamic processes in the organization of species in a saxicolous lichen community. *The Canadian Journal of Botany*, **67**, 2025–2038.

John, E. A. (1990). Fine scale patterning of species distributions in a saxicolous lichen community at Jonas Rockslide, Canadian Rocky Mountains. *Holarctic Ecology*, **13**, 187–194.

John, E. A., & Dale, M. R. T. (1989). Niche relationships amongst *Rhizocarpon* species at Jonas Rockslide, Alberta, Canada. *Lichenologist*, **21**, 313–330.

John, E. A., & Dale, M. R. T. (1990). Environmental correlates of species distributions in a saxicolous lichen community. *Journal of Vegetation Science*, **1**, 385–392.

John, E. A., & Dale, M. R. T. (1995). Neighbor relations within an epiphytic lichens and bryophytes. *The Bryologist*, **98**, 27–37.

Johnson, W. C., Sharik, T. L., Mayes, R. A., & Smith, E. P. (1987). Nature and cause of zonation discreteness around glacial prairie marshes. *Canadian Journal of Botany*, **65**, 1622–1632.

Johnston, C. A., Pastor, J., & Pinay, G. (1992). Quantitative methods for studying landscape boundaries. In *Landscape Boundaries*. eds. A. J. Hansen & F. di Castri, pp.107–125. New York: Springer-Verlag.

Jumars, P. A., Thistle, D., & Jones, M. L. (1977). Detecting two-dimensional spatial structure in biological data. *Oecologia*, **28**, 109–123.

Kanzaki, M. (1984). Regeneration in subalpine coniferous forests. I. Mosaic structure and plant processes in a *Tsuga diversifolia* forest. *The Botanical Magazine*, **97**, 297–311.

Kareiva, P. (1982). Experimental and mathematical analyses of herbivore movement:

quantifying the influence of plant spacing and quality on foraging discrimination. *Ecological Monographs*, **52**, 261–282.

Kareiva, P. (1985). Finding and losing host plants by *Phyllotreta*: patch size and surrounding habitat. *Ecology*, **66**, 1809–1816.

Kareiva, P. (1987). Habitat fragmentation and the stability of predator-prey interactions. *Nature*, **326**, 388–390.

Keddy, P. A. (1991). Working with heterogeneity: an operator's guide to environmental gradients. In *Ecological Heterogeneity*. eds. I. Kolasa & S. T. A. Pickett, pp. 181–200. New York: Springer-Verlag.

Keenan, R. J. (1994). *The Population Structure of Western Redcedar and Western Hemlock Forests on Northern Vancouver Island, Canada.* M.Sc. Thesis. Vancouver: University of British Columbia.

Kendall, M. G., & Stuart, A. (1973). *The Advanced Theory of Statistics*, 3rd edn., Vol. 2. New York: Hafner.

Kenkel, N. C. (1988a). Pattern of self-thinning in jack pine: testing the random mortality hypothesis. *Ecology*, **69**, 1017– 1024.

Kenkel, N. C. (1988b). Spectral analysis of hummock-hollow pattern in a weakly minerotrophic mire. *Vegetatio*, **78**, 45–52.

Kenkel, N. C. (1993). Modeling Markovian dependence in populations of *Aralia nudicaulis*. *Ecology*, **74**, 1700–1706.

Kenkel, N. C., Juhász-Nagy, P., & Podani, J. (1989). On sampling procedures in population and community ecology. *Vegetatio*, **83**, 195–207.

Kenkel, N. C., & Podani, J. (1991). Plot size and estimation efficiency in plant community studies. *Journal of Vegetation Science*, **2**, 539–544.

Kenkel, N. C., & Walker D. J. (1993). Fractals and ecology. *Abstracta Botanica*, **17**, 53–70.

Kent, M., & Coker, P. (1992). *Vegetation Description and Analysis*. Boca Raton, FL: CRC Press.

Kershaw, K. A. (1957). The use of cover and frequency in the detection of pattern in plant communities. *Ecology*, **38**, 291– 299.

Kershaw, K. A. (1958). An investigation of the structure of a grassland community. I. The pattern of *Agrostis tenuis*. *Journal of Ecology*, **467**, 571–592.

Kershaw, K. A. (1959a). An investigation of the structure of a grassland community. II. The pattern of *Dactylis glomerata Lolium perenne*, and *Trifolium repens*. *Journal of Ecology*, **47**, 31–43.

Kershaw, K. A. (1959b). An investigation of the structure of a grassland community. III Discussion and conclusions. *Journal of Ecology*, **47**, 44–53.

Kershaw, K. A. (1960). The detection of pattern and association. *Journal of Ecology*, **48**, 233–242.

Kershaw, K. A. (1961). Association and co-variance analysis of plant communities. *Journal of Ecology*, **49**, 643–654.

Kershaw, K. A. (1962). Quantitative ecological studies from Landmannahellir, Iceland. III The variation of performance of *Carex bigelowii*. *Journal of Ecology*, **50**, 393–399.

Kershaw, K. A. (1963). Pattern in vegetation and its causality. *Ecology*, **44**, 377–388.

Kershaw, K. A. (1964). *Quantitative and Dynamic Ecology*. London: Edward Arnold.

Keuls, M., Over, H. J., & de Wit, C. T. (1963). The distance method for estimating densities. *Statistica Neerlandica*, **17**, 71–91.

Kimball, B. F. (1947). Some basic theorems for developing tests of fit for the case of nonparametric probability distribution functions. *Annals of Mathematical Statistics*, **18**, 540–548.

Kimmins, J. P. (1992). *Balancing Act: Environmental Issues in Forestry*. Vancouver: UBC Press.

Kitayama, K. (1992). An altitudinal transect study of the vegetation on Mount Kinabalu, Borneo. *Vegetatio*, **102**, 149–171.

Koopmans, L. H. (1974). *The Spectral Analysis of Time Series*. London: Academic Press.

Kotliar, N. B., & Wiens, J. A. (1990). Multiple scales of patchiness and patch structure: a hierarchical framework for the study of heterogeneity. *Oikos*, **59**, 253–260.

Krebs, C. J., Boutin, S., Boonstra, R., Sinclair, A. R. E., Smith, J. N. M., Dale, M. R. T., Martin, K., & Turkington, R. (1995). Impact of food and predation on the snowshoe hare cycle. *Science*, **269**, 1112–1115.

Laca, E. A., Distal, R. A., Griggs, T. C., & Demment, M. W. (1994). Effects of canopy structure on patch depression by grazers. *Ecology*, **75**, 706–716.

Lamont, B. B., & Fox, J. E. D. (1981). Spatial pattern of six sympatric leaf variants and two size classes of *Acacia aneura* in a semi-arid region of Western Australia. *Oikos*, **37**, 73–79.

Leduc, A., Drapeau, P., Bergeron, Y., & Legendre, P. (1992). Study of spatial components of forest cover using partial Mantel tests and path analysis. *Journal of Vegetation Science*, **3**, 69–78.

Lee, P. C. (1993). The effect of seed dispersal limitations on the spatial distribution of a gap species, seaside goldenrod (*Solidago sempervirens*). *Canadian Journal of Botany*, **71**, 978–984.

Leemans, R. (1990). Sampling establishment patterns in relation to light gaps in the canopy of two primeval pine-spruce forests in Sweden. In *Spatial Processes in Plant Communities*. eds. F. Krahulec, A. D. Q. Agnew, S. Agnew & H. J. Willems, pp 111–119. Prague: Academia.

Legendre, P. (1993). Spatial autocorrelation: trouble or new paradigm? *Ecology*, **74**, 1659–1673.

Legendre, P., & Fortin, M. J. (1989). Spatial pattern and ecological analysis. *Vegetatio*, **80**, 107–38.

Lepš, J. (1990a). Can underlying mechanisms be deduced from observed patterns? *In Spatial Processes in Plant Communities*. eds. F. Krahulec, A. D. Q. Agnew, S. Agnew, & J. H. Willem, pp. 1–11. Prague: Academia Press.

Lepš, J. (1990b). Comparison of transect methods for the analysis of spatial pattern. In *Spatial Processes in Plant Communities*. eds. F. Krahulec, A. D. Q. Agnew, S. Agnew & J. H. Willem, Prague: Academia Press, pp. 71–82.

Lewis, J. R. (1964). *The Ecology of Rocky Shores*. London: English Universities Press.

Lillesand, T. M., & Kiefer, R. W. (1987). *Remote Sensing and Image Interpretation*, 2nd edn. New York: Wiley.

Ludwig, J. A. (1979). A test of different quadrat variance methods for the analysis of spatial pattern. In *Spatial and Temporal Analysis in Ecology*. eds. R. M. Cormack & J. K. Ord, pp. 289–303. Fairland: International Co-operative Publishing House.

Ludwig, J. A., & Goodall, D. W. (1978). A comparison of paired with blocked quadrat variance methods for the analysis of spatial pattern. *Vegetatio*, **38**, 49–59.

Ludwig, J. A., & Reynolds, J. F. (1988). *Statistical Ecology*. New York: Wiley.

Mahdi, A., & Law, R. (1987). On the spatial organization of plant species in a limestone grassland community. *Journal of Ecology*, **75**, 459–476.

Mandelbrot, B. B. (1982). *The Fractal Geometry of Nature*. San Fransisco: Freeman.

Mandossian, A., & MacIntosh, R. P. (1960). Vegetation zonation on the shore of a small lake. *American Midland Naturalist*, **64**, 301–308.

Manly, B. F. J. (1991). *Randomization and Monte Carlo Methods in Biology*. London: Chapman and Hall.

Mantel, N. (1967). The detection of disease clustering and a generalized regression approach. *Cancer Research*, **27**, 209– 220.

Maslov, A. A. (1989). Small-scale patterns of forest plants and environmental heterogeneity. *Vegetatio*, **84**, 1–7.

Maslov, A. A. (1990). Multi-scaled and multi-species pattern analysis in boreal forest communities. In *Spatial Processes in Plant Communities*. eds. F. Krahulec, A. D. Q. Agnew, S. Agnew, & J. H. Willem, pp. 83–88. Prague: Acadamia.

Matérn, B. (1947). *Metoder att uppskatta nogrannheten vid linje-och provytetaxering*. Meddelanden från Statens Skogforskningsinstitut 36(1). [*Methods to estimate accuracy of line- and area-sampling*. Report from the Government Forest Research Institute.]

Matérn, B. (1979). The analysis of ecological maps as mosaics. In *Spatial and Temporal Analysis in Ecology*. eds. R. M. Cormack & J. K. Ord, pp. 271–287. Fairland: International Co-operative Publishing House.

Matérn, B. (1986). *Spatial Variation*, 2nd edn. New York: Springer- Verlag.

Maubon, M., Ponge, J.-F., & André, J. (1995). Dynamics of *Vaccinium myrtillus* patches in mountain spruce forest. *Journal of Vegetation Science*, **6**, 343–348.

McCoy, E. D., & Bell, S. S. (1991). Habitat structure: the evolution and diversification of a complex topic. In *Habitat Structure: The Physical Arrangement of Objects in Space*. eds. S. S. Bell, E. D. McCoy, & H. R. Mushinsky, pp. 3–21. London: Chapman & Hall.

McCoy, E. D., Bell, S. S., & Mushinsky, H. R. (1991). *Habitat Structure: The Physical Arrangement of Objects in Space*. London: Chapman & Hall.

McGarigal, K., & Marks, B. J. (1995). FRAGSTATS: Spatial pattern analysis program for quantifying landscape structure. General Technical Report PNW-GTR-351. Portland, OR: U.S. Department of Agriculture, Forest Service, Pacific Northwest Research Station.

McGregor, J. R. (1988). Asymptotic normality of the Münch numbers. *Scandinavian Journal of Statistics*, **15**, 299–301.

Mead, R. (1974). A test for spatial pattern at several scales using data from a grid of contiguous quadrats. *Biometrics*, **30**, 965–981.

Mertes, L. A. K., Daniel, D. L., Melack, J. M., Nelson, B., Martinelli, L. A., Forsberg, B. R. (1995). Spatial patterns of hydrology, geomorphology, and vegetation on the flood plain of the Amazon River in Brasil from a remote sensing prospective. *Geomorphology*, **13**, 215–232.

Milne, B. (1991a). Lessons from applying fractal models to landscape patterns. In *Quantitative Methods in Landscape Ecology*. eds. M. G. Turner & R. H. Gardner, pp. 199–235. New York: Springer-Verlag.

Milne, B. T. (1991b). Heterogeneity as a multiscale characteristic of landscapes. In *Ecological Heterogeneity*. eds. J. Kolasa & S. T. A. Pickett, pp. 69–84. New York: Springer-Verlag.

Minchin, P. R. (1989). Montane vegetation of the Mt. Field massif, Tasmania: a test of some hypotheses about properties of community patterns. *Vegetatio*, **83**, 97–110.

Mithen, R., Harper, J. L., & Weiner, J. (1984). Growth and mortality of individual plants as a function of 'available area'. *Oecologia*, **62**, 57–60.

Molofsky, J. (1994). Population dynamics and pattern formation in a theoretical population. *Ecology*, **75**, 30–39.

Montaña, C. (1992). The colonization of bare areas in two-phase mosaics of an arid ecosystem. *Journal of Ecology*, **80**, 315–327.

Mooers, H. D., & Glaser, P. H. (1989). Active patterned ground at sea level, Fourchu, Nova Scotia, Canada. *Arctic and Alpine Research*, **21**, 425–432.

Moore, J. A., Budelsky, C. A., & Schlesinger, R. C. (1973). A new index representing individual tree competitive status. *Canadian Journal of Forestry*, **3**, 495–500.

Moran, P. A. P. (1947). The random division of an interval. *Journal of the Royal Statistical Society Supplement*, **9**, 92–98.

Mordelet, P., Barot, S., & Abbadie, L. (1996). Root foraging strategies and soil patchiness in a humid savanna. *Plant and Soil*, **182**, 171–176.

Morse, D. R., Lawton, J. H., Dodson, M. M., & Williamson, M. H. (1985). Fractal dimension of vegetation and the distribution of arthropod lengths. *Nature*, **314**, 731–733.

Mosby, H. S. (1959). Reconnaissance mapping and map use. In *Wildlife Management Technique*, ed. R. H. Giles, pp. 119–134. Ann Arbor: Edwards Bros.

Moss, E. H. (1983). *Flora of Alberta*, 2nd edn. revised by J. G. Packer. Toronto: University of Toronto Press.

Musik, H. B., & Grover, H. D. (1991). Image textural measures as indices of landscape pattern. In *Quantitative Methods in Landscape Ecology*. eds. M. G. Turner & R. H. Gardner, pp. 77–103. New York: Springer-Verlag.

Navas, M.-L., & Goulard, M. (1991). Spatial pattern of a clonal perennial weed, *Rubia peregrina* (Rubiaceae) in vineyards of southern France. *Journal of Applied Ecology*, **28**, 1118–1129.

Newbery, D. McC., Renshaw, E., & Brunig, E. F. (1986). Spatial pattern of trees in Kerangas forest, Sarawak. *Vegetatio*, **65**, 77–89.

Noy-Meir, I., & Anderson, D. (1971). Multiple pattern analysis or multiscale ordination: towards a vegetation hologram. In Statistical Ecology, Vol. 3. *Populations, Ecosystems, and Systems Analysis*. eds. G. P. Patil, E. C. Pielou, & W. E. Water, pp. 207–232. University Park: Pennsylvania State University Press.

Oden, N. L., & Sokal, R. R. (1986). Directional autocorrelation: an extension of spatial correlograms to two dimensions. *Systematic Zoology*, **35**, 608–617.

Okabe, A., Boots, B., & Sugihara, K. (1992). *Spatial Tessellations: Concepts and Applications of Voronoi Diagrams*. New York: Wiley.

Økland, R. H. (1992). Studies in SE Fennoscandian mires: relevance to ecological theory. *Journal of Vegetation Science*, **3**, 279–284.

Økland, R. H. (1994). Patterns of bryophyte associations at different scales in a Norwegian boreal spruce forest. *Journal of Vegetation Science*, **5**, 127–138.

Økland, R. H., & Eilertsen, O. (1994). Canonical correspondence analysis with variation partitioning: some comments and an application. *Journal of Vegetation Science*, **5**, 117–126.

Oliver, M. A., & Webster, R. (1986). Combining nested and linear sampling for determining the scale and form of spatial variation of regionalized variables. *Geographical Analysis*, **18**, 227–242.

O'Neill, R. V., Gardner, R. H., Milne, B. T., Turner, M. G., & Jackson, B. (1991). Heterogeneity and spatial hierarchies. In *Ecological Heterogeneity*. eds. J. Kolasa & S. T. A. Pickett, pp. 85–96. New York: Springer-Verlag.

Pacala, S. W. (1997). Dynamics of plant communities. In *Plant Ecology*, 2nd edn. ed. M. J. Crawley, pp. 532–555. Cambridge: Blackwell Science, Ltd.

Palmer, M. W. (1988). Fractal geometry: a tool for describing spatial patterns of plant communities. *Vegetatio*, **75**, 91–102.

Palmer, M. W. (1990a). Spatial scale and patterns of species–environment relationships in hardwood forest of the North Carolina Piedmont. *Coenoses*, **5**, 79–87.

Palmer, M. W. (1990b). Spatial scale and patterns of vegetation, flora and species richness in hardwood forests of the North Carolina piedmont. *Coenoses*, **5**, 89–96.

Peart, D. R. (1989). Species interactions in a successional grassland. III. Effects of canopy gaps, gopher mounds, and grazing on colonization. *Journal of Ecology*, **77**, 267–289.

Persson, S. (1981). Ecological indicator values as an aid to the interpretation of ordination diagrams. *Journal of Ecology*, **69**, 71–84.

Peters, R., & Ohkubo, T. (1990). Architecture and development in *Fagus japonica-Fagus crenata* forest near Mount Takahara, Japan. *Journal of Vegetation Science*, **1**, 499–506.

Petersen, C. J., & Squiers, E. R. (1995). An unexpected change in spatial pattern across 10 years in an aspen-white-forest, *Journal of Ecology*, **83**, 847–855.

Phillips, J. D. (1985). Measuring complexity of environmental gradients. *Vegetatio*, **64**, 95–102.

Pielou, E. C. (1974). Competition on an environmental gradient. In *Proceedings of the Conference on Mathematical Problems in Biology*. ed. P. van den Driessche. Berlin: Springer-Verlag.

Pielou, E. C. (1975). Ecological models on an environmental gradient. In *Applied Statistics*. ed. R. P. Gupta, pp. 261–269. The Netherlands: North-Holland Publishing Co.

Pielou, E. C. (1977a). *Mathematical Ecology*. New York: Wiley.

Pielou, E. C. (1977b). The latitudinal spans of seaweed species and their patterns of overlap. *Journal of Biogeography*, **4**, 299–311.

Pielou, E. C. (1978). Latitudinal overlap of seaweed species: evidence for quasi-sympatric speciation. *Journal of Biogeography*, **5**, 227–238.

Pielou, E. C. (1979). On A. J. Underwood's model for a random pattern. *Oecologia*, **44**, 143–144.

Pielou, E. C., & Routledge, R. D. (1976). Salt marsh vegetation latitudinal gradients in the zonation patterns. *Oecologia*, **24**, 311–321.

Platt, W. J., & Strong, D. R. (1989). Gaps in forest ecology. *Ecology*, **70**, 535–576.

Plotnick, R. E., Gardner, R. H., Hargrove, W. H., Prestegaard, K., & Perlmutter, M. (1996). Lacunarity analysis: a general technique for the analysis of spatial pattern. *Physical Review E*, **53**, 5461–5468.

Plotnick, R. E., Gardner, R. H., & O'Neill, R. V. (1993). Lacunarity indices as measures of landscape texture. *Landscape Ecology*, **8**, 201–211.

Podani, J. (1984). Analysis of mapped and simulated patterns by means of computerized sampling techniques. *Acta Botanica Hungarica*, **30**, 403–425.

Powell, R. D. (1990). The role of spatial pattern in the population biology of *Centaurea diffusa*. *Journal of Ecology*, **78**, 374–388.

Prentice, I. C., & Werger, M. J. A. (1985). Clump spacing in a desert dwarf shrub community. *Vegetatio*, **63**, 133–139.

Qinghong, L., & Hytteborn, H. (1991). Gap structure, disturbance and regeneration in primeval *Picea abies* forest. *Journal of Vegetation Science*, **2**, 391–402.

Rapoport, E. H. (1982). *Areography: Geographical Strategies of Species*. Oxford: Pergamon Press.

Ratz, A. (1995). Long-term spatial patterns created by fire: a model oriented towards boreal forests. *International Journal of Wildland Fire*, **5**, 25–34.

Reed, R. A., Peet, R. K., Palmer, M. W., & White, P. S. (1993). Scale dependence of vegetation-environment correlations: a case study of a North Carolina piedmont woodland. *Journal of Vegetation Science*, **4**, 329–340.

Reeve, J. D. (1987). Foraging behaviour of *Aphytis melinus:* effects of patch density and host size. *Ecology*, **68**, 1008–1016.

Remmert, H. (1991). The mosaic-cycle concept of ecosystems – an overview. In *The Mosaic-cycle Concept of Ecosystems*. ed. H. Remmert, pp. 1–21. Berlin: Springer-Verlag.

Renshaw, E., & Ford, E. D. (1984). The description of spatial pattern using two-dimensional spectral analysis. *Vegetatio*, **56**, 75–85.

Reynolds, C. S. (1984). *The Ecology of Freshwater Plankton*. Cambridge: Cambridge University Press.

Ricklefs, R. E. (1990). *Ecology*, 3rd edn. New York: Freeman.

Ripley, B. D. (1976). The second order analysis of stationary point processes. *Journal of Applied Probability*, **13**, 255–266.

Ripley, B. D. (1977). Modelling spatial patterns. *Journal of the Royal Statistical Society, Series B*, **39**, 172–212.

Ripley, B. D. (1978). Spectral analysis and the analysis of pattern in plant communities. *Journal of Ecology*, **66**, 965– 981.

Ripley, B. D. (1981). *Spatial Processes*. New York: Wiley.

Ripley, B. D. (1988). *Statistical Inference for Spatial Processes*. Cambridge: Cambridge University Press.

Ripley, B. D., & Kelly, F. P. (1977). Markov point processes. *Journal of the London Mathematical Society*, **15**, 188–192.

Rohlf, F. J., & Archie, J. W. (1978). Least-squares mapping using interpoint distances. *Ecology*, **59**, 126–132.

Roland, J. (1993). Large-scale forest fragmentation increases the duration of tent caterpillar outbreak. *Oecologia*, **93**, 25–30.

Rossi, R. E., Mulla, D., Journel, A. G., & Franz, E. H. (1992). Geostatistical tools for modeling and interpreting ecological spatial dependence. *Ecological Monographs*, **62**, 277–314.

Rydin, H. (1986). Competition and niche separation in *Sphagnum*. *Canadian Journal of Botany*, **64**, 1817–1824.

Schaefer, J. A. (1993). Spatial patterns in taiga plant communities following fire. *Canadian Journal of Botany*, **71**, 1568–1573.

Schroeder, M. (1991). *Fractals, Chaos, Power Laws*. New York: Freeman.

Shipley, B., & Keddy, P. A. (1987). The individualistic and community-unit concepts as falsifiable hypotheses. *Vegetatio*, **69**, 47–55.

Shmida, A., & Whittaker, R. H. (1981). Pattern and biological microsite effects in two shrub communities, Southern California. *Ecology*, **61**, 234–251.

Shorrocks, B., Marsters, J., Ward, I., & Evennett, P. J. (1991). The fractal dimension of lichens and the distribution of arthropod body lengths. *Functional Ecology*, **5**, 457–460.

Sih, A., & Baltus M.-S. (1987). Patch size, pollinator behavior, and pollinator limitation in catnip. *Ecology*, **68**, 1679– 1690.

Silander, J. A. Jr., & Pacala, S. W. (1985). Neighborhood predictors of plant performance. *Oecologia*, **66**, 256–263.

Silander, J. A. Jr., & Pacala, S. W. (1990). The application of plant population dynamic models to understanding plant competition. In *Perspectives on Plant Competition*. eds. J. B. Grace & D. Tilman, pp. 67–91. San Diego: Academic Press.

Silvertown, J. W., & Lovett Doust, J. (1993). *Introduction to Plant Population Biology*. Oxford: Blackwell.

Skarpe, C. (1991). Spatial patterns and dynamics of woody vegetation in an arid savanna. *Journal of Vegetation Science*, **2**, 565–572.

Skellam, J. G. (1952). Studies in statistical ecology. I. Spatial pattern. *Biometrica*, **39**, 346–362.

Slezak, E., Bijaoui, A., & Mars, G. (1992). Structures identification from galaxy counts – use of the wavelet transform. In *Wavelets and Applications*. ed. Y. Meyer. Masson: Springer-Verlag.

Smith, T. M., & Urban, D. L. (1988). Scale and resolution of forest structural pattern. *Vegetatio*, **74**, 143–150.

Sokal, R. R. (1986). Spatial data analysis and historical processes. In *Data Analysis and Informatics*, IV. Proceedings of the Fourth International Symposium of Data Analysis and Informatics, Versailles, France, pp. 29–43. Amsterdam: North Holland.

Sokal, R. R., & Oden, N. L. (1978a). Spatial autocorrelation in biology. I. Methodology. *Biological Journal of the Linnean Society*, **10**, 199–228.

Sokal, R. R., & Oden, N. L. (1978b). Spatial autocorrelation in biology. II. Some biological applications of evolutionary and ecological interest. *Biological Journal of the Linnean Society*, **10**, 229–249.

Sokal, R. R., & Rohlf, F. J. (1981). *Biometry*. San Fransisco: Freeman.

Sowig, P. (1989). Effects of flowering plant's patch size on species composition of pollinator communities, foraging strategies, and resource partitioning in bumblebees (Hymenoptera: Apidae). *Oecologia*, **78**, 550–558.

Sprugel, D. G. (1976). Dynamic structure of wave-generated *Abies balsamea* forests in the North Eastern United States. *Journal of Ecology*, **64**, 889–911.

Stadt, J. J. (1993). Pinus contorta *community dynamics in Banff and Jasper National Parks*. M.Sc. Thesis. Edmonton, Canada: University of Alberta.

Steinauer, E. M., & Collins, S. L. (1995). Effect of urine deposition on small-scale patch structure in prairie vegetation. *Ecology*, **76**, 1195–1205.

Sterling, A., Peco, B., Casado, M. A., Galiano, E. F., & Pineda, F. D. (1984). Influence of microtopography on floristic variation in the ecological succession in grassland. *Oikos*, **42**, 334–342.

Stowe, L. G., & Wade, M. J. (1979). The detection of small-scale patterns in vegetation *Journal of Ecology*, **67**, 1047–1064.

Strang, G. (1994). Wavelets. *American Scientist*, **82**, 250–255.

Sugihara, G., & May, R. M. (1990). Applications of fractals in ecology. *Trends in Ecology and Evolution*, **5**, 79–86.

Sumida, A. (1995). Three-dimensional structure of an mixed boreal broad-leaved forest in Japan. *Vegetatio*, **119**, 67–80.

Swanson, D. K., & Grigal, D. F. (1988). A simulation model of mire patterning. *Oikos*, **53** 309–314.

Symonides, E., & Wierzchowska, U. (1990). Changes in the spatial pattern of vegetation structure and soil properties in early old-field succession. In *Spatial Processes in Plant Communities*. eds. F. Krahulec, A. D. Q. Agnew, S. Agnew, & H. J. Willems, pp. 201–214. Prague: Academia.

Szwagrzyk, J., & Czerwczak, M. (1993). Spatial patterns of trees in natural forests of East-Central Europe. *Journal of Vegetation Science*, **4**, 469–476.

Taylor, A. H., & Halpern, C. B. (1991). The structure and dynamics of *Abies magnifica* forests in the Southern Cascade Range, USA. *Journal of Vegetation Science*, **2**, 189–200.

ter Braak, C. J. F. (1986). Canonical correspondance analysis: a new eigenvector technique for multivariate direct gradient analysis. *Ecology*, **67**, 1167–1179.

ter Braak, C. J. F. (1987). The analysis of vegetation-environment relationships by canonical correspondance analysis. *Vegetatio*, **69**, 69–77.

ter Braak, C. J. F., & Prentice, I. C. (1988). A theory of gradient analysis. *Advances in Ecological Research*, **18**, 271–317.

Termier, H., & Termier, G. (1963). *Erosion and Sedimentation*. London: Van Nostrand.

Thiéry, J. M., D'Herbès, J.-M., & Valentin, C. (1995). A model simulating the genesis of banded vegetation patterns in Niger. *Journal of Ecology*, **83**, 497–507.

Thompson, J. N. (1956). Distribution of distance to n^{th} nearest neighbor in a population of randomly distributed individuals. *Ecology*, **27**, 391–394.

Thompson, J. N. (1978). Within-patch structure and dynamics in *Pastinaca sativa* and resource availability to a specialized herbivore. *Ecology*, **59**, 443–448.

Thomson, J. D., Weiblen, G., Thomson, B. A., Alfaro, S., & Legendre, P. (1996). Untangling multiple factors in spatial distribution: lilies, gophers, and rocks. *Ecology*, **77**, 1698–1715.

Tilman, D. (1994). Competition and biodiversity in spatially structured habitats. *Ecology*, **75**, 2–16.

Turkington, R., & Harper, J. L. (1979a). The growth, distribution and neighbor relationships of *Trifolium repens* in a permanent pasture. I. Ordination, pattern and contact. *Journal of Ecology*, **67**, 201–218.

Turkington, R., & Harper, J. L. (1979b). The growth, distribution and neighbor relationships of *Trifolium repens* in a permanent pasture. II. Inter- and intra-specific contact. *Journal of Ecology*, **67**, 219–230.

Turner, M. G. (1989). Landscape ecology: the effect of pattern on process. *Annual Review of Ecology and Systematics*, **20**, 171–197.

Turner, M. G., & Bratton, S. P. (1987). Fire, grazing and the landscape heterogeneity of a Georgia barrier island. In *Landscape Heterogeneity and Disturbance*. ed. M. G. Turner, pp. 85–101. New York: Springer-Verlag.

Turner, M. G., & Gardner, R. H. (1991). *Quantitative Methods in Landscape Ecology*. New York: Springer-Verlag.

Turner, M. G., Hargrove, W., Gardner, R. H., & Romme, W. H. (1994). Effects of fire on landscape heterogeneity in Yellowstone National Park, Wyoming. *Journal of Vegetation Science*, **5**, 731–742.

Turner, S. J., O'Neill, R. V., Conley, W., Conley, M. R., & Humphries, H. C. (1991). Pattern and scale: statistics for landscape ecology. In *Quantitative Methods in Landscape Ecology*. eds. M. G. Turner & R. H. Gardner, pp. 17–49. New York: Springer-Verlag.

Umbanhowar, C. E. Jr. (1992). Abundance, vegetation, and environment of four patch types in northern mixed prairie. *Canadian Journal of Botany*, **70**, 227–284.

Underwood, A. J. (1978a). The detection of nonrandom patterns of distribution of species along a gradient. *Oecologia*, **36**, 317–326.

Underwood, A. J. (1978b). A refutation of critical tide levels as determinants of the structure of the intertidal communities on British shores. *Oecologia*, **37**, 261–276.

Upton, G. J. G., & Fingleton, B. (1985). *Spatial Data Analysis by Example*, Vol. I. *Point Pattern and Quantitative Data*. New York: Wiley.

Upton, G. J. G., & Fingleton, B. (1989). *Spatial Data Analysis by Example*, Vol. II. *Categorical and Directional Data*. New York: Wiley.

Usher, M. B. (1975). Analysis of pattern in real and artificial plant populations. *Journal of Ecology*, **63**, 569–586.

Usher, M. B. (1983). Pattern in the simple moss-turf communities of the Sub-Antarctic and Maritime Antarctic. *Journal of Ecology*, **71**, 945–958.

Usher, M. B., Booth, R. G., & Sparks, K. E. (1982). A review of progress in understanding the organisation of communities of soil arthropods. *Pedobiologia*, **23**, 126–144.

Valiente-Banuet, A., & Ezcurra, E. (1991). Shade as a cause of the association between the cactus *Neobuxbaumia tetetzo* and the nurse plant *Mimosa luisana* in the Tehuacan Valley, Mexico. *Journal of Ecology*, **79**, 961–971.

Van der Hoeven, E. C., de Kroon, H., & During, H. J. (1990). Fine-scale spatial distribution of leaves and shoots of two chalk grassland perennials. *Vegetatio*, **86**, 151–160.

van der Maarel, E. (1996). Pattern and process in the plant community: fifty years after A. S. Watt. *Journal of Vegetation Science*, **7**, 19–28.

Veblen, T. T. (1979). Structures and dynamics of *Nothofagus* forests near timberline in South Chile. *Ecology*, **60**, 937–945.

Veblen, T. T. (1992). Regeneration dynamics. In *Plant Succession: Theory and Prediction*. eds. D. C. Glenn-Lewin, R. K. Peet & T. T. Veblen, pp. 152–187. London: Chapman & Hall.

Ver Hoef, J. M., Cressie, N. A. C., & Glenn-Lewin, D. C. (1993). Spatial models for spatial statistics: some unification. *Journal of Vegetation Science*, **4**, 441–452.

Ver Hoef, J. M., & Glenn-Lewin, D. C. (1989). Multiscale ordination: a method for detecting pattern at several scales. *Vegetatio*, **82**, 59–67.

Ver Hoef, J. M., Glenn-Lewin, D. C., & Werger, M. J. A. (1989). Relationship between horizontal pattern and vertical structure in a chalk grassland. *Vegetatio*, **83**, 147–155.

Vince, S. W., & Snow, A. A. (1984). Plant zonation in an Alaskan salt marsh. I. Distribution, abundance, and environmental factors. *Journal of Ecology*, **72**, 651–667.

Washburn, A. L. (1980). *Geocryology: A Survey of Periglacial Processes and Environments*. New York: Wiley.

Watkins, A. J., & Bastow Wilson, J. (1992). Fine-scale community structure of lawns *Journal of Ecology*, **80**, 15–24.

Watt, A. S. (1947). Pattern and process in the plant community. *Journal of Ecology*, **35**, 1–22

Werger, M. J. A., Louppen, J. M. W., & Eppink, J. H. M. (1983). Species performances and vegetation boundaries along an environmental gradient. *Vegetatio*, **52**, 141–150.

Westman, W. F. (1983). Xeric Mediterranean-type shrubland associations of Alta and Baja California and the community/continuum debate. *Vegetatio*, **52**, 3–19.

Wheelwright, N. T., & Bruneau, A. (1992). Population sex ratios and spatial distribution of *Ocotea tenera* (Lauraceae) trees in a tropical forest. *Journal of Ecology*, **80**, 425–432.

White, L. P. (1971). Vegetation stripes on sheet wash surfaces. *Journal of Ecology*, **59** 615–622.

Whittaker, R. H. (1975). *Communities and Ecosystems*, 2nd edn. New York: MacMillan Publishing Co., Inc.

Whittaker, R. H., & Levin, S. A. (1977). The role of mosaic phenomena in natural communities. *Theoretical Population Biology*, **12**, 117–139.

Whittaker, R. H., & Naveh, Z. (1979). Analysis of two-phase patterns. In *Contemporary*

Quantitative Ecology and Related Ecometrics. eds. G. P. Patil & M. Rosenzweig, pp. 157–165. Fairland, MD: International Co-operative Publishing House.

Whittaker, R. J. (1991). Small-scale pattern: an evaluation of techniques with an application to salt marsh vegetation. *Vegetatio*, **94**, 81–94.

Wiens, J. A., & Milne, B. T. (1989). Scaling of 'landscapes' in landscape ecology, or, landscape ecology from a beetle's perspective. *Landscape Ecology*, **3**, 87–96.

Williams, D. G., Anderson, D. J., & Slater, K. R. (1978). Influence of sheep on pattern and process in *Atriplex vesicaria* populations from the Riverine Plain of New South Wales. *Australian Journal of Botany*, **26**, 381–92.

Williamson, G. B. (1975). Pattern and seral composition in an old-growth beech-maple forest. *Ecology*, **56**, 727–731.

Wilson, J. B., & Agnew, A. D. Q. (1992). Positive-feedback switches in plant communities. *Advances in Ecological Research*, **23**, 263–336.

Wright, H. E. Jr., & Bent, A. M. (1968). Vegetation bands around Dead Man Lake, Chuska Mountain, New Mexico. *American Midland Naturalist*, **79**, 8–30.

Wulff, B. L., & Webb, K. L. (1969). Intertidal zonation of marine algae at Gloucester Point, Virginia. *Chesapeake Science*, **10**, 29–35.

Yarranton, G. A. (1966). A plotless method of sampling vegetation. *Journal of Ecology*, **59**, 224–237.

Yarranton, G. A., & Green, W. G. E. (1966). The distributional pattern of crustose lichens on limestone cliffs at Rattlesnake Point, Ontario. *The Bryologist*, **69**, 450–461.

Yarranton, G. A., & Morrison, R. G. (1974). Spatial dynamics of a primary succession: nucleation. *Journal of Ecology*, **62**, 417– 427.

Young, C. G. (1994). *Spatial Distribution of Vascular Plants in High Arctic Sedge Meadow Communities*. M.Sc. Thesis. Edmonton, Canada: University of Alberta.

Zeide, B., & Gresham, C. A. (1991). Fractal dimension of tree crowns in three loblolly pine plantations of coastal South Carolina. *Canadian Journal of Forest Research*, **21**, 1208–1212.

Zhang, W., & Skarpe, C. (1995). Small-scale species dynamics in semi-arid steppe vegetation in Inner Mongolia. *Journal of Vegetation Science*, **6**, 583–592.

Glossary of abbreviations

ANOVA Analysis of variance is a statistical method in which the total variability in data is partitioned according to the source of the variability. For example, given several sets of quadrats from different sites, we can test whether the variance among sites is significantly greater than the variance among quadrats within sites.

BQV Blocked quadrat variance is the original method for analysis of spatial pattern by combining quadrats into blocks. In this approach, the number of quadrats in a block is a power of 2, and for each blocksize any quadrat is included in only one block. The variance among blocks is calculated for each block size and the positions of peaks in the variance are interpreted as reflecting scales of pattern in the data (Eq. 3.3).

CA Correspondence analysis, also referred to as reciprocal averaging, is an ordination technique in which samples and species are ordinated simultaneously. It is accomplished by an iterative process of weighted averages (hence 'reciprocal averaging') or by eigenanalysis.

CCA Canonical correspondence analysis is an extension of correspondence analysis (**CA**) not only to ordinate samples and species simultaneously, but to include environmental variables in the procedure using multiple regression.

CSR Complete spatial randomness is the usual null model for the analysis of spatial point patterns. The points are distributed in the plane independently of each other. Because the number of points per unit area follows a Poisson distribution, this pattern is also referred to as a Poisson forest.

DCA Detrended correspondence analysis. In correspondence analysis (**CA**) (and in other ordination procedures), samples that are arranged linearly on a gradient often form an arch or horseshoe in the ordination diagram. Detrending is a technique to remove this arch effect.

ED — Euclidean distance refers generally to a measure of distance or difference between two samples equal to the square root of the sum of the squares of the differences in density of each of the individual species. In multispecies pattern analysis, the average Euclidean distance for a particular block size is the sum of the **TTLQV** or **3TLQV** values for all species, calculated at that blocksize (Eq. 5.11).

FD — Fractal dimension is a measure of the spatial complexity of an object, based on the concept that it is possible for objects to have noninteger dimensions. For example, when iterated, the Koch curve shown in Figure 1.11 has fractional dimension 1.26. In spatial pattern analysis, fractal dimension is usually calculated from the slope of the semivariogram (Eq. 3.44).

GIS — Geographical Information System: a combination of computer hardware and software for the acquisition, manipulation, display, and analysis of large-scale spatial data.

LANDSAT — This term is sometimes written 'Landsat' and refers to a class of remote sensing satellites and the images acquired by them. The Thematic Mapper (**TM**) sensor records the intensity in seven wavelengths of electromagnetic radiation; this information is then converted into inferences about characteristics of the earth's surface.

LDNT — Least diagonal neighbor triangulation is a method of creating a framework for point pattern analysis by joining the points in pairs to form triangles. This particular triangulation joins pairs of points from the shortest line to the longest, with the provision that at each stage the new line crosses no other.

MSO — Multiscale ordination is one method for analysing multispecies pattern. It combines a blocked quadrat method like **TTLQV** with the ordination method of principle components analysis (**PCA**).

NLV — New local variance is a method based on two-term local quadrat variance (**TTLQV**), that detects the average size of patches or of gaps, whichever is smaller. It calculates the average absolute value of the difference between adjacent **TTLQV** terms for each of a range of block sizes (Eq. 3.35).

NMDS Nonmetric multidimensional scaling is a nonlinear ordination method based on the rank order of measured differences between the sample units. The method reduces the number of dimensions used to describe the data while disturbing the rank order of differences as little as possible.

PCA Principle components analysis is an ordination technique that reduces the number of dimensions used to explain multivariate data. It does so by creating new orthogonal axes that are linear combinations of the original variables and that explain as much as possible of the total variance.

PQC Paired quadrat covariance is a method of joint pattern analysis for two species that uses spaced individual quadrats from a grid or transect. It is based on paired quadrat variance (**PQV**) and calculates covariance for a range of distances, b. Positive and negative peaks in the plot of the covariance, $C_{\mathrm{P}}(b)$, are indicative of scales at which the two species are positively or negatively associated (Eq. 4.4).

PQV Paired quadrat variance is a method of spatial pattern analysis that uses spaced individual quadrats from a grid or transect. It calculates the average squared difference between quadrats at each of a range of distances, b. This version uses all possible pairs of quadrats. Peaks in the plot of the variance, $V_{\mathrm{P}}(b)$, are indicative of scales of pattern in the data (Eq. 3.28).

RPQF Random paired quadrat frequencies is a method of analyzing spatial pattern that counts the number of times that pairs of quadrats with a certain horizontal and vertical displacement contain the same or different species.

RPQV Random paired quadrat variance is a method of spatial pattern analysis that uses spaced individual quadrats from a grid or transect. It calculates the average squared difference between quadrats at each of a range of distances, b. This version uses randomly chosen pairs of quadrats rather than all possible pairs. Peaks in the plot of the variance, $V_{\mathrm{P}}(b)$, are indicative of scales of pattern in the data.

3TLQC Three-term local quadrat covariance analyzes the joint spatial pattern of two species using blocks of adjacent quadrats from a transect of contiguous quadrats. It is based on three term local quadrat variance (**3TLQV**) and calculates covariance for a range of distances, b. Positive and negative peaks in the plot of the covariance, $C_3(b)$, are indicative of scales at which the two species are positively or negatively associated.

3TLQV — Three-term local quadrat variance is a method of spatial pattern analysis that uses blocks of adjacent quadrats from a transect of contiguous quadrats. It calculates the average squared difference between a block and the sum of the two adjacent blocks on either side of it for a range of block sizes, b. Peaks in the plot of the variance, $V_3(b)$, are indicative of scales of pattern in the data (Eq. 3.5).

4TLQV — Four-term local quadrat variance is the two-dimensional equivalent of **TTLQV**, for studying the spatial pattern of a single species, based on square blocks of quadrats from a grid of quadrats. It calculates a variance for each of a range of block sizes and peaks in the variance are interpreted as indicating scales of pattern (Eqs. 6.12 and 6.17).

9TLQV — Nine-term local quadrat variance is the two-dimensional equivalent of **3TLQV**, for studying the spatial pattern of a single species, based on square blocks of quadrats from a grid of quadrats. It calculates a variance for each of a range of block sizes and peaks in the variance are interpreted as indicating scales of pattern (Eq. 6.20).

TM — The Thematic Mapper (TM) sensor of a LANDSAT satellite records the intensity in seven wavelengths of electromagnetic radiation; this information is then converted into inferences about characteristics of the earth's surface.

:QC — Triplet quadrat covariance is a method of joint pattern analysis for two species that uses spaced individual quadrats from a grid or transect. It is based on triplet quadrat variance (**tQV**) and calculates covariance for a range of distances, b. Positive and negative peaks in the plot of the covariance, $C_t(b)$, are indicative of scales at which the two species are positively or negatively associated (Eq. 4.7).

QV — Triplet quadrat variance is a method of spatial pattern analysis that uses spaced individual quadrats from a grid or transect. It is the three term equivalent of **PQV**, calculating the average squared difference between a quadrat and the sum of the two on either side of it at distance b. Peaks in the plot of the variance, $V_t(b)$, as a function of block size are indicative of scales of pattern in the data (Eq. 3.34).

TTLQC — Two-term local quadrat covariance is a method of joint pattern analysis for two species that uses blocks of quadrats from a transect. It is based on two-term local quadrat variance (**TTLQV**) and calculates covariance for a range of distances, b. Positive and negative peaks in the plot of the covariance, $C_2(b)$, are indicative of scales at which the two species are positively or negatively associated.

TTLQV — Two-term local quadrat variance is a method of spatial pattern analysis that uses blocks of adjacent quadrats from a transect of contiguous quadrats. It calculates the average squared difference between pairs of adjacent blocks for a range of block sizes, b. Peaks in the plot of the variance, $V_2(b)$, are indicative of scales of pattern in the data (Eq. 3.4).

List of plant species

*Species**	*Family*	*Reference*
Abies amabalis Dougl. *ex* Forb.	Pinaceae	Keenan 1994
Abies balsamea (L.) Mill.	Pinaceae	Sprugel 1976
Abies magnifica Murr.	Pinaceae	Taylor & Halpern 1991
Acacia aneura F. Muell.	Fabaceae	Lamont & Fox 1981
Acacia ehrenbergiana Hayne	Fabaceae	Greig-Smith & Chadwick 1965
Achillea borealis Bong.	Asteraceae	Dale 1990
Acer spp. L.	Aceraceae	Upton & Fingleton 1985
Acer campestre L.	Aceraceae	Szwagrzyk & Czerwczak 1993
Acer pseudoplatanus L.	Aceraceae	Szwagrzyk & Czerwczak 1993
Adenostoma fasciculatum H., & A.	Rosaceae	Crawley 1986
Agrostis spp. L.	Poaceae	Kershaw 1958
Agrostis castellana Boiss & Reuter	Poaceae	Galiano 1986
Agrostis tenuis Sibth.	Poaceae	Kershaw 1958
Ammophila arenaria (L.) Link	Poaceae	Grieg-Smith 1961b
Arabidopsis thaliana (L.) Heynh	Brassicaceae	Silander & Pacala 1985
Aralia nudicaulis L.	Aralicaceae	Kenkel 1993
Arctagrostis latifolia (R. Br.) Griseb	Poaceae	Dale *et al.* 1993
Arctostaphylos rubra Rehder & Wils.)	Poaceae	Blundon *et al.* 1993; Dale & MacIsaac 1989
Arctostaphylos uvi-ursi (L.) Spreng.	Ericaceae	Schaefer 1993
Arrhenatherum elatius (L.) Beauv. *ex* J., & C. Presl.	Poaceae	Gibson & Greig-Smith 1986

*Species**	*Family*	*Reference*
Ascophyllum nodosum (L.) Le Jol.	Fucaceae	Lewis 1964
Aspicilia cinerea (L.) Körber	Hymeneliaceae	John 1989
Aster spp. L.	Asteraceae	Dale & Powell 1994
Atriplex vesicaria Heward *ex* Benth.	Chenopodiaceae	Williams *et al.* 1978
Betula glandulosa Michx.	Betulaceae	Dale 1990
Brachythecium groenlandicum (C. Jens.) Schljak.	Brachytheciaceae	Blundon *et al.* 1993
Bryum caespiticium Hedw.	Bryaceae	Blundon *et al.* 1993
Calluna vulgaris (L.) Hull	Ericaceae	Diggle 1981
Carex aquatilis Wahleb.	Cyperaceae	Dale *et al.* 1993
Carex bigelowii Torr.	Cyperaceae	Kershaw 1964
Carex flacca Schreber	Cyperaceae	Ver Hoef *et al.* 1989
Carex membranacea Hook.	Cyperaceae	Dale *et al.* 1993
Carex misandra R. Br.	Cyperaceae	Dale *et al.* 1993
Carex nigra (L.) Reichard	Cyperaceae	Gibson & Greig-Smith 1986
Carpinus betulus L.	Betulaceae	Szwagrzyk & Czerwczak 1993
Carya spp. Nutt.	Junglandaceae	Upton & Fingleton 1985
Centaurea diffusa Lam.	Asteraceae	Powell 1990
Centaurea jacea L.	Asteraceae	Ver Hoef *et al.* 1989
Chamaedaphne calyculata (L.) Moench	Ericaceae	Kenkel 1988b
Chamaerhodos erecta	Rosaceae	Dale (unpublished)
Chondrus crispus Stackh.	Fucaceae	Dale 1979
Cladonia cariosa (Ach.) Spreng.	Cladoniaceae	Blundon *et al.* 1993
Cladonia pyxidata (L.) Hoffm.	Cladoniaceae	Dale & MacIsaac 1989
Clematis fremontii S. Wats.	Ranunculaceae	Ricklefs 1990; Silvertown & Lovett Doust 1993
Cornus canadensis L.	Cornaceae	Schaefer 1993
Cupressus pygmaea (Lemmon) Sarg.	Cupressaceae	Whittaker & Levin 1977
Dactylis glomerata L.	Poaceae	Kershaw 1959a
Ditrichum spp. Hampe	Ditrichaceae	Blundon *et al.* 1993
Ditrichum flexicaule (Schwaegr.) Hampe	Ditrichaceae	Blundon *et al.* 1993
Drepanocladus uncinatus (Hedw.) Warnst.	Amblystegiaceae	Blundon *et al.* 1993
Dryas drummondii Richards	Rosaceae	Blundon *et al.* 1993; Dale and MacIsaac 1989
Dryas integrifolia M. Vahl	Rosaceae	Dale *et al.* 1993

*Species**	*Family*	*Reference*
Dryas octopetala L.	Rosaceae	Blundon *et al.* 1993
Epibolium angustifolium L.	Onagraceae	Dale 1990
Equisetum variegatum Schleich.	Equisetaceae	Gibson & Greig-Smith 1986
Eriogonum ovalifolium Nutt.	Polygonaceae	Day & Wright 1989
Eriophorum angustifolium Honck.	Cyperaceae	Kershaw 1964
Eriophorum scheurchzeri Hoppe	Cyperaceae	Dale *et al.* 1993
Eriophorum triste (Th. Fr.) Hadac & Löve	Cyperaceae	Dale *et al.* 1993
Erythronium grandiflorum Pursh	Liliaceae	Thomson *et al.* 1996
Fagus crenata Blume	Fagaceae	Peters & Ohkubo 1990
Fagus japonica Maxim.-	Fagaceae	Peters & Ohkubo 1990
Fagus sylvatica L.	Fagaceae	Szwagrzyk & Czerwczak 1993
Festuca spp. L.	Poaceae	Kershaw 1958
Festuca altraica Trin.	Poaceae	Dale 1990
Festuca pratensis Huds.	Poaceae	Ver Hoef *et al.* 1989
Festuca rubra L.	Poaceae	Kershaw 1964
Fraxinus angustifolia Vahl	Oleaceae	Szwagrzyk and Czerwczak 1993
Fucus serratus L.	Fucaceae	Dale 1979
Fucus spiralis L.	Fucaceae	Dale 1979
Fucus vesiculosus L.	Fucaceae	Dale 1979
Gigartina stellata (Stackh. in With.) Batt.	Gigartinaceae	Dale 1979
Hedysarum boreale var. *mackenzii* (Nutt.) Rich.	Fabaceae	Blundon *et al.* 1993
Hedysarum mackenzii Richards	Fabaceae	Dale & Blundon 1991
Hydrocotyle vulgaris L.	Apiaceae	Gibson & Greig-Smith 1986
Hypericum perforatum L.	Hypericaceae	Ver Hoef *et al.* 1989
Knautia arvensis (L.) Coulter	Dipsicaceae	Ver Hoef *et al.* 1989
Kochia scoparia (L.) Schrad.	Chenopodiaceae	Franco & Harper 1988
Lagarostrobus franklinii (Hook. f.) Quinn	Pinaceae	Gibson & Brown 1991
Lecidea auriculata Th. Fr.	Lecideaceae	Dale 1995
Lecidea paupercula Th. Fr.	Lecideaceae	John 1989
Leontodon hispidus L.	Asteraceae	Ver Hoef *et al.* 1989
Lepraria neglecta (Nyl.) Erichsen	Leprariaceae	John 1989
Linum catharcticum L.	Linaceae	Ver Hoef *et al.* 1989

*Species**	*Family*	*Reference*
Lolium perenne L.	Poaceae	Kershaw 1959a
Melanelia granulosa (Lynge) Essl.	Parmeliaceae	John 1989
Melanelia sorediata (Ach.) Goward & Ahti	Parmeliaceae	John 1989
Melanelia stygia (L.) Essl.	Parmeliaceae	John 1989
Mercurialis perennis L.	Euphorbiaceae	Hutchings 1979
Mimosa luisana Brandeg	Fabaceae	Valiente-Banuet & Ezcurra 1991
Neobuxbaumia tetetzo (Weber) Backeberg	Cactaceae	Valiente-Banuet & Ezcurra 1991
Nepeta cataria L.	Lamiaceae	Sih & Baltus 1987
Nothofagus betuloides (Mirb.) Bl.	Fagaceae	Veblen 1979
Nothofagus pumilio (Poepp.et Endl.) Krasser	Fagaceae	Veblen 1979
Ocotea tenera Mez & J.D. Smith *ex* Mez	Lauraceae	Wheelwright & Bruneau 1992
Oxalis spp. L.	Oxalidaceae	Persson 1981
Pastinaca sativa L.	Apiaceae	Thompson 1978
Phaeophyscia sciastra (Ach.) Moberg	Physciaceae	John & Dale 1989
Picea abies (L.) Karst.	Pinaceae	Szwagrzyk & Czerwczak 1993
Picea engelmannii Parry *ex* Englm.	Pinaceae	Blundon *et al.* 1993; Dale and MacIsaac 1989
Picea glauca (Moench) Voss.	Pinaceae	Dale 1990
Pinus banksiana Lamb.	Pinaceae	Kenkel 1988a
Pinus contorta Loudon	Pinaceae	Stadt 1993
Pinus muricata D. Don.	Pinaceae	Whittaker & Levin 1977
Pinus ponderosa Dougl.	Pinaceae	Getis & Franklin 1987
Pinus strobus	Pinaceae	Petersen & Squiers 1995
Pistacia lentiscus L.	Anacardiaceae	Whittaker & Naveh 1979
Plantago lanceolata L.	Plantaginaceae	Galiano 1986
Poa bulbosa L.	Poaceae	Galiano 1986
Polygonum viviparum L.	Polygonaceae	Dale *et al.* 1993
Polysiphonia lanosa (L.) Tandy	Rhodomelaceae	Lewis 1964
Populus grandidentata Michx.	Salicaceae	Petersen & Squiers 1995
Populus tremuloides Michx.	Salicaceae	Thompson 1978
Pseudephebe pubescens (L.) Choisy	Usneaceae	John 1989
Quercus nigra L.	Fagaceae	Upton & Fingleton 1985
Quercus rubra du Roi	Fagaceae	Upton & Fingleton 1985
Rhacomitrium spp. Brid.	Grimmiaceae	Kershaw 1964

*Species**	*Family*	*Reference*
Rhinanthus serotinus (Schöenheit) Oborny	Scrophulariaceae	Ver Hoef *et al.* 1989
Rhizocarpon bolanderi (Tuck.) Herre	Lecideaceae	John 1989
Rhizocarpon disporum (Nägeli ex Hepp) Mull. Arg.	Lecideaceae	John 1989
Rhizocarpon eupetraeoides (Nyl.) Blommb & Forss.	Lecideaceae	John 1989
Rhizocarpon geographicum (L.) DC.	Lecideaceae	John 1989
Rhizocarpon grande (Flörke *ex* Flotow) Arnold	Lecideaceae	John 1989
Rhizocarpon polycarpum (Hepp) Th. Fr.	Lecideaceae	Dale 1995
Rhizocarpon superficiale (Schaer.) Vain	Lecideaceae	John & Dale 1989
Rhododendron macrophyllum D. Don *ex* G. Don	Ericaceae	Whittaker & Levin 1977
Rubia peregrina L.	Rubiaceae	Navas & Goulard 1991
Rubus fruticosus (L.)	Rosaceae	Hutchings 1979
Salix arctica Pall.	Salicaceae	Dale *et al.* 1993
Salix barclayi Anderss.	Salicaceae	Blundon *et al.* 1993
Salix glauca L.	Salicaceae	Dale 1990
Salix vestita Pursh	Salicaceae	Blundon *et al.* 1993
Saxifraga oppositifolia L.	Saxifragaceae	Dale *et al.* 1993
Schaereria tenebrosa (Flowtow) Hertel & Poelt	Lecideaceae	John 1989
Sequoia sempervirens (D. Don.) End	Taxodiaceae	Whittaker & Levin 1977
Setaria incrassata (Hochst.) Hack	Poaceae	Carter & O'Connor 1991
Shepherdia canadensis (L.) Nutt.	Eleagnaceae	Dale & MacIsaac 1989
Solidago canadensis L.	Asteraceae	Dale & Powell 1994
Solidago multiradiata Ait.	Asteraceae	Dale 1990
Solidago sempervirens L.	Asteraceae	Lee 1993
Sphagnum spp. Dum.	Sphagnaceae	Kenkel 1988b
Sphagnum angustifolium C. E. O. Jensen	Sphagnaceae	Gignac & Vitt 1990
Sphagnum fuscum (Schimp.) Klinggr.	Sphagnaceae	Gignac & Vitt 1990
Sphagnum megellanicum Brid.	Sphagnaceae	Gignac & Vitt 1990
Spilonema revertens Nyl.	Coccocarpiaceae	John 1989

*Species**	*Family*	*Reference*
Themeda triandra Forsk.	Poaceae	Carter & O'Connor 1991
Thuja plicata D. Don	Cupressaceae	Keenan 1994
Tilia cordata Mill.	Tiliaceae	Szwagrzyk & Czerwczak 1993
Tortella inclinata (Hedw.) Limpr.	Pottiacaeae	Blundon *et al.* 1993
Tortella tortuosa (Hedw.) Limpr.	Pottiacaeae	Blundon *et al.* 1993
Trifolium repens L.	Fabaceae	Kershaw 1964
Tsuga canadensis L. Carr.	Pinaceae	Legendre & Fortin 1989
Tsuga heterophylla (Raf.) Sarg.	Pinaceae	Keenan 1994
Typha domingensis Pers.	Typhaceae	Grace 1987
Typha latifolia L.	Typhaceae	Grace 1987
Ulmus glabra Huds.	Ulmaceae	Szwagrzyk & Czerwczak 1993
Umbilicaria hyperborea (Ach.) Hoffm.	Umbilicariaceae	John 1989
Umbilicaria torrefacta (Lightf.) Schrader	Umbilicariaceae	John 1989
Umbilicaria vellea (L.) Hoffm.	Umbilicariaceae	John 1989
Vaccinium myrtillus L.	Ericaceae	Maubon *et al.* 1995
Vaccinium vitis-idea L.	Ericaceae	Schaefer 1993

*Compiled from, among many sources, Egan 1987, Ireland *et al.* 1980; Moss 1983, and through the Missouri Botanical Garden species list, http://mobot.mobot.org/Pick/Search/pick.html.

Index

Numbers in bold refer to glossary definition

aggregation 193–8, 231–7, 279, 284
allelopathy 27
analysis of variance (ANOVA) 56, **314**
anisotropy 37, 39, 77, 168, 174–7, 186–97, 227–31, 282–8
association 25–7, 31, 42, 45–7, 100–3, 117–24, 134–5, 147–8, 198, 277ff
autocorrelation 23–4, 46, 113, 123–4, 151, 160–5, 170–4, 226, 235, 262–71, 284

bivariance 231–41, 279
blocked quadrat variance (BQV) 56–8, 104–6, 168–74, **314**
bryophytes 7, 9, 127, 143, 162, 274

canonical correspondence analysis (CCA) 139, 203, **314**
cellular automata 25
classification 100, 204, 243
clonal 6, 25, 32, 38, 51, 295
clumped 20–1, 206–7, 220–32, 261–6, 283
combinatorics 96, 248
competition 2, 9, 24, 26–7, 34, 104–6, 120–4, 165, 220–1, 245, 259, 274
complete spatial randomness (CSR) 20, 209, 229, 239, 281, **314**
contiguity hypothesis 249–54
contingency table 123–4, 147–8, 277
correlation 26, 72, 106–9, 114–16, 119–20, 139, 170–2, 198–200, 241, 278
correspondence analysis (CA) 137ff, **314**
covariance 72, 104–14, 128–33, 193–205
see also: paired quadrat covariance; three-term local quadrat covariance; triplet quadrat covariance; two-term local quadrat covariance

detrended correspondence analysis (DCA) 138, 198, **314**
Dirichlet domain 14, 221, 282
dispersal 19, 165
dispersion 9, 19–21, 37, 95, 197, 207–26, 275, 281
disturbance 8, 31, 53, 231, 293
diversity 3, 9, 147, 165

edge effect 209, 214, 233, 280
eigenanalysis 129–35, 202
eigenvalue 129ff
environment 3, 6ff, 10, 26, 44–51, 120, 127, 139, 148, 198–200, 203, 242–5, 286–8, 293
epiphyte 26, 280
Euclidean distance (ED) 139–46, **315**

fire 8, 10, 25, 44
forest 4–5, 7–8, 30, 32, 50, 53, 78, 95, 140, 146–7, 169, 176, 191, 197, 204, 216, 220, 237–8, 291–4
four-term local quadrat variance (4TLQV) 178–95, **317**
fractal 28, 90–8, 136–46, 200–1, 296
fractal dimension (FD) 28, 90–1, 136–42, 286, **315**

gap 1, 5, 8, 12, 14, 16–19, 50–4, 96, 176, 206–10, 216, 257, 270, 279, 291, 293
Geographic Information System (GIS) 44, **315**
geostatistics 31, 71, 108, 136, 190, 198, 286
goodness-of-fit 163, 187–8, 226
gradient 30–2, 37–8, 48–9, 138, 242–52
grassland 6, 38, 47, 77, 126, 137–9, 146, 156, 274
Gruyère model 245, 289

habitat 3, 280
herbivory 4–10, 30, 197

intensity 16, 35, 50, 61–8, 83–4, 114, 133, 140, 209
intertidal 242–8

join counts 161–2, 234–5

K function 214–20, 228, 236

lacunarity 201
LANDSAT 43, 168, 184–5, **315**
landscape 10, 43–4, 183, 200–5
Least Diagonal Neighbor Triangulation (LDNT) 221–6, 235, 240, 282, **315**
lichen 25, 135, 159–62, 186–8, 286

map(s) 1, 14, 33–4, 39, 43–9, 193–7, 206–7, 237–40, 280–5
Markov (models) 24–5, 51, 102, 123, 151
microclimate 5
microhabitat 26, 160
microtopography 12, 120, 138–9, 143
Monte Carlo 123, 151, 161, 216, 229, 262, 269
mortality 8, 9, 207, 220, 280
mosaic 1–8, 14–16, 43, 100–1, 119, 126–7, 166, 186–8, 200–6, 242
multiscale ordination (MSO) 128, 135–46, 156, 166, 202, **315**
multispecies pattern 17, 125–46, 156, 202
multivariate analysis 43, 100, 273

neighborhood 3, 24, 26, 46–7, 103–4, 157–62, 171–3, 207–9, 220–1, 231–40
new local variance (NLV) 78–83, 97–9, 138, 203, **315**
nine-term local quadrat variance (9TLQV) 178, **317**
nonmetric multidimensional scaling (NMDS) 143, **316**
nucleation 9, 47–8, 103–4
nutrient(s) 7, 9, 53, 221, 243, 274

ordination 128–46, 198, 203–5, 288
overlap 249–61

paired quadrat covariance (PQC) 106–17, 193, **316**
paired quadrat variance (PQV) 71–8, 86, 117, 119, 136, 169–76, 285, **316**
patch 1–9, 14–19, 28–30, 50–5, 78–86, 96–100, 125–7, 135, 155–61, 196–207, 220, 245
phase (of cycle) 8, 112, 116, 121, 143–5
phase (of mosiac) 1, 14–17, 77, 98, 100–1, 119, 126–7, 166
phytosociological structure 6, 27, 100
point(s) 12–14, 19–21, 24–5, 32–3 46–9, 157–62, 206–19, 280–6
point-contact 46, 47, 147, 157–63
presence/absence 41, 54, 91, 103–4, 111, 115, 117, 123, 124, 135, 140, 143, 181, 248–52, 270–2, 280, 285
principal components analysis (PCA) 128ff, **316**

quantification 3, 35, 46, 101, 200–2, 277, 286

random paired quadrat frequency (RPQF) 186–90, **316**
random paired quadrat variance (RPQV) 71–8, **316**
randomization 77–8, 196–7, 233–4, 269, 279
randomness 1, 12–13, 19–21, 32, 37, 71–4, 157–63, 197, 206–10, 214–16, 225–9, 274–6, 280–4
resonance 67, 74–7, 81, 91–5, 108–9, 117, 132, 173, 181–2, 189

sampling 31–49, 103, 157, 247, 277–85
segregation 110, 165, 193–8, 205–6, 231–7, 284
self-thinning 48, 220
semivariogram 72, 90, 136, 190, 202, 286, 287
shrub 5, 9, 41–2, 47, 126, 140, 143, 146, 293
spectral analysis 91–8, 121, 143, 174–7, 283–90
standardized residuals 149, 159, 186, 285
stationarity 18, 51, 83, 177, 288
succession 8–9, 12, 26, 47–8, 55, 78

tessellation 220–1, 234, 241, 281–2
Thematic Mapper (TM) 43, **317**
three-term local quadrat covariance (3TLQC) 113–17, 128, 317
three-term local quadrat variance (3TLQV) 58–61, 68–70, 82, 85, 95, 115, 128, 132, 138, 144–5, 184–5, 277, 293
tides/tidal 30, 244, 248, 259, 261
topography 177, 204
trees 5, 8, 30, 32, 47, 50, 53, 95, 98, 137, 146, 169–73, 176, 191, 194, 197, 203, 220, 234, 237–40, 275
triplet quadrat covariance (tQC) 109–11, 117, 128, 139, 317
triplet quadrat variance (tQV) 75–6, 86, **317**
two-term local quadrat covariance (TTLQC) 113–17, 121, 317
two-term local quadrat variance (TTLQV) 58–71, 83–6, 97–9, 113–19, 136–9, 181–8, 202–5, **318**

unicornian distribution 226

variance
see: blocked quadrat variance; four-term local quadrat variance; new local variance; nine-term local quadrat variance; paired quadrat covariance; random paired quadrat variance; three-term local quadrat covariance; triplet quadrat covariance; two-term local quadrat covariance

variogram 71–3, 90–1, 108–9, 136–7, 190–2, 198–200, 286–8

wavelet 290–3
wetland 5, 10, 34, 45, 53, 267, 272

zonation 30, 45, 242, 254, 270

Printed in the United States
By Bookmasters